Topics in Current Physics 4

Topics in Current Physics Founded by Helmut K.V. Lotsch

Volume 1 **Beam-Foil Spectroscopy**
Editor: S. Bashkin

Volume 2 **Modern Three-Hadron Physics**
Editor: A. W. Thomas

Volume 3 **Dynamics of Solids and Liquids by Neutron Scattering**
Editor: S. W. Lovesey and T. Springer

Volume 4 **Electron Spectroscopy for Surface Analysis**
Editor: H. Ibach

Electron Spectroscopy for Surface Analysis

Edited by H. Ibach

With Contributions by
J. D. Carette B. Feuerbacher B. Fitton
H. Froitzheim M. Henzler H. Ibach
J. Kirschner D. Roy

With 123 Figures

Springer-Verlag Berlin Heidelberg New York 1977

Professor Dr. Harald Ibach

Institut für Grenzflächenforschung und Vakuumphysik,
Kernforschungsanlage Jülich GmbH, D-5170 Jülich 1

ISBN 3-540-08078-3 Springer-Verlag Berlin Heidelberg New York
ISBN 0-387-08078-3 Springer-Verlag New York Heidelberg Berlin

Library of Congress Cataloging in Publication Data. Main entry under title: Electron spectroscopy for surface analysis. (Topics in current physics ; v. 4). 1. Electron spectroscopy. 2. Surface chemistry. I. Ibach, H., 1941 – II. Series. QC454.E4E45.   539.7'2112.   76-51772.

This work is subject to copyright. All rights are reserved, whether the whole or part of the material is concerned, specifically those of translation, reprinting, re-use of illustrations, broadcasting, reproduction by photocopying machine or similar means, and storage in data banks. Under § 54 of the German Copyright Law where copies are made for other than private use, a fee is payable to the publisher, the amount of the fee to be determined by agreement with the publisher.

© by Springer-Verlag Berlin Heidelberg 1977
Printed in Germany.

The use of registered names, trademarks, etc. in this publication does not imply, even in the absence of a specific statement, that such names are exempt from the relevant protective laws and regulations and therefore free general use.

Offset printing and bookbinding: Konrad Triltsch, Graphischer Betrieb, Würzburg.   2153/3130-543210

Preface

The development of surface physics and surface chemistry as a science is closely related to the technical development of a number of methods involving electrons either as an excitation source or as an emitted particle carrying characteristic information. Many of these various kinds of electron spectroscopies have become commercially available and have made their way into industrial laboratories. Others are still in an early stage, but may become of increasing importance in the future. In this book an assessment of the various merits and possible drawbacks of the most frequently used electron spectroscopies is attempted. Emphasis is put on practical examples and experimental design rather than on theoretical considerations. The book addresses itself to the reader who wishes to know which electron spectroscopy or which combination of different electron spectroscopies he may choose for the particular problems under investigation.

After a brief introduction the practical design of electron spectrometers and their figures of merit important for the different applications are discussed in Chapter 2. Chapter 3 deals with electron excited electron spectroscopies which are used for the elemental analysis of surfaces. Structure analysis by electron diffraction is described in Chapter 4 with special emphasis on the use of electron diffraction for the investigation of surface imperfections. For the application of electron diffraction to surface crystallography in general, the reader is referred to Volume 4 of "Topics in Applied Physics". Chapter 5 discusses phonon excited electron spectroscopies and Chapter 6 is devoted to electron-loss spectroscopy. This technique has found rather important applications recently for the investigation both of electronic transitions and surface vibrations.

Specific surface systems are discussed in Chapters 3 to 6; however, only in connection with the experimental technique used and only insofar as the question of the specific imformation provided by the technique is addressed. For further details of the physical interpretation, the reader is referred to the literature. It is hoped that this book may serve as a guide through the embarrassing number of different electron spectroscopies that are in practical use.

Jülich, January 1977 H. Ibach

Contents

1. *Introduction* By H. Ibach 1
   1.1 Electron Spectroscopy and Its Importance in Surface Science 1
   1.2 The Information Depth 4
   1.3 Electron Spectroscopies 8
   References 11

2. *Design of Electron Spectrometers for Surface Analysis.* By D. Roy
   and J.D. Carette 13
   2.1 Specific Requirements of the Various Electron ....
       Spectroscopies for Surface Analysis 13
   2.2 General Principles and Characteristics of Electron Spectrometers .. 16
       2.2.1. Basic Principles of Energy Analysis 16
       2.2.2. Characteristics Related to Energy Resolution and
              Sensitivity 17
       2.2.3. Figures of Merit 18
   2.3 Design Principles of Electron Spectrometers 19
       2.3.1 Electron Optics for Energy Analysis 20
       2.3.2 Optimization of Electron Monochromators and Analyzers 22
       2.3.3 Detection Methods and Data Processing 28
       2.3.4 Further Practical Considerations 29
   2.4 Description of the Electron Spectrometers and the Methods of
       Energy Analysis for Surface Studies 31
       2.4.1 Electrostatic Deflection Spectrometers 31
       2.4.2 Magnetic Deflection Spectrometers 43
       2.4.3 Crossed-Field Deflection Spectrometers 43
       2.4.4 Retarding Potential Spectrometers 44
       2.4.5 Other Techniques 48
   2.5 Comparison of Electron Spectrometers 48
   2.6 List of Abbreviations and Acronyms 51
   References 52

3. *Electron-Excited Core Level Spectroscopies.* By J. Kirschner 59
   3.1 Basic Processes 60
       3.1.1 Free Atoms 60
       3.1.2 Surface Atoms 68
   3.2 Threshold Spectroscopies 71
       3.2.1 Oberserving the Excitation: Disappearance Potential
             Spectroscopy (DAPS) 74
       3.2.2 Observing the Deexcitation................................. 80
   3.3 Ionization Loss Spectroscopy (ILS) 87
   3.4 Auger Electron Spectroscopy (AES) 92
       3.4.1 Influence of the Atomic Environment 92
       3.4.2 Quantitative Auger Analysis 95
       3.4.3 Auger Microanalysis 99
       3.4.4 Combined Auger/X-Ray Microanalysis 102
   3.5 Comparisons 104
       3.5.1 Threshold Spectroscopies Inter Se 104
       3.5.2 Threshold Spectroscopies Versus ILS 107
       3.5.3 Elemental Analysis 108
   References 111

4. *Electron Diffraction and Surface Defect Structure.* By M. Henzler 117
   4.1 Principles of Defect Detection by Electron Diffraction 118
       4.1.1 Validity of the Kinematical Approximation 118
       4.1.2 Construction and Calculation of the Ideal LEED Pattern .. 119
       4.1.3 Instrumental Limitations 120
       4.1.4 Diffraction Pattern of Simple Defect Structures 121
       4.1.5 The Kind of Information in the Diffraction Pattern 123
   4.2 Point Defects 124
       4.2.1 Variation of Scattering Factor 124
       4.2.2 Variation of Atom Position 126
   4.3 Atomic Steps 130
       4.3.1 Regular Step Arrays on Primitive Lattices 130
       4.3.2 Irregular Step Arrays 134
       4.3.3 Nonprimitive Lattices 138
       4.3.4 Examples of Stepped Surfaces 139
   4.4 Domains and Facets 141
       4.4.1 Superstructures and Domains 141
       4.4.2 LEED Patterns of Domain Structures 142
       4.4.3 Quantitative Description of LEED Patterns 143
       4.4.4 Facets 145

		4.5 The Interpretation of a LEED Pattern	146
		4.5.1 Parameters to be Observed	146
		4.5.2 Interrelation of Defects and Effects	148
References			149

5. *Photoemission Spectroscopy*. By B. Feuerbacher and B. Fitton 151
    5.1 Principles of Photoemission 152
        5.1.1 Parameters and Ranges 154
        5.1.2 Basic Processes 156
    5.2 Instrumentation 159
        5.2.1 Light Sources 159
    5.3 Theoretical and Practical Aspects 162
        5.3.1 Electron Excitation and Emission 163
        5.3.2 Surface Sensitivity 168
        5.3.3 Relaxation and Chemical Shift 171
        5.3.4 Photoemission from Adsorbates 174
    5.4 Measurement Methods 179
        5.4.1 Energy-Resolved Spectroscopy 180
        5.4.2 Angle-Resolved Photoemission 183
        5.4.3 Yield Spectroscopies 189
        5.4.4 Spin-Polarized Photoemission 194
References 197

6. *Electron Energy Loss Spectroscopy*. By H. Froitzheim 205
    6.1 Definition of ELS 206
    6.2 Theory of Inelastic Scattering 207
        6.2.1 The Classical Theory (Concept of the "Dielectric Theory") 207
        6.2.2 Quantum Mechanical Description of the Dielectric Theory 210
        6.2.3 The Excitation of Optical Surface Phonons in Infrared-Active Material 213
        6.2.4 Excitation of Optical Surface Phonoms on Noninfrared-Active Substrates 214
        6.2.5 Excitation of Plasma Waves 216
        6.2.6 Electronic Surface Transitions 217
        6.2.7 Data Reduction 217
        6.2.8 Anisotropic Effects of ELS 218
    6.3 Experimental Studies of Surface Vibrations (Clean Surfaces)..... 219
        6.3.1 The Apparatus 219
        6.3.2 Infrared Active Material 220
        6.3.3 Noninfrared Active Material 222

6.4	Vibrational Modes on Gas-Covered Surfaces	223
	6.4.1 Apparatus	223
	6.4.2 Information	223
	6.4.3 Oxygen Adsorption on Si(111) 2 × 1	226
	6.4.4 Adsorption of Hydrogen on Si(111) 2 × 1	228
	6.4.5 Hydrogen Adsorption on W(100)	229
	6.4.6 Adsorption of Oxygen on W(100)	230
	6.4.7 The Adsorption of CO on Tungsten (100)	232
6.5	Experimental Studies of Electronic Transitions	233
	6.5.1 The Apparatus	233
	6.5.2 Relationship Between the Spectrometer and the Interpretation of the Loss Spectra	233
	6.5.3 Excitations of Electronic Transitions at Clean Silicon Surfaces	234
	6.5.4 Electronic Excitations at Ge(111) Surfaces	238
	6.5.5 Gallium Arsenide	239
	6.5.6 Selection Rule Effects Observed at Ge and GaAs	242
	6.5.7 Electronic Transitions at SiO and SiO_2	243
6.6	Conclusion	245
References		246
Subject Index		251

List of Contributors

FEUERBACHER, BERNDT
FITTON, BRIAN

European Space Agency,
Domeinweg, Noordwijk, The Netherlands

FROITZHEIM, HERMANN

Institut für Grenzflächenforschung und Vakuumphysik,
Kernforschungsanlage Jülich GmbH, Postfach 365,
5170 Jülich 1, Fed. Rep. of Germany

HENZLER, MARTIN

Institut B für Experimentalphysik,
Technische Universität Hannover, Appelstraße
3000 Hannover 1, Fed. Rep. of Germany

IBACH, HARALD

Institut für Grenzflächenforschung und Vakuumphysik,
Kernforschungsanlage Jülich GmbH, Postfach 365,
5170 Jülich 1, Fed. Rep. of Germany

KIRSCHNER, JÜRGEN

Institut für Grenzflächenforschung und Vakuumphysik,
Kernforschungsanlage Jülich GmbH, Postfach 365,
5170 Jülich 1, Fed. Rep. of Germany

ROY, DENIS
CARETTE, JEAN DENIS

Laboratoire de Physique Atomique et Moléculaire, C.R.A.M.,
Dept. de Physique, Université Laval, Québec, G1K 7P4, Canada

1. Introduction

H. Ibach

With 2 Figures

1.1 Electron Spectroscopy and Its Importance in Surface Science

Modern technologies frequently make use of the interaction of gases and fluids with solids and the properties of interfaces or thin films. Still our knowledge about fundamental processes at surfaces and interfaces is rather limited. The great difficulties in the understanding of heterogeneous catalysis, corrosion protection, semiconductor and thin-film technology have spurred many scientists to develop new tools for the study of surfaces and to learn more about the chemical and physical nature of the solid in its outermost atomic layers. Concurrently, the commercial availability of surface analysis instruments and the mature stage of ultrahigh-vacuum equipment in general is already beginning to have a major impact on further progress of the applied sciences.

The use of ultrahigh vacuum is in general considered as an advantage since it permits the investigation of surfaces under static and stable conditions. However, for many questions of technical importance especially those in the field of catalysis and corrosion, the high vacuum required by the surface analytical technique itself may actually be a drawback. The availability of surface analytical tools that operate while the surface is in contact with a liquid or high gas pressures remains the major need in surface science. While such a technique would allow analysis of the surface under the conditions to which they are exposed in reality, the present tools are limited to investigations under static low gas pressure conditions. Possible major differences in composition and structure of surfaces under these two conditions are the reason for a "credibility gap" between pure surface science and applied technology.

Surface analytical tools mostly make use of particles such as photons, electrons, atoms, molecules or ions. As in material science in general, the most valuable information is obtained by various kinds of spectroscopies. Surface spectroscopies, however, encounter the difficult problem of being selective to a rather small number of surface atoms compared to bulk atoms. The natural way to circumvent this problem is to reduce the number of contributing bulk atoms by using particles of appropriate energy that probe only a few Å into the solid. Still, separation between surface and bulk properties may remain difficult.

Among the available particles for surface spectroscopies electrons as the carriers of the specific information have found by far the widest application for a number of reasons:

1) Electrons have an inelastic mean free path of a few Å depending on the energy. Energy and momentum of an electron are therefore characteristic of elementary excitations near the surface.

2) Electrons are easily focused into beams and the energy may be varied by applying appropriate potentials.

3) Electrons are efficiently detected and counted.

4) Electrons may be analyzed with respect to angular and energy distribution using electrostatic lenses and deflection systems.

5) Another major advantage of electrons which one comes to appreciate, if one has ever worked with atoms, molecules, or ions as probing particles, is that electrons disappear from the vacuum system after being used for the surface analysis.

Electrons offer a wide range of different spectroscopies with different type of information: surface structure, elemental composition, chemical bond and surface vibrations may be investigated on clean or gas-covered surfaces. Single crystal surfaces are, in general, not required however widely used in fundamentally oriented studies. The various kinds of electron spectroscopies may be divided into two subgroups. In the first group of spectroscopies, the solid is excited by an electron beam and either the backscattered electrons or secondary electrons are analyzed. In the second group, electrons are generated by photon excitation. Other sources of electron excitation are possible, however, but not in wide-spread use so far.

Although electrons have been so successfully applied in surface science and the overwhelming part of our knowledge about surfaces stems from one or the other kind of electron spectroscopy, it should not be concealed that the use of the electron as the carrier of surface information has certain inherent limitations.

As already mentioned, rather good high vacuum conditions are required for the normal operation of the electron spectroscopic methods. Unfortunately this high vacuum requirement makes electron spectroscopic studies of surfaces in technical environments impossible. Even the analysis of such surfaces *in situ* after evacuating the gas is of limited value as evacuation may change the surface condition. Nevertheless this last variant is to be preferred to the transportation of the surface in air. The requirement of good vacuum is automatically overfulfilled in fundamental surface studies because in this case the partial pressure of reactive gases as hydrogen, oxygen, carbon monoxide, and nitrogen must be kept at least in the low 10^{-10} mbar range. Because of the much lower sticking coefficient, 10^{-9} mbar may be tolerated for clean semiconductor surfaces.

Another limitation arises from the fact that electron spectroscopies are in general not nondestructive. This holds especially for those spectroscopies where electron beams of high current are used as a primary source of excitation. The excitation or ionization of the surface atoms of the substrate and of adsorbed molecules or atoms provides enough energy for dissociation, desorption, and chemical reactions. Cross sections for electron-stimulated desorption of absorbed gases range from 10^{-15} cm^2 down to unmeasurable small values [1.1]. The highest values have been found for adsorbed halides, which are therefore rather difficult to detect by electron spectroscopies. Smaller values in the 10^{-18} cm^2 range and below apply to more stable adsorption systems. These are sufficiently low to allow an electron-beam excited surface analysis, provided that one works with a moderately focused beam. Highly focused beams of a few microns diameter as used, for instance, in scanning Auger spectroscopy still may cause problems and one has to check for electron stimulated desorption effects on any particular system under investigation. Only for substrate atoms, electron-stimulated desorption may be neglected. In addition to electron-stimulated desorption, the electron beam may cause surface chemical reactions. Frequently a cracking of hydrocarbons and carbon monoxide adsorbed on the surface is observed. Light products may desorb and carbonaceous layers are deposited on the surface. Carbon deposition may proceed at a rather rapid rate in poor vacuum.

Another source of continuous concern and confusion is the information depth of electron spectroscopies, i.e., the thickness of the surface layer that is actually probed in the particular experiment. The information depth depends on the electron energy and the material; in some cases it is even difficult to define what one may consider as the information depth. This most important question warrants a more detailed discussion which is given in the next paragraph. In any case electron spectroscopies are tools for a true surface analysis only in a limited sense. Even the smallest obtainable information depth of about 5 Å is sufficiently high so that not only true surface properties but also properties of bulk atoms are investigated simultaneously. This causes the problem of discriminating between properties of surface atoms and bulk atoms for which a general solution has not yet been found. For adsorbate systems one may consider the difference between the clean and the adsorbate-covered surface as being the property of the adsorbate layer. For the investigation of the properties of surface substrate atoms, no such method exists. Substantial confusion may arise from this fact, especially in alloy substrates where the chemical composition of the surface layer may be different from the second and deeper layers. Surface analytical techniques other than electron spectroscopies (e.g., ion backscattering) should be used to solve such questions.

1.2 The Information Depth

For most electron spectroscopies (with a few exceptions) the characteristic surface information is contained in the energy E and momentum k of the electron escaping the material. Therefore the information is lost after elastic or inelastic scattering events that the electron may encounter on its way between the point where it was generated and the surface. Inelastic scattering processes arise from electron-electron or electron-phonon interactions. With the present experimental resolution the energy loss in electron-phonon scattering is negligibly small. Large-angle phonon scattering events that would affect the angular distribution have a relatively small cross section corresponding to a mean free path of several hundred Å and are therefore not considered in the following.

For bulk material the inelastic scattering probability is proportional to the path length in the solid. The flux of electrons of certain energy and momentum, therefore, decays exponentially

$$I = I_0 \, e^{-x/\lambda} \tag{1.1}$$

where λ is the mean free path of the electron. The mean free path is inversely proportional to the imaginary part of the electron self-energy [1.2,3] which may be calculated from the dielectric response function $1/\varepsilon(q,\omega)$

$$\lambda^{-1} = \frac{me^2}{\pi\hbar E} \int \frac{1}{q} \, dq \, \frac{1}{2} \int_0^{\omega_{max}} d\omega \, \mathrm{Im}\left\{\frac{-1}{\varepsilon(q,\omega)}\right\} \theta(E-E_{min}-\hbar\omega) \tag{1.2}$$

with E the energy of the electron and q and $\hbar\omega$ the momentum transfer and energy involved in a transition, respectively. E_{min} is the Fermi energy or the conduction band edge E_c in case of an insulator. The step function θ and the upper integration limit

$$\omega_{max} = \frac{\hbar}{2m} [k^2 - (k-q)^2]$$

takes care of energy conservation. As the response function represents the electronic excitation spectrum of the solid which is different for different materials, the mean free path of the electron depends on the material under investigation. For an insulator, for example, $\mathrm{Im}[-1/\varepsilon(q,\omega)]$ becomes considerable only when $\hbar\omega$ exceeds the band-gap energy E_g. Because of energy conservation (the θ function in (1.2)) the electronic contribution to λ^{-1} becomes rather small for electron energies $E<E_c+E_g$ (dashed line in Fig.1.1d). Impurity scattering may set a lower limit. Thus for low electron energies the mean free path for an insulator is much higher than for a metal.

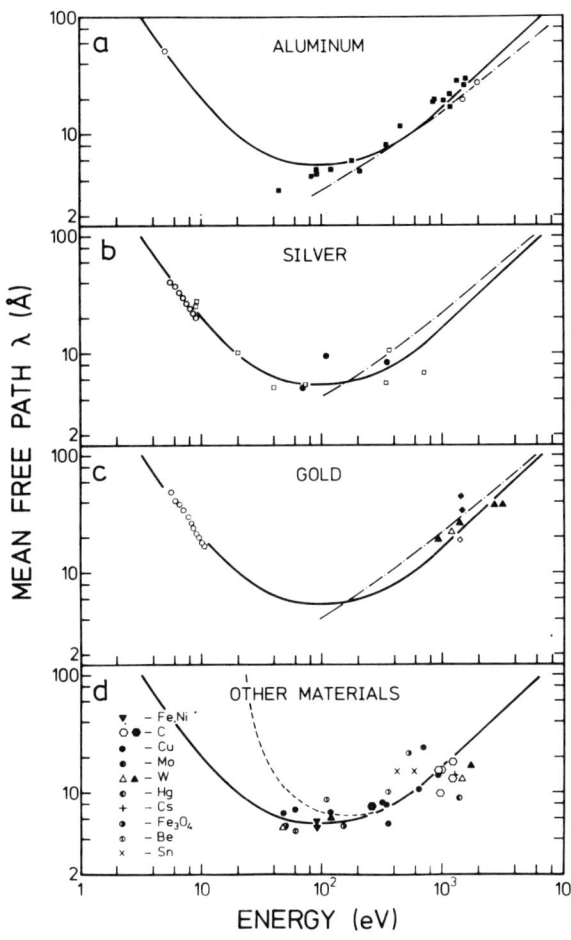

Fig.1.1a-d Electron mean free path of various materials. The solid line is an average of all experimental values. The dashed line is the expected mean free path for a wide band gap ($E_g \sim 10$ eV) insulator. The dashed-dotted line represents a calculation according to (1.6)

(a) Aluminum	(b) Silver	(c) Gold	(d) Other materials
TRACY [1.6]	KANTER [1.7]	KANTER [1.7]	RIDGWAY [1.15]
KANTER [1.7]	SEAH [1.8]	HENKE [1.11]	SEAH [1.8]
	EASTMAN [1.9]	EBEL [1.12]	STEINHARDT [1.16]
	PALMBERG [1.10]	BAER [1.13]	JACOBI [1.17]
		KLASSON [1.14]	TARNG [1.18]
			CARLSON [1.19]
			BRUNDLE [1.20]
			FRASER [1.21]
			NEEDHAM [1.22]
			SEAH [1.23]

A compilation of the available experimental data for the mean free path of various materials, mostly metals, is shown in Fig.1.1a-c. The experimental data were obtained either by measuring the reduction in a characteristic electron emission from a different substrate after the deposition of an overlayer of the material, or by direct transmission experiments in a few cases. Additional values obtained indirectly from a certain theoretical evaluation of photoemission data [1.4] are not included, since they yield much lower values as one would estimate from (1.2). As immediately evident from the large scattering in the experimental points, reliable data aré rather difficult to obtain. The reason is that the overlayer technique requires a strictly two-dimensional growth of a preferably amorphous layer in order to avoid errors by holes in the overlayer or diffraction effects. Fig.1.1 shows that the differences between different materials are of the same order as the scattering between different data for a single material. This has encouraged many authors to assume a single master curve for the mean free path. The full line in Fig.1.1 may represent such a master curve. It should, however, be pointed out once more that there is no theoretical reason for assuming a single curve for all materials, and large deviations are expected especially for low electron energies. Note that no values for insulators are available in this energy range. On the other hand, it is easy to see from (1.2) why the differences between different materials are small for high energies.

Representing the materials by a free electron gas of proper density, one may see that at high electron energies it is only the plasmon excitation which contributes to the integral in (1.2) significantly and thus the response function may be replaced by

$$\text{Im}\left\{\frac{-1}{\varepsilon(q,\omega)}\right\} = \frac{\pi}{2} \omega_p \delta(\omega-\omega_p) \tag{1.3}$$

with ω_p the plasma frequency.
One obtains then for λ^{-1}

$$\lambda^{-1} = \frac{e^2 \omega_p}{\hbar v^2} \ln\left(\frac{q_c}{q_{min}}\right) \tag{1.4}$$

where v is the electron velocity, $q_{min} = \omega_p/v$ the lower integration limit derived from energy conservation

$$\frac{\hbar^2 k^2}{2m} - \frac{\hbar^2 (k-q_{min})^2}{2m} = \hbar\omega_p \tag{1.5}$$

and $q_c = \omega_p/v_F$ the cut-off wave vector where the plasmon merges into the single electron excitation spectrum. Introducing the average electron radius in atomic units r_s which is simply calculated from the free electron density, one obtains

$$\lambda^{-1} = \sqrt{3}\, a_0\, \frac{R}{E}\, r_s^{-3/2}\, \ln(\frac{4}{9\pi})^{2/3}\, \frac{E}{R}\, r_s^2 \tag{1.6}$$

with $a_0 = 0.53$ Å the radius and $R = 13.6$ eV the Rydberg constant. The two factors containing r_s partly compensate each other. This together with the fact that the values for r_s fall in a limited range explains the small variation in the mean free path. In Fig.1.1a-c the mean free path calculated according the (1.6) is plotted as a dashed-dotted line. The agreement is rather satisfactory and (1.6) may therefore be used for an estimate of the mean free path when no experimental data are available. It should be pointed out that for the tetrahedrally coordinated semiconductors the free electron gas model provides a very acceptable description of the response function. A more sophisticated procedure for the estimation of the mean free path that includes core excitations was suggested by POWELL [1.5]. The agreement with the experimental results, however, seems to be not substantially improved compared to (1.6).

A question that has been raised frequently is a possible influence of characteristic surface excitations (surface plasmons) associated with the two boundaries of an overlayer. If these excitations played an important role for the electron self-energy, a mean free path could not be defined and the electron escape depth for a substrate material could not be measured by transmission through an overlayer of this material on top of another substrate. Four terms contribute [1.24] to the characteristic losses an electron experiences during transmission through a thin layer. One is the already mentioned bulk loss P proportional to $\text{Im}\{-1/E(q,\omega)\}$. The three others are the so-called "Begrenzungsterm" Q_B and the terms arising from the coupled surface excitation of the two boundaries, Q_+ and Q_- (surface plasmons for the free electron gas). Since, however, the bulk loss alone exhausts the sum rule [1.24] the latter three terms compensate each other and the integral

$$\int \omega (Q_B + Q_+ + Q_-)\, d\omega = 0 \tag{1.7}$$

vanishes independent of the thickness of the layer. Therefore the surface- and Begrenzungsterm need not be considered for the mean free path in the high-energy limit. Apart from experimental difficulties, the results obtained from the overlayer technique may therefore be considered as reliable and characteristic for layers as well as for substrate materials. Due to surface electronic transitions, the ω,q dependence of the response function may differ considerably for thin layers and substrate materials (Subsec.6.5.3). Since for small electron energies the integral in (1.6) is taken only over a limited energy range, substantial deviations in the mean free path are expected that depend directly on the surface conditions itself. This may be a reason why the mean free path for overlayers of alkali and earth alkali metals [1.4] was found to be smaller than expected.

The mean free path of an electron is not always equivalent to the information depth. In disappearance-potential spectroscopy (Subsec.3.3.1) and energy-loss spectroscopy of bulk losses (Subsec.6.2.1), the characteristic electron is the primary electron itself. It penetrates the surface layer twice and therefore the information depth is half the mean free path. For an electron leaving the surface at some angle θ from the normal, the information depth is $\propto \lambda \cos \theta$. Finally for surface excitations in electron energy-loss spectroscopy, the information depth is not related to the mean free path at all but is determined by the electron energy, the reflection angle, and the angular resolution of the spectrometer (Subsec.6.2.2).

It is evident from the above consideration that the information depth is a most uncertain quantity in electron spectroscopy and one has to be extremely aware of this problem if quantitative work is desired. In general one may say that the information depth becomes more uncertain the higher the surface sensitivity of the specific electron spectroscopy is. For quantitative work, high-energy electrons should therefore be used at the expense of a high surface sensitivity.

1.3 Electron Spectroscopies

In this chapter a brief view on the various kinds of electron spectroscopies is given together with a short assessment of the inherent potentialities and limitations. References are made to the chapters of this volume where a more detailed description is presented. Two electron spectroscopies are not treated in this volume, namely, field emission spectroscopy and ion neutralization spectroscopy. Both have not found a widespread use so far and have also been reviewed recently [1.25,26] A schematic overall spectrum of electrons emitted from a solid subject to irradiation by a 2 keV electron beam is shown in Fig.1.2. Electrons of different origin are observed in the four energy ranges (named I-IV in Fig.1.2). The relative intensity of various contributions to the spectrum depends on primary energy, angle of incidence and angle of emission. An actual spectrum obtained under particular conditions may therefore differ significantly in shape from Fig.1.2. The narrow peak on the right side (range I) is made up of elastically diffracted electrons, which contain the structural information of the bulk and surface. Energy losses due to phonon excitations (see insert on the right) are resolved only by more sophisticated analysers (Subsec.6.4.1). They are followed in a range II by characteristic energy losses due to electronic excitations and ionization losses. The dashed line in Fig.1.2 represents the secondaries. Their characteristic energies are related to zero kinetic energy, not to the energy of the exciting beam. Most of the secondary electrons especially those in the large hump at low energies (range IV) result from a cascade process and are therefore of little analytical use except in yield spectroscopies and in the scanning microscope. Secondary electrons generated in Auger processes are indicated in the insert on the left.

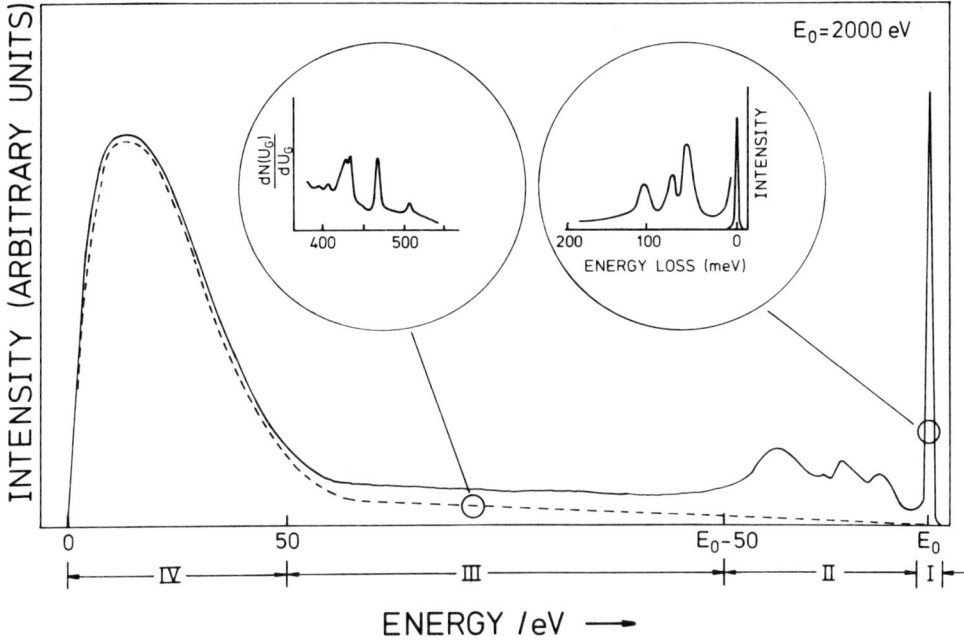

Fig.1.2 Spectrum of inelastic electrons. The angular distribution of the various contributions is quite different. The relative intensities therefore depend on the acceptance angle. The inserts show Auger transitions and phonon losses, respectively

Low- (High-) Energy Electron Diffraction (L(H)EED). Especially LEED has found a widespread use for the investigation of surface structures. Yet a full structure analysis of surface lattices is still impeded by the considerable theoretical difficulties in the intensity evaluation. Present interest is focused on a further development of surface crystallography but also on the use of LEED for the investigation of surface imperfections such as steps, kinks and adatoms. These studies are not limited to single crystal surfaces and therefore have found some interesting technical applications. Especially this aspect of LEED is stressed in Chapter 4.

Inelastic Low-Energy Electron Diffraction (ILEED). In ILEED electrons are analyzed that have experienced one or more elastic and inelastic scattering process. ILEED has been used to study the plasmon dispersion. Because of the sophisticated theoretical background involved in its interpretation, ILEED has not found many applications with the exception of *Electron Energy Loss Spectroscopy (ELS)* where the inelastic electrons are collected either in the vicinity of a diffracted beam or by a wide angular aperture (Chap.6). With ELS surface electronic transition and surface vibrations of clean and gas-covered surfaces have been investigated. ELS provides information about the spectrum of unfilled electron states near the surface in combination with valence and core level spectroscopies. The application to surface vibrations may be even more important in the future. Indeed, the localized vibrations of adsorbates contain collective and structural information about the degree of dissociation, binding energies, binding sites, and lateral interactions. These studies may therefore contribute significantly to catalytic research. As mentioned the information depth of ELS is not simply determined by the electron mean free path but by the wavelength of the charge density fluctuation near the surface. Sensitivities of 10^{-1} and 10^{-2} monolayers have been reported for electronic and vibrational excitations, respectively.

Auger Electron Spectroscopy (AES). A core hole generated by the primary electron may be filled by another core or valence electron. In the Auger process the transition energy is transmitted radiationless to a second electron, whose kinetic energy is measured for analysis (Sec.3.5). AES is used preferably for analysis of elements and has found the widest application of all surface analytical techniques. In the form of the scanning Auger microprobe, a direct imaging of the lateral distribution of elements has become possible. The surface sensitivity is 10^{-3} to 10^{-2} of a monolayer and not very different for the elements. However small signals may sometimes be difficult to separate when they overlap substrate peaks. Supplementary information for the elemental analysis may be obtained by other electron-excited spectroscopies such as *Ionization Loss Spectroscopy (ILS), Disappearance Potential Spectroscopy (DAPS), Auger Electron Appearance Potential Spectroscopy (AEAPS)* and *Soft X-ray Appearance Potential Spectroscopy (SCAPS)*. In general these methods are less sensitive and the sensitivity differences between elements are larger than with AES. A detailed assessment of these methods and their relative merits is given in Chapter 3.

Photon-excited electron spectroscopies have attracted much attention in the past. The interest became even more stimulated with the availability of the continuous synchrotron light source. As most experiments are still carried out with conventional noncontinuous sources, one distinguishes, as shown in Chap.5 between *Ultraviolet Photoemission Spectroscopy (UPS)* for photon energies up to 40.8 eV (He II line) and *X-ray Photoemission Spectroscopy (XPS)* for photon energies in the keV-range (typically Al-K_α and Mg-K_α with 1486.6 eV and 1253.6 eV, respectively). Both spectroscopies probe the electronic spectrum with an information depth determined by the kinetic energy of the excited electron. Surface sensitivity ranges from 10^{-1} to 10^{0} monolayer for UPS and XPS, respectively. Due to matrix element effects, XPS is more sensitive to core levels than to valence-band electrons. The influence of the chemical environment of an atom and the chemical bond is reflected in the "chemical shift" of the core level and in changes in the overall shape of the spectra in the valence-band region. With UPS these changes may be also observed for surfaces covered with adsorbed layers and may be related to the electronic structure of the surface adsorbate bond. With recent further developments such as angle-resolved photoemission spectroscopy and partial yield spectroscopy even the two-dimensional surface-band structure and the unoccupied surface states may be studied.

Each of the described electron spectroscopies has proved to be a powerful tool for surface investigations. The broadest physical and chemical information, however, is obtained by a suitable combination of several spectroscopies. This book may assist the reader in making the proper choice.

References

1.1 T.E. Madey, J.T. Yates: J. Vac. Sci. Technol. $\underline{8}$, 525 (1971)
1.2 L. Kleinman: Phys. Rev. $\underline{B3}$, 2982 (1971)
1.3 J.C. Ashley, R. Ritchie: Phys. Stat. Sol. $\underline{62}$, 253 (1974)
1.4 I. Lindau, W.E. Spicer: J. Elect. Spectros. $\underline{3}$, 409 (1974)
1.5 C.J. Powell: Surface Sci. $\underline{44}$, 29 (1974)
1.6 J.C. Tracy: General Motors Corp. Research Publication GMR-1502 (1973)
1.7 H. Kanter: Phys. Rev. $\underline{B1}$, 2357 (1970)
1.8 M.P. Seah: Surface Sci. $\underline{32}$, 703 (1972)
1.9 D.E. Eastman: In Proc. 32nd Phys. Electr. Conf., Albuquerque, New Mexico (1972)
1.10 P.W. Palmberg: Anal. Chem. $\underline{45}$, 549A (1973)
1.11 B.L. Henke: Phys. Rev. $\underline{A6}$, 94 (1972)
1.12 H. Ebel, M.F. Ebel: X-ray Spectros. $\underline{2}$, 19 (1973)
1.13 Y. Baer, P.F. Heden, H. Hedman, M. Klasson, C. Nordling: Solid State Comm. $\underline{8}$, 1479 (1970)
1.14 M. Klasson, J. Hedman, A. Berndtsson, R. Nilsson, C. Nordling, P. Melnik: Phys. Scripta $\underline{5}$, 93 (1972)
1.15 J.W.T. Ridgway, D. Haneman: Surface Sci. $\underline{24}$, 451 (1971); $\underline{26}$, 683 (1971)
1.16 R.C. Steinhardt, J. Hudis, M.L. Perlman: Phys. Rev. $\underline{B5}$, 1016 (1972)
1.17 K. Jacobi, J. Hölzl: Surface Sci. $\underline{26}$, 54 (1971)
1.18 M.L. Tarng, G.K. Wehner: J. Appl. Phys. $\underline{44}$, 1534 (1973)
1.19 T.A. Carlson, G.E. McGuire: J. Elect. Spectros. $\underline{1}$, 162 (1972)
1.20 C.R. Brundle, W.M. Roberts: Chem. Phys. Lett. $\underline{18}$, 380 (1973)
1.21 W.A. Fraser, J.V. Florio, W.N. Delgass, W.D. Robertson: Surface Sci. $\underline{36}$, 661 (1973)

1.22 P.B. Needham, T.J. Driscoll: J. Vac. Sci. Technol. $\underline{11}$, 278 (1974)
1.23 M.P. Seah: J. Phys. $\underline{F3}$, 1538 (1973)
1.24 H. Boersch, J. Geiger, W. Stickel: Z. Phys. $\underline{212}$, 130 (1968)
1.25 E.W. Plummer: In *Topics in Applied Physics*, Vol. 4, ed. by R. Gomer (Springer, Berlin, Heidelberg, New York 1975) p. 143
1.26 H.D. Hagstrum: In *Metals*, Vol. 6, ed. by R.F. Bunshaw (Wiley, New York 1972) p. 309

2. Design of Electron Spectrometers for Surface Analysis

D. Roy and J. D. Carette

With 11 Figures

The purpose of this chapter is to act as a guideline for the choice and design of the electron spectrometers for surface analysis. First the characteristics of the various types of electron spectroscopies for surface analysis are presented with their specific requirements regarding angular and energy resolutions, and minimum sensitivity. The basic operating principles and general characteristics of electron spectrometers are then presented. This is followed by a discussion of the factors which determine their optimum performance and sensitivity, and a presentation of practical design principles. Afterwards a critical review of the various types of electron spectrometers and their electron optics is presented, with full description of typical designs. The emphasis is placed on electrostatic instruments which are more widely used and more compatible with surface analysis. Finally the different electron spectrometers are compared and an attempt is made to determine which spectrometers are the most appropriate for each type of spectroscopy.

2.1 Specific Requirements of the Various Electron Spectroscopies for Surface Analysis

For those methods of surface analysis in which electrons are the particles to be collected and analyzed, Table 2.1 summarizes the characteristics of these spectroscopies and their specific requirements dealing with the handling and analysis of electrons. The description of these methods may be found in the following chapters of this book. This classification given in Table 2.1 is not proposed as universal. Some of the distinctions between techniques may be somewhat arbitrary (mainly for the energy loss spectroscopies) and have been introduced to simplify the classification of the various types of electron spectrometers for surface analysis. Most of the data in this table are taken from a review paper of HOBSON [2.1] on the relationship of solid surfaces to vacuum science and technology. Data given by BALLU [2.2] have also been helpful in filling this table.

Table 2.1 Characteristics of the various electron spectroscopies for surface analysis

Name	Incident particle	Incidence energy	Incidence energy resolution	Angle of Incidence[a]	Incidence angular definition	Incident current or flux	Incident beam diameter or area
XPS (ESCA)	Photon (X-ray)	High (1–10 keV)	Medium (0.5–1.5 eV)	70 – 85°	Unimport.	Low	1 – 3 mm
UPS	Photon (UV)	Low (4–40 eV)	High (10–25 meV)	0 – 85°	Unimport.	High	0.01 – 0.1 cm^2
LEED (ELEED)	Electron	15 – 500 eV	Medium	Usually ~0°	High	10^{-8}–10^{-5} A	0.2 – 1 mm
ILEED	Electron	1 – 200 eV	Medium (~0.5 eV)	Usually ~0°	High (~2°)	10^{-8}–5×10^{-6} A	1 – 2 mm
AES	Electron	100 – 5000 eV	Unimport.	0 – 75°	Unimport.	0.1 – 500 μA	0.1 – 1 mm
SAES	Electron	3 – 15 keV	Unimport.	~60°	High spatial resolution	10^{-9}–10^{-6} A	3×10^{-4} – 0.1 mm
AEAPS	Electron	100 – 2000 eV	Medium	~0°	Unimport.	~10^{-4} A	~10 mm
IS (ILS)	Electron	250 – 1000 eV	Medium	~0° or 45°	Unimport.	250 – 500 μA	~3 mm
LEELS	Electron	Low (1–50 eV)	High (10–50 meV)	45 – 85°	High (1 – 5°)	10^{-8}–10^{-10} A	~2 mm
TELS (CELS)	Electron	High (20–60 keV)	Med. or ultrahigh (400–5 meV)	0°	Ultrahigh (~10^{-4} rad)	High or low (~10^{-5}–10^{-9} A)	~0.05 mm
FES	–	–	–	–	–	–	100 – 1000 Å
INS	Ion$^+$ (He$^+$, Ne$^+$)	4 – 10 eV	Medium	0 – 45°	Unimport.	10^{-9}–10^{-10} A	5 mm

[a] With respect to the normal
[b] Large angular acceptance implies low angular resolution
[c] Amplitudes of the measured features with respect to the background
[d] With peak perpendicular to incident beam
[e] Except for angular distribution measurement (~5°)

Table 2.1 (continued)

Emitted particle	Emitted energy	Analysis energy resolution	Angle of exit[a]	Angular resolution or acceptance[b]	Exiting current	Peak/background[c]	Used spectrometers
Electron	0 - 10 keV	Med. to high (~1eV- 0.1%)	0 - 90°[d]	Usually large accept.[e]	10^{-19}-10^{-16} A	0.1 - 100	SDA, MDA, DFA, SGRPA, CMA, TOFA
Electron	0 - 30 eV	Med. to high (50-200 meV)	0 - 90°	Usually large accept.[e]	10^{-15}-10^{-9} A	Very high or ~ 1	CDA, SDA, RPA, PMA LEEDOA, CMA, TOFA
Incident electron	Within 0.5 eV of prim. energy	Medium	Specific direction 0-90°	Large accept.	10^{-4}-10^{-1} × inc. current	Very high	LEEDOA
Incident electron	Losses 5 - 50 eV	Medium (~ 0.5 eV)	As in LEED	High resolution (2 - 6°)	~ 10^{-3}× inc. current	0.1 - 2	RPA, CDA, LEEDOA
Electron	20 - 2000 eV	Med. to high (~ 0.1eV-0.5%)	0 - 75°	Usually large accept.[e]	10^{-5}-5×10^{-3} × inc. current	10^{-1} - 10^{-2}	LEEDOA, CMA, SDA CDA, SGRPA
Electron	10 - 1000 eV	Medium (~1%)	~ 0 - 60°	Usually large accept.[e]	~ 10^{-4} × inc. current	10^{-1} - 10^{-2}	CMA, LEEDOA
Electron	50 - 1400 eV	Total detection	All angles	Medium accept.	~10^{-4} × inc. current	10^{-1} - 10^{-2}	IEMM
Incident electron	100 - 500 eV	Medium	All angles	Usually large accept.	10^{-7} A	10^{-1} - 10^{-2}	CMA, LEEDOA
Incident electron	Losses 0.02 - a few eV	High (10-50 meV)	Near specular	High resolution (1 - 5°)	~ 10^{-3} × inc. current	0.1 - 10	TSDS, TCDS
Incident electron	Losses 0.005 - a few eV	Med. or ultrahigh (200-5meV)	Forward	Ultrahigh resolution (~ 10^{-4} rad)	~ 10^{-3}× inc. current	0.1 - 10	WFA, FLA, MLA, SDA, CDA, RPA, MDA
Electron	~ 2 keV	High (~ 20 meV)	Forward with wide spread	Very high resolution (~ 5×10^{-3} rad)	10^{-4} - 10^{-8} A	10^{-1} - 10^{-2}	RPA, SDA, CDA
Electron	0 - 16 eV	Medium (~ 0.1 eV)	0 - 90°	Large accept.	10^{-10} - 10^{-9} A	Very high or ~ 1	RPA

For each electron spectroscopy described, the characteristics of both the incident beam and the emitted particle are given. For the latter, which is to be analyzed, the emission energy, direction, and flux are presented with the required energy and angular resolutions for the analysis. A tentative appreciation of the minimum sensitivity required for the measurements of the features of interest is given. This figure is the approximate ratio of the amplitudes of the features with respect to the physical background on which they are superimposed in the experimental measurement. The reciprocal of the figure gives a rough estimate of the minimum signal-to-noise ratio (S/N) required for each type of measurement. The electron spectrometers used so far are also listed for each spectroscopy. The various abbreviations and acronyms used in Table 2.1 and throughout the text are listed in Section 2.6.

In most of these techniques for surface studies the measurements are mainly "integral type" measurements performed with large angular acceptance analyzers. However, for many of these studies, angular dependence measurements are now also carried out since they provide more detailed information (see, for example, [2.3]). Hence for some applications, one must prefer devices which provide facilities for angular measurements. Further, as outlined by HOBSON [2.1], surface analysis generally requires the conditions of pressure (about 10^{-9}-10^{-11} Torr) and cleanliness of ultrahigh vacuum. Thus the electron spectrometers or the techniques of energy analysis must also be compatible with the UHV conditions.

2.2 General Principles and Characteristics of Electron Spectrometers

2.2.1 Basic Principles of Energy Analysis

Fundamentally, there are at least four basic principles which have been used in experimental physics to achieve energy analysis of charged particles. Listed in order of increasing importance, they are: a) The use of a collision phenomenon which is sharply resonant as the electron attachment to a molecule [2.4] or the resonant formation of a temporary negative ion [2.5]. b) The measurement of the time taken by a particle to travel through a given distance between two points, i.e., the time-of-flight analysis [2.6]. c) The use of a retarding potential in front of a collector so that only particles having energy higher than this barrier can be collected [2.7]. This is the principle of a "high-pass filter". d) The deflection of the particle beam in a magnetic [2.8] or electrostatic [2.9] field, or even crossed fields [2.10], which disperses them in energy, so that a narrow energy band can be filtered by a slit ("band-pass filter").

Though elegant, since it is based on physical phenomena, principle a) has severe inherent limitations and is hardly compatible with surface analysis requirements since its application requires high pressures. Principle b) may be seen as compatible a priori but so far very few applications of it have been made in surface analysis since this principle is mainly efficient for very low energy electrons. However the two other principles (c and d) have been used for a long time in many research fields and are quite compatible with surface analysis. They are the basis of a great variety of models of energy analyzers.

2.2.2 Characteristics Related to Energy Resolution and Sensitivity

The energy resolution of analyzers is usually defined as the ratio $\Delta E/E_0$, where ΔE is the full width at half maximum (FWHM) of the energy distribution after the analysis of a monochromatic beam, and E_0 the energy at which the analyzer is tuned (i.e., the pass energy within the analyzer). The ratio $\Delta E/E_0$ expresses the resolution in its relative form and therefore an absolute value cannot be given except for a particular value of energy. The resolution $\Delta E/E_0$ is often expressed in percentage and sometimes with the reciprocal $\rho = E_0/\Delta E$ (called resolving power) In calculations, it is often easier to determine the base resolution ΔE_B, instead of ΔE, but it is generally assumed that for a well-designed analyzer one will have $\Delta E \simeq \Delta E_B/2$.

For the deflection analyzers which exhibit linear energy dispersion and at least first- order focusing, the base resolution may be expressed [2.11] as a function of the geometrical parameters in this general form:

$$\Delta E_B/E_0 = A\Delta S + B\alpha^n + C\beta^2 , \qquad (2.1)$$

where α and β are the semiangular apertures, respectively, in the plane of deflection and perpendicular to this plane. ΔS is the aperture or slit width for equal widths at the entrance and exit; for unequal widths, ΔS should be replaced by $(\Delta S_1 + \Delta S_2)/2$. This relation is obtained from expansions of the equations describing the electron trajectories, keeping only the low-order terms. The coefficients A, B, C and the exponent n are constants particular to the type of instruments; they are given in Table 2.2 for the most commonly used deflection energy analyzers. R_0 is the radius of curvature of the central trajectory (usually the mean radius of electrodes in electrostatic instruments), and ℓ_0 is the distance between the slits. According to RUDD [2.16], the same constants could be used to evaluate $\Delta E/E_0$ with this equation:

$$\Delta E/E_0 = \frac{1}{2} A\Delta S + \frac{1}{4} B\alpha^n + \frac{1}{4} C\beta^2 . \qquad (2.2)$$

But this is very approximate and the use of this relation is recommended only if dimensions are such that one has $B\alpha^n \simeq C\beta^2 \simeq A\Delta S/2$. More accurate determinations of the energy resolution require the calculation of energy distribution profiles from exact (or second order) trajectory equations [2.17]. Such a method is described in Subsection 2.3.2.

Table 2.2 Coefficients of the relation giving the base energy resolution of the most commonly used deflection energy analyzers

Analyzer	A	B	C	n
CDA - 127° [a]	$2/R_0$ [b]	4/3	1	2
SDA - 180° [a]	$1/R_0$ [b]	1	0	2
PMA - 45° [a]	$2/\ell_0$ [c]	2	1	2
PMA - 30° [d]	$3/\ell_0$ [c]	9.2	1	3
CMA - 42° [e]	$2.2/\ell_0$ [c]	5.55	0	3
MDA - 180° [a]	$2/R_0$ [b]	1	1	2
WFA [a]	$2\pi/\ell_0$ [c]	3	1	2

[a] Data from [2.11,12]
[b] R_0 is the radius of curvature of the central trajectory
[c] ℓ_0 is the distance between the slits
[d] Data from [2.13]
[e] Data from [2.14,15]

Other quantities have been defined in order to describe and characterize the sensitivity of the energy analyzer, namely: the transmission T, used by most authors as the entrance solid angle Ω relative to 4π (i.e., $T=\Omega/4\pi$); the *etendue* ε, defined by HEDDLE [2.18] as the product of entrance area S and entrance solid angle (i.e., $\varepsilon=S\Omega$); the luminosity L, which is the transmission integrated over the entrance aperture (i.e., L=ST); the energy dispersion D, defined as the change in image position r of a trajectory for a unit fractional change in the energy of the particle (i.e., $D=E\partial r/\partial E$, usually $D=2/A$); the trace width TW, which is the spread of the image of a monoenergetic point source caused by the angular aberrations and related to the angular acceptance of the analyzer; and the effective resolution η, which is equivalent to the energy resolution $\Delta E/E_0$ expressed in terms of angular aberrations only. SAR-EL [2.14] uses this quantity as being 1.26 times the angular aberration terms in the expression of $\Delta E_B/E_0$, (2.1).

2.2.3 Figures of Merit

In order to characterize the overall electron optical quality of the energy analyzers and to allow comparisons between them, many authors have proposed "figures of merit". First PURCELL [2.19] used the "reduced dispersion", defined as the ratio of energy dispersion to trace width (D/TW). This criterion was re-used by HAFNER et al. [2.20] in its reciprocal form as the "reduced aberration". Since one type of energy analyzer (SMA) is free of angular aberration, SAR-EL [2.14] argued that the above criterion cannot be generalized and rather proposed the effective resolution η as a function of the luminosity $L/(OP)^2$ (normalized to the optical path OP of the particle). This criterion was based on the works of HAYWARD [2.21] in beta spectroscopy.

Arguing that criteria based on luminosity are rather for spectrometers with extended sources, AKSELA et al. [2.22] proposed the ratio transmission to resolution $T/(\Delta E/E_0)$. For the same type of applications, HEDDLE [2.18] compared the *etendue* of the analyzers. ROY and CARETTE [2.17] also compared the performances of most of the electrostatic spectrometers; they calculated the energy distributions transmitted by these devices (with normalized dimensions) and compared their resolution and transmitted intensity as well as the ratios of these quantities.

Another figure of merit was proposed by WANG [2.23]: that expression of the optimum transport current of charged particle as a function of the resolution and the geometrical parameters of the analyzers. In their work on improvement of the plane mirror analyzer, SCHMITZ and MEHLHORN [2.24] used the criterion base resolution as a function of transmission. Recently READ et al. [2.15] presented an extensive work on optimization of electron monochromators. They compared devices on the basis of their capabilities of offering the best compromise current-resolution. Finally GELLENDER and BAKER [2.25] determined the general expression of the maximum ideally attainable transmission in energy analyzers, as a function of their energy resolution. This too could probably be used for the comparison of energy analyzers. In Section 2.5, Table 2.3 summarizes the rankings of the energy analyzers according to these various criteria.

2.3 Design Principles of Electron Spectrometers

A schematic representation of the various components of an idealized instrument for surface analysis is presented in Fig.2.1. Each component plays an important role, but essentially the heart of electron spectrometer is the energy analyzer. This section gives the principles for its optimum design and operation.

However it is evident that the sensitivity of an electron spectrometer for the analysis of a surface also depends on the characteristics of the incident particle beam. For a given technique of analysis, the sensitivity can be enhanced if the incident beam is used in the optimum conditions with respect, for example, to the following characteristics: incidence energy and resolution, incidence direction and angular definition, intensity and density of the bombarding beam. The potential in the region of the sample and the position and orientation of the latter relative to the analyzer entrance also play an important role. These conditions, specific for each method of analysis, may be found in the other chapters of this book.

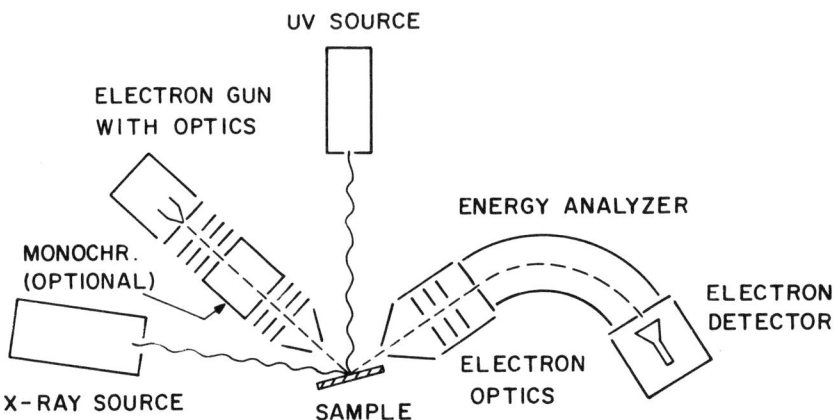

Fig.2.1 Schematic representation of the main components of an idealized instrument for surface analysis: three alternative excitation sources, the sample, an energy analyzer with electron optics, and the electron detector. The electron gun includes an optional electron monochromator

2.3.1 Electron Optics for Energy Analysis

The Use of Electron Optics

With energy analyzers other than LEED optics and analyzers of this type, a high enhancement of sensitivity and collection efficiency may be obtained by the use of properly designed electron optics. This makes easier the definition of the acceptance angle and of the viewed area, and also the achievement of angular measurements. The analyzer being more distant, the vicinity of the sample is less crowded and more easily shielded against stray electric fields. The optical properties and practical realization of electron lenses were reviewed by MULVEY and WALLINGTON [2.26]. Useful data on electrostatic lenses may be found in [2.27,28].

Pre-Retardation for Energy Analysis

The usefulness of electron lenses may even be increased if they are combined with the use of the "pre-retardation" technique, which was shown to enhance significantly the sensitivity of electron spectrometers. Actually the principle of pre-retardation is based on an essential property of deflection spectrometers: according to (2.1) the absolute energy resolution ΔE is proportional to E_0, the tuning energy of the analyzer; thus the absolute resolution of an analyzer may be increased if the electrons are decelerated and analysis achieved at lower energy. This property has been exploited for a long time for energy analysis [2.29,30], and later for electron monochromatization [2.31-33].

A new aspect of this property was presented by HELMER and WEICHERT [2.34]. They showed that when electron of energy E are decelerated to E_0 for the analysis, the luminosity of the spectrometer may be improved by a factor $(E/E_0)^2$ if the aperture dimensions and acceptance angles are increased so as to keep the same absolute resolution ΔE. This compensates the loss of brightness and the transmitted current is then increased by a factor E/E_0. HEDDLE [2.18] later showed that in a general manner if the *etendue* of a given spectrometer is proportional to $(\Delta E/E_0)^n$, the pre-retardation improves this figure by a factor $(E/E_0)^{n-1}$. The advantages of the pre-retardation principle are exploited in many instruments, and extensive discussions are available on its effects on the transmission of spectrometers [2.35-39].

Sweeping Modes for Energy Analysis

It is evident that the principle of pre-retardation is really efficient if it is used with properly designed electron optics whose function is to keep the electron focused during their deceleration [2.36]. However the properties of these lenses must be suitable for the sweeping mode used to achieve the energy analysis. Two main sweeping modes must be distinguished [2.16,36,40]: Mode I, the constant relative-resolution mode ($\Delta E/E_0$=const, or constant-deceleration mode) in which the electrons to be analyzed always enter the analyzer with their own energy, or an energy decreased by a constant amount of pre-retardation, while the pass energy E_0 of the analyzer is swept from one end of the spectrum to the other; and Mode II, the constant-absolute-resolution mode (ΔE=const, or constant-energy mode) in which the electrons to be analyzed are decelerated by a variable amount so as to adapt their energy to the fixed pass energy E_0. Schemes of electrical connections are given by RUDD [2.16] for these two modes.

As pointed out by NÖLLER et al. [2.36], with Mode I it is easier to design proper electron optics since the transmission conditions between the sample and the analyzer are constant. But this method also has important drawbacks: since the absolute resolution ΔE is not constant, this is not convenient at all for quantitative spectrum analysis, and in the low energy region of an extended spectrum, the resolution becomes too high and hence the signal-to-noise ratio is spoiled.

Mode II offers better conditions for quantitative analysis, since the absolute resolution remains constant and optimum for the whole spectrum. However the difficulty lies in the requirement of electron optics properly designed so as to control the varying transmission conditions between the sample and the analyzer. This may be achieved by the use of zoom lenses whose function is to focus at a fixed position an object of fixed energy and position, in spite of the fact that the energy of the image is varying.

Usually the zoom lenses are effective for a rather limited energy range; however by following the basic works on the simple arrangements [2.41-44], it is possible to design multi-element arrangements (corresponding to many lenses in series) and to obtain properly extended energy range. KUREPA et al. [2.45] and NÖLLER et al. [2.36] have both proposed four-element tube lenses which are quite convenient for the use of pre-retardation with the analysis Mode II.

2.3.2 Optimization of Electron Monochromators and Analyzers

Optimization Principles of Energy Analyzers

As mentioned in Subsections 2.2.2 and 2.2.3 the properties and characteristics of electron spectrometers (or energy analyzers) are usually described and compared by means of electron optical characteristics, as aberrations, dispersion, luminosity, etc.,or their ratios. But additional information may be obtained by the determination of the transmission function (or energy distribution) of these devices [2.46-50]. The useful method proposed by DELAGE and CARETTE [2.49] has thus been used to determine the optimum conditions of electron spectrometers by the calculations of the energy distributions for various values of the parameters involved [2.51-55].

This method proceeds via the calculation of a great number of electron trajectories, each electron having a different set of initial conditions (position, direction and energy). The transmission function is determined by taking into account the filtering action of both the deflection electrodes and the exit aperture. Since the angular acceptance is then only limited by the position of the deflection electrodes, the angular aberrations strongly manifest their influences and the energy distributions exhibit "wings". The analysis of trajectories [2.52] reveals that, in spite of the symmetrical positions of the electrodes, they actually define an asymmetrical angular acceptance aperture and this seriously distorts the transmission function [2.17].

Further calculations showed that angular acceptance may be symmetrized, and the distribution profiles optimized, by changing internal components of the spectrometers, as the deflection angle or the positions of the slits or electrodes [2.52]. In general this optimization is achieved without losses in the transmitted intensity. On the other hand for devices in which the angular divergence of the electron beam is symmetrically defined before the entrance aperture (via the use of electron optics), it has been verified that the distribution profile is then optimum at the focusing angle, provided that the positions of the deflection electrodes are such that the initial divergence is not affected [2.56].

Nevertheless the shape of the energy distribution may be optimized by establishing a compromise between the contributions of the apertures and of the angular aberrations. By means of an analytical treatment, RUDD [2.16] has studied this pro-

blem; for the case of the electrostatic spectrometers with first-order focusing
(i.e., CDA, SDA, PMA), he established that for a high-resolution spectrometer, the
distribution profile is optimum if the angle terms in (2.1) are equal, and each
equal to one-half the aperture term:

$$B\alpha^n = C\beta^2 = \frac{1}{2} A\Delta S \quad . \tag{2.3}$$

For the case of a spectrum with sharp and narrow peaks to be analyzed at low resolution, instead of one-half, Rudd proposed one-third. By means of the method of calculation described above, we calculated a large number of energy distributions for various dimensions of the SDA-180° and the CMA-42°, and compared them with the characteristics of the energy distributions. The optimum conditions thus determined were in good agreement with those given by (2.3) and showed that this relation can be used as a practical guideline for the design of the electrostatic deflector energy analyzers.

On the other hand, WANG [2.23] also proposed relations for optimization of energy analyzers. Using the method of Lagrange multipliers, he determined the conditions for the optimum transport current, and he gave the expressions for the optimum values of α and ΔS as a function of $\Delta E/E_0$. According to his results, the angle α must be equal to 0.4 $(\Delta S/R_0)^{1/2}$ for the SDA, and 0.4 $(\Delta S/\ell_0)^{1/3}$ for the CMA. The coefficient 0.4 must be compared to 0.71 and 0.58 obtained from Rudd's relation (2.3) for the SDA and CMA, respectively. Since Wang's approach does not take into account the transmission function profile while this directly affects the spectra to be analyzed, we think that Rudd's relation is to be preferred.

In the case of the analyzers of the retarding-potential type, the conditions which determine their optimum performances and sensitivity will be discussed in Subsections 2.3.3 and 2.4.4; often the optimization of this type of analyzers is rather a question of detection method and data processing.

Design Principles of Electron Monochromators

In most of the applications, a thermionic cathode is used for the production of the bombarding electron beam. The beam thus produced exhibits Maxwell energy distribution whose FWHM is given in eV by [2.11]:

$$\Delta E_K = 2.54 \, k_B T \quad , \tag{2.4}$$

where k_B is the Boltzmann constant (11600^{-1} eV.K^{-1}) and T the cathode temperature (in K degrees). In practice, this means an energy resolution in the range 0.3-0.6 eV (additional energy spreads are possible [2.11]). While this is quite satisfactory for many applications, LEELS and TELS do need a far better energy resolution, thus requiring the use of an electron monochromator.

Actually the most current electron monochromators are electrostatic energy analyzers; hence the optimization principles presented in the preceeding subsection for analyzers are fully applicable to the design of monochromators. The additional aspect is that a compromise must be established between the current and the resolution, and therefore the low operation energy of the monochromator (which determines its resolution, see *Pre-Retardation for Energy Analysis*) must be matched with the space charge.

KUYATT and SIMPSON [2.57] proposed to achieve this compromise by using the properties of space-charge limited electron beams, as outlined by PIERCE [2.58]. Under these conditions (see Fig.2.2), the maximum electron current I_m (in µA) which may be focused in a cylindrical volume at an energy E_0 (in eV), with a half-angular divergence α, is given by:

$$I_m = 38.5 \, E_0^{3/2} \tan^2\alpha \quad . \tag{2.5}$$

Under the action of the mutual repulsion of electrons, the beam exhibits a finite minimum diameter which may be matched to the entrance aperture ΔS of the monochromator. This is achieved by using a collimator with a diameter $a = 2.35 \, \Delta S$ located at a distance z, so that the half-angular divergence of the beam is given by (Fig.2.2):

$$\tan \alpha = a/2z = 1.17 \, \Delta S/z \quad . \tag{2.6}$$

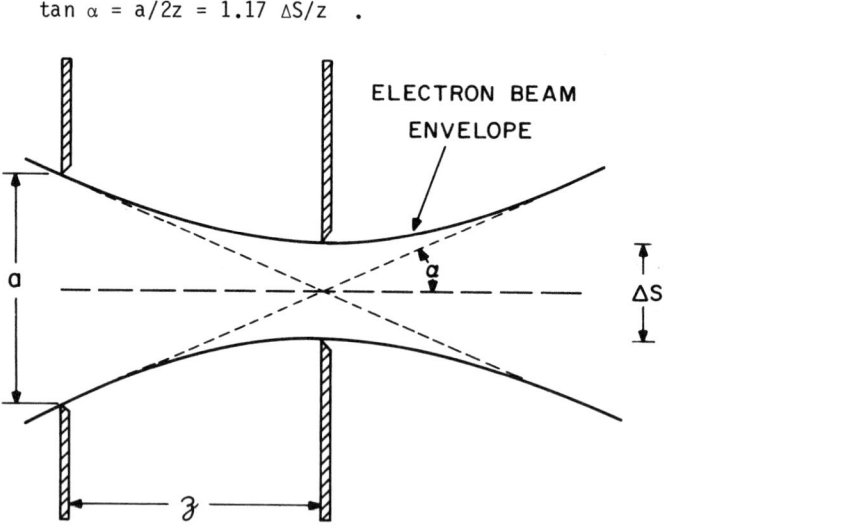

Fig.2.2 Schematic representation of the injection of a space-charge limited electron beam in a monochromator

It is assumed that proper electron optics is used between the electron source and the collimator in order to focus the beam at the entrance of the monochromator as illustrated in Fig.2.2. The relations (2.3,6) are thus the practical guidelines for the optimization of electron monochromators. They were used for the design of a high-resolution tandem cylindrical spectrometer [2.56] which is described in *Cylindrical Deflector Analyzer*. According to READ et al. [2.15], even though (2.5,6) were established for beams and apertures with circular symmetry, they are also valid for slit apertures; the current density has the same limitations, but the total amount of current transmitted through the slit may be larger.

In this study on optimum transport current in energy analyzers, WANG [2.23] extended his treatment to the conditions of space charge limitation. Since the transport current may be further increased by means of pre-retardation at low energy, the maximum current given by (2.5) then becomes the limit current. By replacing α in (2.5) by its optimum value (in terms of $\Delta E/E_0$), Wang was able to propose relations and graphs giving the optimum operation energy (E_0) and current, as a function of the energy resolution ΔE. His data may be used for the design of electron monochromators.

In their study on the optimization of electron monochromators READ et al. [2.15] analyzed the roles of all the various components of the selection system. Instead of (2.5), they used more general equations for the maximum current I_m, valid for the range $0.0035 \leq \Delta S/a \leq 1.0$. They gave the following relation for the determination of the usable current I_t which may be transmitted by the monochromator:

$$I_t = I_m(\Delta E/k_B T)(E_c/k_B T) \exp(-E_c/k_B T) [1 - \exp(-E_c/k_B T)]^{-1} \quad , \tag{2.7}$$

where E_c is the electron energy at the collimator (usually $E_c = E_0$). They presented a comprehensive optimization procedure which was applied to CMA and SDA. However this procedure is too complex to be presented as practical guideline easily applicable, in comparison with the simple principles described above.

When one desires to measure the energy distribution or transmission function of an electron monochromator (or any spectrometer), this may be achieved by using a second instrument, similar to the first, which acts as analyzer [2.31,33,48,50,59]. The observed energy resolution ΔE_{Obs} is then given by (the subscripts m and a refer to the monochromator and analyzer, respectively):

$$\Delta E_{Obs} = (\Delta E_m^2 + \Delta E_a^2)^{1/2} = \sqrt{2}\ \Delta E \quad , \tag{2.8}$$

assuming gaussian shapes and equal widths ΔE for both distributions. However according to calculations simulating such an analysis [2.40,49], this measurement is valid provided that the angular aperture of the instruments is symmetric and not larger than about $10°$.

Fringing Field Shielding

One of the important factors which may disturb the ideal operation of electron spectrometers is the fringing fields near the ends of the electrodes and in the region of the apertures. In the case of mirror type instruments, the problem was investigated by many authors [2.60-62], showing that these effects could be minimized. In practice, since the beam is injected through an equipotential, grids are used to minimize distortions when the apertures are large. At the ends of deflection electrodes, a few guard electrodes are sufficient to retrieve ideal conditions [2.62]; some experimenters use end plates coated with semiconductor films (see for example [2.63]).

The case of deflector type (or prism type) analyzers is quite different since the electron beam is injected through an equipotential plate which lies normal to the natural equipotential lines of the condenser. BRYCE et al. [2.64] recently showed that this disturbs the electron trajectories in this region (in the CDA-127°) and pointed out the importance of minimzing these effects; for the same purpose BALLU [2.48] proposed to design apertures with a conical shape.

HERZOG [2.65] proposed devices for the correction of fringing field effects, which WOLLNIK and EWALD [2.66] generalized and presented in practical forms. These Herzog corrections are now currently used in SDA [2.9,33,67]. Their application is illustrated in Fig.2.3

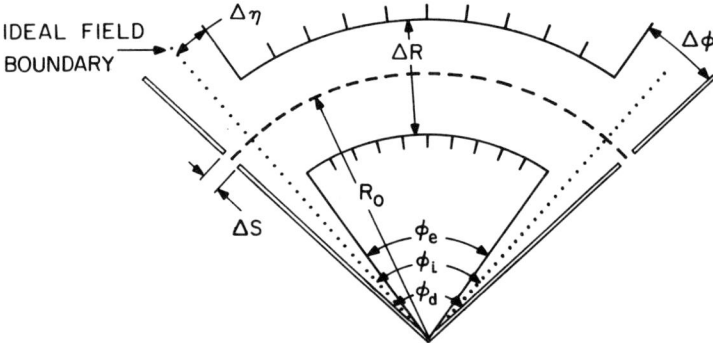

Fig.2.3 Arrangement proposed by WOLLNIK and EWALD [2.66] for the application of Herzog corrections. The aperture plates act as thin shielding diaphragms. The dotted line is the ideal field boundary, and the dashed line the main path of the beam

If a shielding diaphragm (which may be the entrance or exit aperture plate) is properly placed at a distance $R_0 \Delta\phi$ from the ends of the electrodes, it may be assumed in first order approximation that the deflection field stays homogeneous as far as an "ideal field boundary", at a distance $R_0 \Delta\eta$ from the electrodes along the electron path, beyond which it becomes negligible. The potential of the diaphragm must be that of the mean equipotential. The values of $\Delta\eta$ and $\Delta\phi$ are functions of ΔS and ΔR (the electrode separation), and may be determined by means of the nomograph presented in Fig.2.4, reproduced from [2.66]. These values are valid for a "thin" diaphragm, but values are also available for a thick diaphragm [2.66].

Since initially Herzog calculated these corrections by assuming a parallel-plate condenser with a uniform field, Wollnik and Ewald suggested to place the diaphragms oblique as in Fig.2.3 rather than parallel to the ends of the electrodes in order to keep the radial inhomogeneity of the field also in the fringing field. In this arrangement, the angle ϕ_d between the two diaphragms is:

$$\phi_d = \phi_e + 2\Delta\phi \qquad (2.9)$$

where ϕ_e is the sector angle of the electrodes. Actually it is believed that this arrangement is appropriate mainly when the sector deflector is used with finite object and image distances, and with wide apertures.

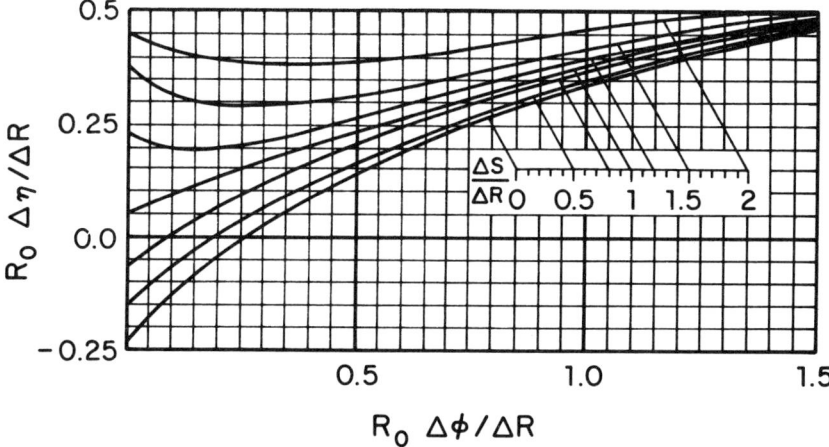

Fig.2.4 Nomograph giving the values of $\Delta\eta$ and $\Delta\phi$ as a function of ΔS and ΔR, for the achievement of fringing field shielding by means of a thin diaphragm, according to Herzog corrections (data taken from [2.66])

The angle ϕ_i of the ideal field (Fig.2.3) is then the actual deflection angle of the spectrometer. When the object and image distances are zero (as in the CDA-127° and SDA-180°), the width ΔS of the apertures is usually much smaller than the electrode separation (see Fig.2.3), and the diaphragms may then as well be placed parallel to the ends of the electrodes. However, with any of these two arrangements (diaphragms parallel or oblique), when the object and image distances are zero, the effective deflection angle of the spectrometer is actually the angle ϕ_d between the two apertures. This was shown by calculations of the energy distributions [2.52] simulating these conditions and taking into account additional second-order corrections (given in [2.66]) to be brought to the trajectories at the ideal field boundary. This means that for the 180°-spherical and 127°-cylindrical spectrometers, the angle of the electrodes should be (in degrees):

$$\phi_e(°) = (180° \text{ or } 127.3°) - 2\Delta\phi(°) \tag{2.10}$$

According to the work of WOLLNIK and EWALD [2.66], the Herzog corrections are also applicable in magnetic sector spectrometers.

2.3.3 Detection Methods and Data Processing

The sensitivity of electron spectrometers may also be enhanced at the stages of detection and data processing. The technique of pulse counting (by detection of individual electrons) with repetititve energy sweeps is now used in many fields of electron spectroscopy. Its advantages lie in the fact that, when the experiment does not depend on time-dependent phenomena, ultrahigh resolution measurements are achievable at low counting rate, and very high signal-to-noise ratios are accessible at normal counting rate (for N accumulated counts, the noise is $2N^{1/2}$). In electron impact ionization in gases for example, very small features may be detected with S/N ratio of about 10^5 [2.69]. Many types of small electron multipliers are available and are quite compatible with the requirements of surface analysis. When the analyzer has a well-defined focal surface, even multidetector arrays consisting in multichannel electron multipliers may be used to further increase the rate of data acquisition [2.35,70]. On the other hand the digitalization of data offers further advantages for their recording and processing [2.71] so that even measurements performed at high current (as with the retarding-potential type analyzers) should be digitalized by means of an analog-to-digital converter in order to make possible not only the improvement of S/N ratio by repetitive sweeps and data accumulation, but also the processing of these data by numerical methods such as those mentioned below.

In AES and IS, since the features are small and superimposed on the large background of secondary electrons, it is usual to enhance the detection sensitivity by modulating the analyzing energy and synchronously detecting the output current [2.35]. The first derivative of the collected electron current I(E) is then directly detected. With a band-pass analyzer (as the CMA), such a detection scheme allows the observation of only the interesting features since the slow varying background is eliminated through the derivative. However in the case of a high-pass analyzer (as the RPA), the collector is given by:

$$I(E) = \int_E^{E_P} N(E')dE' \tag{2.11}$$

where E is the energy corresponding to the retarding potential, E_p the primary energy, and N(E) the energy distribution under investigation. Thus the detection at the fundamental frequency ω only gives N(E), the energy distribution of secondary electrons, and the lock-in amplifier must be tuned at 2ω in order to obtain its derivative N'(E). The properties and consequences of these modulation techniques have been investigated by many authors [2.72-79].

However the fact that the data are differentiated has inherent drawbacks; it makes it difficult to localize the actual energy of the levels involed, to distinguish overlapping features, and to determine the actual peak amplitudes or areas (as required for quantitative analysis). Moreover the use of such technique usually leads to a loss in the S/N ratio. That is why we think that for many applications it would be preferable to digitalize the measured signal and further process it by means of a proper numerical treatment (see for example [2.80-83]). The method of "straightening through smoothings" proposed by BOLDUC et al. [2.69], and analyzed by CARBONNEAU et al. [2.84] and DE CELLES [2.85], is very efficient for background removal.

For some measurements and analysis the background subtraction may be insufficient or inadequate since the energy resolution always remains a limitation. The problem of data deconvolution from the instrumental resolution function has often been treated [2.86,87], but the process is not currently used: it has the reputation of leading often to nonphysical results since the solutions can be shown to be nonunique. However these methods are in progress and they are applied in the fields of AES [2.81], LEED [2.88], UPS [2.89], and XPS [2.90].

2.3.4 Further Practical Considerations

The ideal operation of an electron spectrometer depends in a large part on the shielding of stray magnetic fields. Even though this sensitivity to magnetic fields may be more or less important for the various types of spectrometers, a general relation may be proposed [2.91] for the evaluation of the required degree of annulment. For an electron beam of energy E (in eV) and path length PL (in cm), the maximum field H_m to be tolerated in given in gauss by:

$$H_m = 6.74 \ E^{1/2} \ d/(PL)^2 \ , \tag{2.12}$$

where d is the allowed maximum beam deviation (in cm). For a CDA having a mean radius of 3cm and a tolerable beam deviation of 0.05mm, the maximum field would be about 1 mG at 1 eV and 10 mG at 100 eV.

The shielding of magnetic fields may be achieved by means of two types of devices: either by placing the apparatus inside a system of Helmholtz coils, or by enclosing it in a high-permeability magnetic shield (mumetal). RUDD [2.92] presented a survey of the shielding properties of Helmholtz coils and gave the design principles. The equivalent information about mumetal shielding may be found in the article of WADEY [2.93].

On the other hand it is recommended to enclose each part of an electron spectrometer in boxes acting as electrostatic shields against stray electrons and electric fields. The electrodes must be designed so that the electron beam cannot see any insulators which accumulate charges and create strong perturbing fields. On the other hand, in addition to the care required by UHV, further precautions are needed in the cleaning and handling of electrodes; residual organic impurities form insulating films under electron bombardment [2.94] and this spoils the operation of electron spectrometers. The vacuum heating at high temperature of all the pieces of the spectrometer is a good way to achieve complete cleaning. Further efficient cleaning procedures for various materials may be found in the work of ROSEBURY [2.95]. Alumina is very good (as beads or hollow cylinders) for the insulation in UHV. For the screws, it is recommended to drill them through the center, and to slit down the threads.

Equally important is the choice of metal for the construction of the electron spectrometers [2.96]. The perturbing effects which are to be minimized are: secondary emission, surface potential variations, surface properties affected by gas adsorption or baking, residual magnetic fields near the edges and in the apertures, etc. So far many types of materials (gold, Advance, stainless, steel, copper) or coatings (platinum black, soot, graphite, electron velvet) have been used, and it appears that molybdenum is probably the most desirable surface, even though this solution is not the least expensive nor the easiest. However there is a lack of objective data on the reflectivity and potential uniformity of these surfaces for low-energy electrons. Some measurements were reported by MARMET and KERWIN [2.31] but these should now be carried out in UHV conditions for a more extended choice of materials and electron energies. Such data would be extremely valuable particularly for the construction of electron monochromators in which very low-energy electrons must be handled to achieve high resolution. In these devices, the electron reflection is usually minimized by using gridded electrodes [2.31], "open" electrodes [2.97], or corrugated electrodes [2.98]. But it is not clear which solution is the best since all have drawbacks and limitations. It is not superfluous to mention here that, when one wishes to use grids, the same material should be used for their supports if severe heat treatments are expected.

Finally a very accurate alignment of the various components of an electron spectrometer is a must for optimum performance. Hence when designing such a device, it is recommended to foresee and achieve further adjustability of the critical components with respect to the others, unless a very high precision machining is guaranteed for the whole system. A small laser source is a convenient tool to achieve the ultimate alignment of the apparatus.

2.4 Description of Electron Spectrometers and the Methods of Energy Analysis for Surface Studies

In this section, we present a review of the electron spectrometers and the methods of energy analysis used in the field of surface analysis. Other review have already been published from a more general point of view by KLEMPERER [2.99], RUDD and SEVIER [2.16], and more recently by STECKELMACHER [2.13]. The electron monochromators were surveyed by SIMPSON [2.11] and by KERWIN et al. [2.100], while the electron spectrometers used in XPS and UPS were reviewed by WANNBERG et al. [2.70].

2.4.1 Electrostatic Deflection Spectrometers

Cylindrical Deflector Analyzer

The basic theory describing the properties of the 127° cylindrical deflector analyzer was presented by HUGHES and ROJANSKY [2.101], and a prototype was experimented by HUGHES and McMILLEN [2.102]. The use of this instrument as electron monochromator was first reported by CLARKE [2.103], but its resolution was limited to about 0.3 eV. MARMET and KERWIN [2.31] then improved its performances by using gridded electrodes with collector plates behind to reduce electron reflections and space charge. This allowed low-energy operation and energy resolution better than 0.1 eV. This improved design led to a breakthrough in the field of electron impact physics [2.100] and, although minor modifications have been proposed, the original design is still used [2.69].

The basic calibration equation which relates the pass energy E_0 to the potential difference ΔV between the electrodes in the CDA is:

$$E_0 = e\Delta V / 2\ln(R_2/R_1) \; , \tag{2.13}$$

where R_2 and R_1 are the radii of the outer and inner electrodes respectively. The base resolution is given by (2.1). Numerous theoretical and experimental studies were carried on this instrument in relations with space charge [2.104,105], calibration [2.48,106], and energy distributions [2.17,48,52,107]. The comparison of the CDA-127° with the other types of analyzers, according to the usual criteria (see Subsec. 2.2.3), does not favor it mainly because it does not achieve space focusing. However it offers many practical advantages: the production or acceptance of a beam with rectangular cross section is quite suitable for some applications; moreover its simple geometry makes its construction easier and its alignment less critical, and allows the use of gridded electrodes contrary to other designs.

In surface analysis, HARROWER [2.108] and HARRIS [2.109] were probably the first to use the 127° cylindrical deflector for the analysis of Auger and secondary electrons. It is also used in other fields of surface analysis [2.110]. In Fig.2.5, we present a scale drawing of a typical tandem cylindrical deflector spectrometer as currently used in LEELS. Such a spectrometer was used for the observation of vibrational states of adsorbed molecules, surface phonons, interband transitions, or plasmon losses by PROPST and PIPER [2.111], IBACH [2.112], ADNOT et al. [2.113], BALLU and LECANTE [2.114], and FROITZHEIM and IBACH [2.68]. This type of instrument is also currently used in electron scattering spectroscopy in gases [2.56,59].

This tandem spectrometer (Fig.2.5) [2.56,115] involves an electron gun, an electron monochromator, electron optics, a collision chamber with a device for heating the sample, a rotatable energy analyzer, and an electron multiplier. For the monochromator and analyzer themselves the fundamental design proposed by MARMET and KERWIN [2.31] with gridded electrodes and low operating energy was kept. But special attention was given to achieve its optimization according to the principles outlined in Section 2.3.

The electron source is a thoria-coated tungsten filament, folded in hairpin shape. The piece 3 acts as a repeller to favor forward emission. A three-element electrostatic lens (7,8,9) [2.44], incorporating deflection electrodes (5), is used to achieve the focusing into the monochromator entrance (11). According to the device shown in Fig.2.2, the collimator (10) defines the angular divergence of the beam at $\pm 5°$ and yields the injection of the maximum space charge (*Design Principles of Electron Monochromators*). The monochromator slit widths are 0.254 mm, and the electrode separation is 6.0 mm with a mean radius of 30.0 mm. The angle between the two slit plates is 127.3°, while the angle of the electrodes is 121.6°, according to energy distribution calculations and Herzog corrections (*Fringing Field Shielding*). The purpose of the plates (13) behind the grids is the collection of stray electrons. Since the grid transparency is very high (90%), the attractive potential applied to the plates must be low in order to avoid the distortion of the deflection field between the grids. Alternative solutions have been proposed in order to avoid this problem [2.97,98] (see Subsec.2.3.4).

The electron optics are three-element zoom lenses (18,19,20 and 28,29,30) designed according to the theory given by READ [2.44]. They include deflection electrodes (17,31) for an optimum alignment, and collimators (22,26) for a proper definition of the investigated area. The sample region is surrounded by a hemi-cylindrical wall (23) properly slotted at -45° relative to the normal for the incident beam and from 35° to 90° for the scattered beam. The energy analyzer is identical to the monochromator; its position relative to the sample may be varied in order to achieve angular measurements. At the exit of the analyzer, a grid (37) and a plate with a conical aperture (38) are used to prevent field penetration within and at exit of the analyzer. The grids and the electrodes with narrow slits are all made of molybdenum, while the others are in gold-plated copper.

The detector is an electron multiplier of the continuous dynode type, preferred for its small size and high gain. The pulses are led to a high voltage vacuum feedthrough by a short shielded cable with low capacitance. Then they are injected in

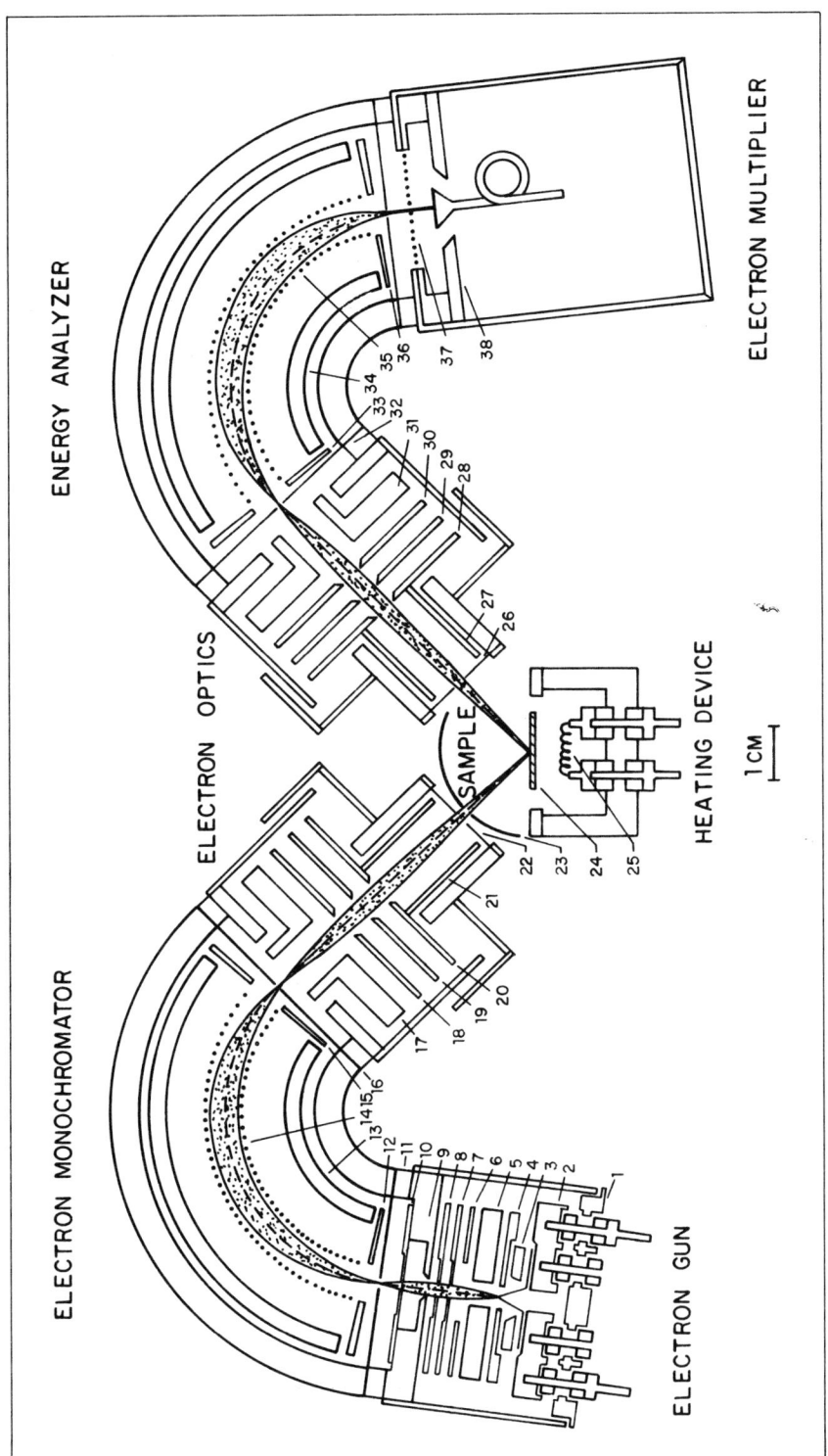

Fig.2.5 Scale drawing of a tandem cylindrical deflector spectrometer used by ADNOT and CARETTE [2.115] for LEELS; the entrance and exit slits of the monochromator and the analyzer (0.254mm) are not to scale

a preamplifier, amplified, discriminated, and shaped. The data are accumulated in a minicomputer operated as a multichannel analyzer, in the multiscale mode. The channel address is synchronized with the sweep voltage of the spectrometer. This data processor is programmed to carry out many operations such as usual arithmetical operations, integration, differentiation, slope subtraction, and curve straightening (Subsec.2.3.3).

All the components of the spectrometer are surrounded by walls or boxes acting as electrostatic shields against stray field and electrons. The elimination of magnetic fields required for a proper operation of the instrument is achieved by means of mumetal shields. The UHV system is capable of an ultimate vacuum of 2×10^{-11} Torr with the beam on.

According to (2.2), the theoretical energy resolution of each of the two components of the electron spectrometer is about 0.01. With the monochromator operated at about 2 eV and the analyzer at a lower value, current measurements are carried out with a resolution of 20 meV with a usable electron current of about 5×10^{-9} A. Further details on this spectrometer may be found in [2.56] where was described the original version of this instrument developed for electron scattering in gases. A few modifications might be suggested for the further improvement of the versatility of this instrument: a rotatable sample manipulator, the increase of the angular path of the analyzer, additional elements to the lenses in order to extend their optimum application range, a more generalized use of molybdenum. FROITZHEIM and IBACH also described a high-resolution spectrometer with promising characteristics [2.68]. Many of its features are identical to that described above; but they preferred the use of copper with graphite coating to minimize electron reflections. In a subsequent work [2.98] they proposed the use of corrugated electrodes.

Spherical Deflector Analyzer

The focusing and energy dispersing properties of the spherical deflector where first investigated by PURCELL [2.19]. Many comparisons of the SDA with other types of analyzers, according to various criteria [2.14,15,17,18,20,23,24], showed its powerful properties. READ et al. [2.15] even claimed that it offers the best compromise current-resolution as monochromator, while HEDDLE [2.18] showed that its *etendue* becomes superior to that of the CMA when it is used with pre-retardation.

The basic calibration equation for the SDA is:

$$E_0 = e\Delta V / [(R_2/R_1) - (R_1/R_2)] \quad . \tag{2.14}$$

The base resolution is given by (2.1). Further studies on this instrument are available on the space charge effects [2.116], the usable current [2.23,104], the energy distributions [2.17,52], and the effects of pre-retardation [2.35-38].

SIMPSON [2.33] was the first to exploit the properties of the spherical deflector for the development of a low-energy electron spectrometer, including a high-resolution monochromator and an analyzer in tandem, and KUYATT contributed to the improvement of his design [2.57]. The main characteristics of the Simpson-Kuyatt design is that it includes a full system of tube lenses, expressly developed for this purpose, which defines virtual slits in the condensers and minimizes the drawbacks of electron reflection and secondary production. This design became very popular and, like the version of the CDA proposed by Marmet, it was adopted by many laboratories [2.117]. In the field of surface analysis, only POWELL and MANDL [2.118] used this design for LEELS and AES measurements, while ARMSTRONG [2.119] joined two condensers of this type in series to make an electron monochromator for LEED measurements.

The spherical condenser is also used as analyzer for surface studies. The Uppsala group [2.35,70] designed a SDA-180° with a special feature: the dispersion compensation scheme. The purpose of this scheme is the exploitation of the full width of the dispersed X-ray line; by an electron lens with variable magnification, the dispersion of the X-ray radiation can be matched to the dispersion of the spectrometer in such a way that they exactly cancel each other.

Good electron optics suitable for pre-retardation and analysis with constant absolute resolution in conjunction with a SDA may be proposed. NÖLLER et al. [2.36] described a setup featuring constant magnification for all energies and retarding factors. It involves two einzel-tube lenses with an homogeneous retarding field between. However this electron optics involves at least the use of three grids, which may affect the sensitivity of this instrument in some conditions. The four-tube zoom lens system proposed by KUREPA et al. [2.45] is also quite suitable for this type of application. Both the image position and magnification are constant for a wide range of energy sweep, without the use of any grid. The voltages to be applied to electrodes during the sweep are mutually linear functions so that no special analog circuits are needed to run the experiment. Thus this lens system is also highly recommended for coupling with electrostatic spectrometers.

The spherical condenser as a sector analyzer is also widely used, and actually it is probable that the first measurements of Auger peaks in spectra of secondary electrons were carried out by LANDER [2.120] with a SDA-90°. The Uppsala group has developed a new apparatus for XPS incorporating a SDA with a deflection angle of 158° and a mean radius of 36 cm [2.9,35,70,121]. This analyzer, capable of a resolution of 5×10^{-4}, is coupled with the X-ray monochromatization scheme which is characterized by the fine focusing and the rotating anode; with this device the X-rays which reach the sample are monochromatic to better than 0.2 eV.

A group at La Trobe University, which constructed two distinct photoelectron spectrometers including a SDA-90°, proposed a useful device to carry out angular measurements without displacing either the sample, or the analyzer [2.122]. A cross section of this device is illustrated in Fig.2.6. The 90° spherical analyzer has a full 360° symmetry around the axis. Since the sample surface makes a 45° angle relative to the axis of the spectrometer, one may see that in the absence of the conical electrode S_0 the analyzer can accept, in the annular aperture S_1, electrons emitted in the range 0-90° with respect to the normal to the surface. If a conical electrode (S_0) with a hole defining a small angular aperture (about 6°) is inserted between S_1 and the sample, it is possible to restrict the electrons entering the analyzer to a small angular range. By the continuous rotation of the electrode S_0 around the symmetry axis, the angular distribution may be measured from 0 to 90°. The conical electrode S_2 is designed to correct the fringing fields according to Herzog specifications (*Fringing Field Shielding*). The interest of this simple device for angular measurements is that it may be used in the spectrometers which have axial symmetry and achieve collection following a cone with apex on the sample; this is thus applicable in the cylindrical mirror and in the plane mirror 30° (fountain-type).

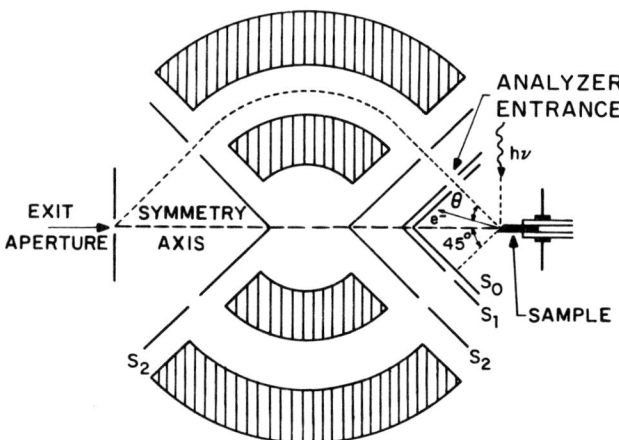

Fig.2.6 Schematic axial section of the device proposed by POOLE et al. [2.122] for angular measurements with 90° spherical analyzer with full 360° symmetry

In the field of FES a few years ago only retarding potential analyzers were used [2.123]. Some of them achieved satisfactory resolution but their sensitivity was seriously limited in the low-energy portion of the distribution by the noise carried by the high-energy electrons. KUYATT and PLUMMER [2.124] then applied the design principles of deflector analyzers to the construction of a SDA capable of overcoming these limitations and compatible with the rigorous UHV requirements. The whole spectrometer is illustrated in Fig.2.7. A deflection angle of 135° was chosen in order to obtain adequate spacing between the detector and the emitter.

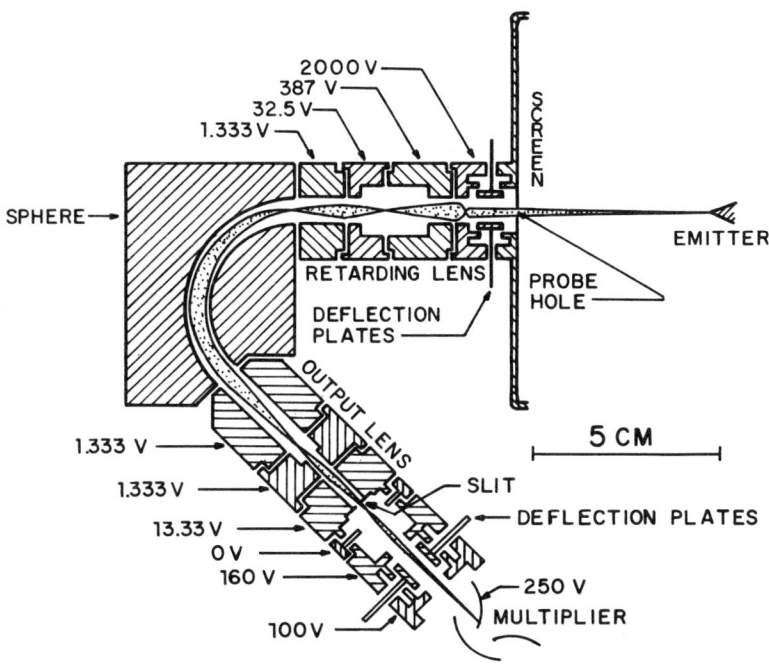

Fig.2.7 Adaptation of the 135° spherical deflector analyzer to a field emission source, achieved by KUYATT and PLUMMER [2.124]. The potentials on the lenses are the design voltages for a 2000 V emitter and 0.020 eV resolution

The basic requirements were a resolution of about 0.020 eV for electrons emitted at 2000 eV. Thus the principle of analysis at low energy following pre-retardation was kept, and this allowed small dimensions to be retained. The mean radius is 2.54 cm; the semi-angular divergence of the incident beam is about 5°, while the effective aperture widths are 0.38 mm at the entrance and 0.46 mm at the exit. With an analyzing energy of 1.33 eV the 0.020 eV resolution is achieved. At the output, as in the previous designs [2.33,57], a lens is first used to image the beam onto the actual exit aperture whose width is 0.31 mm; another one then focuses it onto the first dynode of an electron multiplier. For the deceleration from 2000 to 1.33 eV at the entrance, a special three-stage lens was expressly designed by means of a least-squares computer program to adjust the various parameters involved. This decelerating lens allows the analyzer to be used over a wide range of electron energies and resolutions. The detailed drawings of these lenses are given in [2.124] and are even available from these authors. The chosen materials were mainly copper while molybdenum was preferred for the small apertures.

Plane Mirror Analyzer

The plane mirror analyzer is the simplest device for the analysis of charged particle by deflection in an electrostatic field. Its analyzing properties are not the most powerful [2.17,24] but it offers the greatest simplicity of construction since its uniform field only requires parallel plates. A schematic diagram given in Fig. 2.8 presents the PMA in its most general form. An electron source located at ΔS_1, corresponding to a beam with an energy E_0, an incidence angle θ_0, and a divergence 2α, is imaged at ΔS_2. The object and the image are separated by a distance ℓ_0 given by:

$$\ell_0 = (h_1 + h_2) \cotan\theta_0 + (2E_0 d/e\Delta V) \sin 2\theta_0 \quad , \tag{2.15}$$

where d is the plate separation, and h_1 and h_2 the position of the entrance and exit slits, respectively, relative to the lower plate of the condenser. The condition of first-order focusing for any θ_0 applies on h_1 and h_2, and is given by this relation [2.125]:

$$h_1 + h_2 = (4E_0 d/e\Delta V) \cos 2\theta_0 \sin^2\theta_0 \quad . \tag{2.16}$$

For a given geometry, the relation between the transmitted energy E_0 and the applied difference of potential ΔV is:

$$E_0 = e\Delta V \ell_0 / (8d \sin\theta_0 \cos^3\theta_0) \quad . \tag{2.17}$$

The maximum penetration of the beam into the analyzer (Fig.2.8) is given by:

$$y_m = h_1 + (E_0 d/e\Delta V) \sin^2\theta \quad , \tag{2.18}$$

where $\theta = \theta_0 + \alpha$. The base resolution is given by (2.1). For $\theta = 45°$, one obtains $h_1 + h_2 = 0$; this corresponds to the well-known PMA-45° whose properties were first described by PIERCE [Ref. 2.58, Chap. 3] and in a more comprehensive manner by HARROWER [2.126]. But GREEN and PROCA [2.125] later showed that for $\theta_0 = 30°$, the focusing is of second-order.

Since to date, few uses of this energy analyzer in the field of surface analysis were reported [2.127,128], the properties of this device will not be described in more details. Further information may be found in the following references [2.17, 24,52,107,125,129,130].

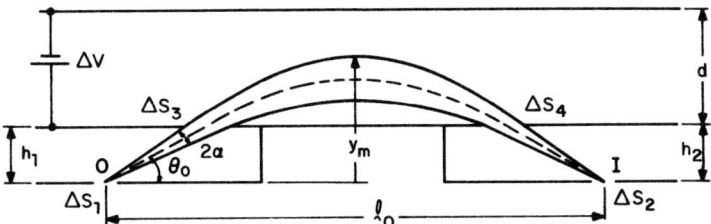

Fig.2.8 Cross section of a plane mirror analyzer in the more general configuration. For the special case of the PMA-45°, the incidence angle θ_0 is 45°, and the separations (h_1 and h_2) between the lower plate and the entrance and exit apertures are zero. For $\theta_0=30°$, the focusing is of second-order and a further improvement may be achieved by displacing the exit slit ΔS_2 toward the plane of minimum trace width (h_2 is then smaller than illustrated)

Cylindrical Mirror Analyzer

The energy analyzer which is probably the most currently used in the field of surface analysis is the cylindrical mirror analyzer. As shown in Fig.2.9, the cross section of this device is quite similar to that of the PMA presented in Fig.2.8. Here the deflection is caused by the potential differences ΔV between two cylinders of radii R_1 and R_2. There is a full 360° symmetry around the cylinder axis, which may be exploited for the improvement of the transmission and sensitivity. The positions of the entrance and exit apertures relative to the inner cylinder are noted d_1 and d_2, respectively. Following this general scheme, it is useful to define a parameter m, as:

$$m = (d_1 + d_2)/R_1 \,, \tag{2.19}$$

since the drift spaces within the inner cylinder are equivalent (usually m=2, sometimes 1 or 0). All the other symbols in Fig.2.9 have the same meaning as in the case of the PMA (Fig.2.8).

The separation ℓ_0 of the source O and its image I is given by [2.22]:

$$\ell_0 = (d_1 + d_2)\cotan\theta_0 + 2R_1(K_0\pi)^{1/2}\cos\theta_0 \exp(K_0 \sin^2\theta_0) \\ \text{erf}(K_0^{1/2} \sin\theta_0) \,, \tag{2.20}$$

where:
$$K_0 = (E_0/e\Delta V)\ln(R_2/R_1) \tag{2.21}$$

is related to the calibration equation, and erf(x) is the error function given by:

$$\text{erf}(x) = (2/\sqrt{\pi}) \int_0^x \exp(-t^2)dt \,. \tag{2.22}$$

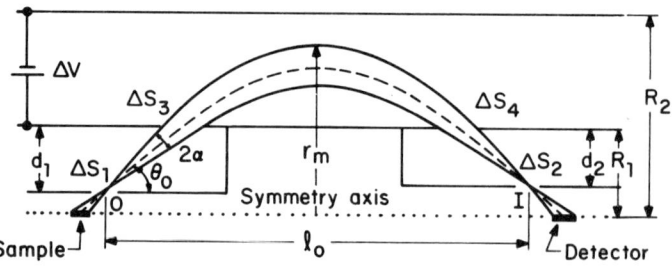

Fig.2.9 Cross section of a cylindrical mirror analyzer in its general geometry. A full 360° symmetry around the axis is possible. Usually both the source O and image I are on the axis ($d_1 = d_2 = R_1$) and $\theta_0 = 42.3°$ for second-order focusing. A further improvement is possible by locating the collector slit ΔS_2 at the minimum trace width (d_2 is then smaller than R_1)

The maximum penetration (or apogee) of the beam into the deflecting field is given by:

$$r_m = R_1 \exp(K_0 \sin^2\theta) \quad , \tag{2.23}$$

where $\theta = \theta_0 + \alpha$.

The first uses of a cylindrical mirror were reported by BLAUTH [2.131] and by MEHLHORN [2.132], respectively, for the analysis of secondary electrons and Auger electrons in gases (m=0, θ_0=54.5°). But the basic theory of this instrument was first given by ZASHKVARA et al. [2.133] and later by SAR-EL [2.134]. They showed a source located on the axis (d_1=R_1) may be imaged on this axis (d_2=R_1, m=2) with first-order focusing for any value of θ_0 between zero and 90°, with a proper choice of the value of K_0 (to each value of K_0 correspond to possible values of θ_0). In addition they found that for the special case of θ_0=42.3° and K_0=1.31 the focusing is actually of second order and the analyzing properties are then improved; some works show that in this geometry the CMA is superior to other analyzers [2.17,23]. For this case ℓ_0=6.13 R_1 and the base resolution may be evaluated by means of (2.1).

ZASHKVARA and RED'KIN [2.135] and, little later AKSELA et al. [2.22] re-investigated the properties of this device and found that, if the position of the source and image are generalized, as with (2.19), second-order focusing is even possible for various values of θ_0, from 30.1° to about 45° if K_0 and m are properly chosen. RISLEY [2.136] has computed and tabulated these values for easy access. For example, second-order focusing may be achieved for m=1 if θ_0=38.8°, K_0=0.94, and ℓ_0=3.6 R_1; this may correspond to a source located on the axis with annular focusing at the surface of the inner cylinder. This setup may be very useful in AES and SAES for example, as discussed by HARRIS [2.137], since the space within the inner cylinder is then unobstructed and fully usable.

In addition to second-order focusing, HAFNER et al. [2.20] showed that further improvement of the CMA (for $\theta_0 = 42.3°$ and m=2) is possible if the collector is displaced at the height of the minimum trace width which occurs before the focus. This improvement is such that the CMA then becomes superior to the other currently used analyzers, according to the criteria of reduced aberration [2.20], luminosity as a function of resolution [2.14], *etendue* [2.18], and transmission-to-resolution ratio [2.24]. SAR-EL [2.14] deduced relations to determine the height r_c above the axis (or above d_2 if $d_2 \neq R_1$) where the minimum trace width occurs, and its width Δz, as a function of α:

$$r_c = 11.66\ R_1\ \alpha^2\ \sin(\theta_0 + \alpha)\ \sin(\theta_0 - \alpha/2) \quad , \quad \text{and} \tag{2.24}$$

$$\Delta z = 3.88\ R_1\ \alpha^3 [3\ \sin(\theta_0 + \alpha) - \sin\theta_0]/\sin\theta_0 \quad . \tag{2.25}$$

The reduction factor of the trace width is about 3 to 4. BISHOP et al. [2.138] proposed more concise expressions valid for α smaller than 6°. Further data were also given by AKSELA [2.139] who stated that in fact the minimum trace width may be as well obtained at the position of second-order focus by properly increasing the value of K_0; a graph presented in [2.139] gives this modified K_0 as a function of α. However the fact that the minimum trace width lies above the axis may be advantageous for some applications since the detecting device may thus have a finite radius (see for example [2.140,141]).

The fringing field in the CMA was described by BOSI [2.60]. RENFRO and FISCHBECK [2.62] investigated both its effects on the focusing properties, and its shielding by means of guard rings. GERLACH [2.63] chose to use alumina end pieces coated by uniform semiconductor films. DIETZ et al. [2.142] used printed circuit plates with ten concentric conductive rings connected by resistors soldered onto the copper rings from the back of the substrate. But the latter solution should require some adaptation to be compatible with UHV requirements. Further mounting details for a CMA may be found in [2.141,143], while interesting data were presented in [2.144, 145] on the influence of the positioning of the object and image on the resolution, transmission and calibration of this device. Further data are available on the space charge effects [2.146], optimum transport current [2.20], and energy distributions [2.17,52,139,147]. READ et al. [2.15] studied its possibilities as an electron monochromator. A few experimenters have described setup by which they achieved angular measurements with the CMA in gases [2.148-150]. The device of Fig.2.6 could also be incorporated in a CMA for angular measurements.

Since the realizations of the first prototypes [2.134,151], a great number of well-designed instruments have been proposed, among which many were manufactured commercially. PALMBERG et al. [2.140] constructed a high transmission CMA for AES which exploited the 360° symmetry, the second-order focusing ($\theta_0=42.3°$) and the minimum trace width. This analyzer is very compact ($R_1=1.27$ cm) and offers good resolution (0.7% for a semi-angular aperture of 6°). As discussed by CHANG [2.152] and TAYLOR [2.153], the major improvement realized in this instrument for AES is its high sensitivity in comparison with the retarding potential analyzers. Since this CMA combines a high transmission and a narrow energy window, a greatly improved S/N ratio is attained. This even makes possible the display of the 0-1000 eV Auger spectrum on an oscilloscope at a scanning rate of 20 kV/s when using a primary beam current of 50 μA. If this current is reduced to 10^{-8} A, the scanning rate is that typical of other instruments (~ 2V/s) while the damage at the sample surface is minimized.

More recently a combined XPS-AES spectrometer based on a double-pass CMA was proposed by PALMBERG [2.154]. It involves two CMA's in series with spherical retarding grids and adjustable defining apertures at the second-order foci. For the Auger spectra, the usual operation mode (without retardation) is kept, and the required resolution is achieved by the use of narrow internal apertures. For the XPS measurements, the larger apertures are used and suitable pre-retardation is applied at the grid stage. For the Auger measurements, it is possible to compare the performances of this double-stage CMA to those of a single-stage one [2.140]. This reveals a loss in the S/N ratio of about 30%, caused by the loss of transmission in the supplementary grids of the double-stage device. We do not think that this loss in S/N ratio is very important, but multiple passages of electrons through gridded electrodes always remain a drawback in a high sensitivity spectrometer.

Other Models

Among the "ideal" instruments which were described in literature there is one which deserves special attention: the spherical mirror. SAR-EL [2.155] showed that a source located inside two concentric spherical shells is imaged on the same diameter without any spherical aberration (exact focusing). Such a property would allow the design of an analyzer with very high luminosity, but unfortunately this device does not exhibit linear energy dispersion. Sar-El suggested the placing of a suitable baffle at the mid-path (where a first-order focus with energy dispersion is) so that the properties of this mirror could be exploited. According to the criteria of optimum transport current [2.23] and luminosity as a function of resolution [2.14], the spherical mirror could thus be superior to the CMA. But so far no construction of this device has been reported because from a practical point of view the positions of sample and detector inside spherical shells must be considered as serious drawbacks. Furthermore the practical installation of a baffle within the radial field, designed so as to respect the ideal operation, is an important technical problem.

2.4.2 Magnetic Deflection Spectrometers

Many magnetic deflection analyzers capable of high performances were developed in the field of β spectroscopy. The properties of the various versions of these magnetic instruments were reviewed by HAYWARD [2.21] and SIEGBAHN [2.156]. When electron spectroscopy appeared, the first instruments used were magnetic devices and good spectrometers were expressly designed. For example, NORDBERG et al. [2.8] constructed an iron-free double-focusing magnetic spectrometer suitable for XPS measurements from samples in the solid, liquid and gaseous phases. FADLEY et al. [2.157] also presented a comprehensive study of the design of an instrument of this type for XPS. MARTON and SIMPSON [2.30] used an iron-core magnetic analyzer with pre-retardation for TELS measurements. A more comprehensive review is presented in the work of SEVIER [Ref. 2.16, pp. 6-17].

However, now that various types of electrostatic spectrometers have been developed, the magnetic instruments are much less used since the former are more easy to construct and in general more versatile. For example, the principle of pre-retardation is hardly applicable in magnetic analyzers, and the energy reading is not a linear function since they are essentially momentum analyzers. The high-resolution magnetic instruments, which are very sensitive to stray magnetic fields, are very difficult to shield since mumetal cannot be used. In addition, the requirement of bakeability of UHV systems makes more complex the construction and use of magnetic devices. Hence the magnetic deflection analyzers are not recommended for surface analysis and they are not systematically reviewed in the present work.

2.4.3 Crossed-Field Deflection Spectrometers

In crossed-field spectrometers, the energy dispersion occurs under the action of mutually perpendicular electric and magnetic fields. Two main types may be distinguished: the Wien filter, in which the electron motion is transverse to the direction of both fields, and the trochoidal analyzer, in which the motion is parallel to the magnetic field.

BOERSCH et al. [2.32] designed a high-resolution spectrometer with two Wien filters in tandem for TELS measurements. SCHRÖDER and GEIGER [2.10] reported a resolution as high as 4 meV in the observation of phonon energy losses in TELS with this instrument. Since this technique falls outside the scope of this book, this type of spectrometer will not be described in more detail.

So far the trochoidal analyzer has been used mainly as an electron monochromator for the study of gases [2.4,54], in conjunction with the method of incidence energy modulation [2.158] (IEEM, see Subsec. 2.4.5). As a backscattering spectrometer [2.55,159], this device proved to be useful for the detection of resonances in the scattering of electrons by gases. But it could be very efficient in surface analysis too, particularly for the observation of surface-state resonances in the elastic reflection from single crystal surfaces [2.160].

2.4.4 Retarding Potential Spectrometers

Classical Models of Retarding Potential Analyzers

As given in Subsection 2.2.1, the basic principle of a retarding potential analyzer is that of a high-pass filter: only particles having energy higher than the potential barrier applied in front of the collector can be detected. Actually the value of this potential barrier corresponds to the electron momentum in the direction perpendicular to the equipotential lines, rather than to its energy, and this limits the obtainable resolution [2.161]. As discussed in Subsection 2.3.3, the measured current is the integral of the energy distribution, and the derivative, i.e., N(E), may be directly obtained by potential modulation and synchronous detection (2.11).

SIMPSON [2.161] has presented a comprehensive study on the design of retarding potential analyzers. Four types were considered: plane parallel plate, spherical condenser, Faraday cage, and filter lens. For the simplest case, the RPA with parallel plates, it was shown that the limit resolution is:

$$\Delta E/E = \sin^2(r_0/4d) \quad , \tag{2.26}$$

where r_0 is the radius of the entrance aperture and d the separation of the retarding electrode from the collector aperture. The case of the RPA with concentric spheres was also extensively studied and it was shown that high resolution is possible with this device.

Many versions of standard RPA's are used in the various fields of electron spectroscopy for surface analysis. However the great simplicity of the standard models of RPA's is partially counterbalanced by many inherent drawbacks, mainly caused by space charge and secondary emission, which sometimes lead to spurious results. This was shown by WEI and KUPPERMAN [2.162] who studied the instrumental affects of a RPA in measurements of energy loss spectra in helium. Many of these problems are related to the use of gridded electrodes, as they showed in their second study on this subject [2.163]. This point is discussed in the following subsection. In addition DISTEFANO and PIERCE [2.164] studied the influence of a magnetic field on the analysis by means of spherically symmetric retarding field, as it is currently done in UPS. They found that this leads to shift and broadening of the features contained in the energy distribution curves. Nevertheless reliable results may be obtained with this type of analyzer if care is taken in order to minimize these effects, and if they are used only when both medium resolution and sensitivity are required.

In the field of TELS, the usual energy analyzers (other than the WFA) are lenses incorporating a retarding potential. This includes the filter lens analyzer, as that proposed by SIMPSON and MARTON [2.165] consisting of two short-focus immersion lenses symmetrically placed back to back about the retarding plane. There is also the MÖLLENSTEDT lens [2.166] which is a three-electrode symmetrical lens whose chromatic aberrations may be exploited for energy analysis if electrons are injected off-axis.

LEED Optics Used as Retarding Potential Analyzer

A schematic diagram of a typical LEED optics analyzer is presented in Fig.2.10 with the associated electronics. This is a four-grid LEED optics with a normal incidence electron gun. The inner and outer grids are both grounded respectively to provide a unipotential sample region and to shield the collector screen (mainly against capacitive effects). Other schemes are possible [2.152]. The retarding potential V_r is applied on both middle grids and a small modulated voltage ΔV_p is superimposed on it. As mentioned in Subsection 2.3.3, the tuning of the lock-in amplifier to the frequency of the modulating voltage gives the energy distribution $N(E)$, while the tuning to the double frequency provides its derivative $N'(E)$. Further improvement of the sensitivity of this device for AES measurements was achieved by PALMBERG [2.167] by incorporating a grazing incidence electron gun. Other optimizations were discussed by SKINNER and WILLIS [2.168]. The detection method is also discussed in Subsection 2.3.3.

There are many factors limiting the performances of the LEEDOA [2.152,153,164, 169], such as the residual magnetic field, improper alignment, grid warping, etc. TAYLOR [2.169] discussed them extensively and concluded that the resolution limitation comes mainly from field penetration and trajectory deviation at the grids. That is why the use of two instead of one retarding grids may improve the resolution from about 3% to 0.3%. WEI et al. [2.163] also investigated the effects of the retarding grids. They observed that the LEEDOA was efficient when the energy losses are small compared to the beam energy; at large energy losses, however, field penetration and scattering from grids contribute considerably to the measured energy distributions. These results were confirmed by CROSS [2.170]. This indicates that good performance with the LEEDOA cannot be obtained except with great care in its design, its construction and its operation.

It is usually believed that this type of analyzer has high sensitivity since its geometry allows a high transmission. As a matter of fact, the use of grid partially limits this characteristic and comparable transmission may be attained in the CMA with the full 360° symmetry [2.153]. However the main drawback arises from the basic fact that retarding potential analyzers allow all electrons of energy greater than the potential barrier to reach the collector. The shot noise thus introduced seriously limits the sensitivity and the S/N ratio in comparison with the band-pass analyzers. Therefore better performances may be obtained with a CMA even though it is more sensitive to magnetic fields and source dimensions. However, for the experimenter who has a LEED apparatus, it is a great advantage to have the possibility to carry out AES, ILEED and LEED measurements with the same instrument. LEED optics are available from many manufacturers.

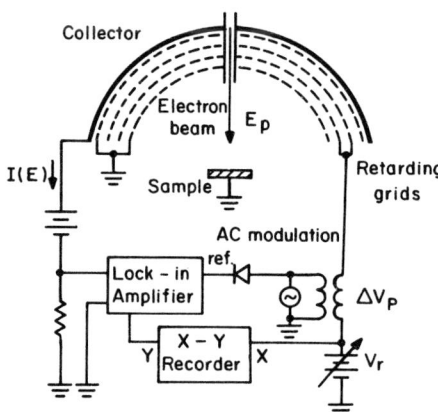

Fig.2.10 Representation of a typical four-grid LEED optics used as energy analyzer, with the associated electronics

Spherical Grid Retarding Potential Analyzer

Since the main limitation on the performances of the retarding analyzers with spherical grids (as the LEEDOA) comes from the shot noise, HUCHITAL and RIGDEN [2.171, 172] proposed to add a setup in order to limit its band-pass and to improve the S/N ratio. In a RPA actually the only electrons of interest are those which pass the retarding grid with a minimum of kinetic energy; then this new setup operates as an effective "post monochromator" by focusing these slow electrons onto a small isolated collector while all fast electrons are collected elsewhere. A sketch of this SGRPA with post monochromator is presented in Fig.2.11. A set of only two grids is used, defining a full acceptance angle of 32°. The purpose of the central stop is to decrease the background by preventing fast electrons from going straight through to the collector. The sidewall potential V_s is generally set at 0-5 V. The collector is run with several hundred volts positive bias.

This analyzer is presented as suitable for XPS and AES. An energy resolution as high as 0.06% was reported for electron energy 200-4000 eV [2.172] with a transmission of about 20% [2.171]. This type of analyzer seems to be ideal since it combines the qualities of both band-pass and high-pass analyzers. However its S/N ratio is not as high as expected because of a great exposure to stray electrons since the collector cannot be properly shielded. A comparison of Auger spectra obtained with this instrument [2.171] and with the CMA [2.140] confirmed that the latter remains superior for AES. Such a conclusion is supported by the tests carried out by HARRIS et al. [2.173] who concluded that this analyzer was more suited for photoelectron spectroscopy where background currents are less important.

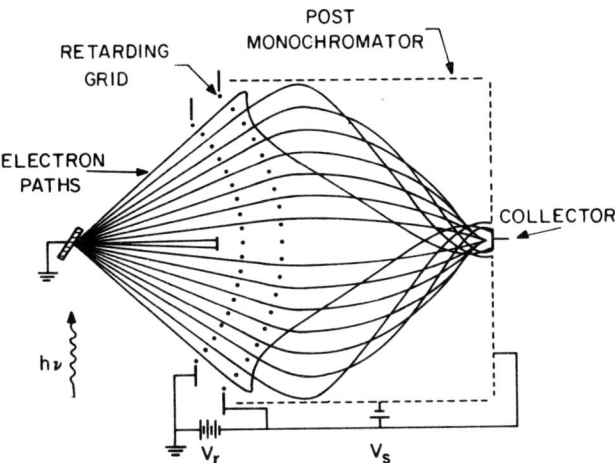

Fig.2.11 Diagram of the spherical grid retarding potential analyzer incorporating a post monochromator, as proposed by HUCHITAL and RIGDEN [2.172]

On the basis of the focusing properties of concentric spherical grids, STAIB [2.174] added to a two-grid RPA a supplementary set of grids with inverted curvature. This second stage acts as post monochromator and focusing device as in the instrument proposed by HUCHITAL and RIGDEN [2.172]. HARRIS et al. [2.173] compared both instruments and reported that Staib's instrument is even more affected by stray electrons because of the electron passages through the additional set of grids.

Double Filter Analyzer

The inherent problem of the retarding potential analyzers may also be overcome by the combination of a low-pass and a high-pass filter in a single instrument, as did LEE [2.175,176]. In this device, the electrons are first slowed down by a retarding field to match the analyzer energy. A grid lens focuses them into a broad-band "prefilter" (a PMA-45°) mainly acting as deflector to lead the electrons towards the low-pass filter. This filter consists in a spherical mirror which reflects the low-energy electrons towards the high-pass filter, which is a conventional RPA with spherical grids. Thus only a narrow slice of energy is detected. The performances in XPS measurements reported by LEE [2.175] indicate a powerful instrument superior to most of the other XPS-spectrometers commercially available [2.177].

2.4.5 Other Techniques

The time-of-flight analysis consists in measuring the time taken for a pulsed beam of particles to traverse a sufficiently long drift space. It has severe limitations since the extremely high velocity of the electron (0.2% of the speed of light at 1 eV) requires either inconveniently long paths or extremely high time resolution [2.11]. So far in the field of surface studies, only one application was reported. This is the TOF spectrometer proposed by BACHRACH et al. [2.6] for photoelectron spectroscopy by means of synchrotron radiation.

It has been mentioned in Subsection 2.4.3 that the method of incidence energy modulation with synchronous detection [2.158] is a powerful technique for the detection of energy-dependent features. In the field of surface analysis, GERLACH [2.178] used such a method for the detection of threshold ionization features. The derivative of the secondary yield with respect to the primary electron energy was detected by measuring the current at the sample in phase with the modulation. KIRSCHNER and STAIB [2.179] improved the method by rather detecting the derivative of the elastically backscattered electrons with a two-grid RPA. This technique was also used by HOUSTON and PARK [2.180] for AEAPS measurements and recently KLUGE [2.181] described a versatile instrument for such measurements. This spectrometer, which incorporates a hemi-cylindrical collector with three grids, makes use of the IEMM and may as well detect electrons, photons, and ions. The results indicate good sensitivity for the three types of measurements.

In addition to the IEMM, other methods and techniques have been proposed for threshold detection of slow electrons. But methods such as the electron attachment [2.182] and the trapped-electron method [2.183] are not compatible with surface analysis. However the threshold UV-photoelectron detector constructed by SPOHR et al. [2.184] proved to be very efficient in gases and could be adapted to the requirements of a solid target.

2.5 Comparison of Electron Spectrometers

As a conclusion, it could be interesting to rank the various types of electron spectrometers in view of their specific applications to surface analysis. Table 2.3 summarizes the results of the rankings of the electrostatic deflection analyzers according to the various criteria and figures of merit defined in Subsection 2.2.3. It is evident from these data that the CMA exhibits inherent characteristics superior as a whole to other analyzers. But among all the instruments discussed in this work, very few appear in this table and therefore the interest and significance of these comparisons are rather limited. This indicates that the usual criteria have a limited degree of generality and applicability. On the other hand the fact that, according to some criteria, the SMA may be classed as superior in spite of its inherent practical drawbacks (see *Other Models*) clearly shows the rather too theoretical character of these criteria. That is why the points of view involved in the following discussion are much more practical.

Among the electron spectrometers used in photoemission spectroscopy (UPS and XPS), the optimized instrument constructed by the Uppsala group [2.9] surely has the higher degree of versatility combined with the best performances. But this spectrometer is very complex and expensive if one considers the point of view of laboratory construction. Quite satisfactory results may as well be obtained with more modest instruments as the SDA, CMA or DFA, if they are constructed with care.

In Auger spectroscopy and related fields, the LEEDOA represents the simplest and most economical approach for the experimenter who already has LEED optics. In addition it is a great convenience to be able to carry out AES, LEED and ILEED measurements with the same instrument. However the CMA is a more powerful and reliable instrument since the developments introduced by PALMBERG [2.140]: it combines high resolution, high sensitivity, and rapidity of data acquisition. Its versatility also allows its coupling with scanning microscopy and other techniques.

For accurate quantitative Auger analysis, the best instrument is probably the spherical deflector analyzer. In addition to high resolution and good sensitivity, it offers all the required conveniences for accurate angular measurements, and for a full exploitation of the advantages of pre-retardation and analysis at constant absolute resolution by means of properly designed electron optics. The cylindrical deflector analyzer is also suitable for this application, but electron optics expressly adapted to its rectangular geometry is far more difficult to design in comparison with the cylindrically symmetric optics involved in the SDA. The geometry of the CMA does not allow the proper design of electron optics. This is overcome by using spherical retarding grids; but this must be considered as a drawback with respect to the SDA whose optics may be designed with accurate control of the transmission without using any grid.

For the measurements of energy losses at low energy and high resolution (LEELS), the existing tandem spectrometers (both TCDS and TSDS) seem to give excellent performances. In spite of the fact that the electron optics used with the cylindrical deflector analyzers are not as accurate as those with the spherical ones, this application makes full use of the advantages of the rectangular beams driven by the cylindrical instruments.

In the field of TELS, the main requirement is the ability to handle high energy and high resolution electron beams. From this point of view, the Wien filters first designed by BOERSCH's group [2.32] have given impressive performances. For field emission measurements, the $135°$-spherical spectrometer proposed by KUYATT and PLUMMER [2.124] is by far the major contribution to the FES instrumentation. Finally

Table 2.3 Summary of the rankings of the electrostatic deflection analyzers according to the various criteria and figures of merit.

Criterion \ Rank	Reduced aberration[a]	Luminosity-resolution[b]	Etendue-resolution[c]	Distribution characteristics[d]	Base resolution-transmission[e]	Optimum transport[f]	Current-resolution[g]
1	CMA-42° (MTW)[h]	SMA (baffles)	CMA-54°	CMA-42°	PMA-30° (MTW)	SMA	SDA-180°
2	CMA-42°	CMA-42° (MTW)	CMA-42°	PMA-30°	CMA-42° (MTW)	CMA-42°	CMA-42°
3	SDA-180°	CMA-42°	SDA-180°	SDA-180°	SDA-180°	SDA-180°	
4		SMA (slits)		CDA-127°	PMA-30°		
5		SDA-180°		PMA-45°	CMA-42°		
6		SMA (grids)			PMA-45°		

[a] [2.20]
[b] [2.14]
[c] [2.18]
[d] [2.17]
[e] [2.24], with 360° symmetry
[f] [2.23]
[g] [2.15]
[h] Minimum trace width

other types of analyzers could be proposed for INS in order to achieve further accuracy in measurements; a cylindrical mirror designed with surface focus and the inner cylinder free for the ion beam would be quite appropriate for this application.

This discussion illustrates the fact that the choice of electron spectrometers does not depend only on their inherent qualitites, but often on practical considerations. We hope that these concluding remarks, with the whole content of the chapter, will be useful guidelines for the choice, the design and the operation of electron spectrometers for surface analysis.

2.6 List of Abbreviations and Acronyms

AEAPS	:	Auger electron appearance spectroscopy
AES	:	Auger electron spectroscopy
CDA	:	Cylindrical deflector analyzer
CELS	:	Characteristic energy loss spectroscopy
CMA	:	Cylindrical mirror analyzer
DFA	:	Double filter analyzer
ELEED	:	Elastic low-energy electron diffraction (sometimes used for LEED)
ELS	:	Energy loss spectroscopy
ESCA	:	Electron spectroscopy for chemical analysis (sometimes used for XPS)
FES	:	Field emission spectroscopy
FLA	:	Filter lens analyzer
FWHW	:	Full width half maximum
IEMM	:	Incidence energy modulation method
ILEED	:	Inelastic low-energy electron diffraction
ILS	:	Ionization loss spectroscopy (sometimes used for IS)
INS	:	Ion neutralization spectroscopy
IS	:	Ionization spectroscopy
LEED	:	Low-energy electron diffraction
LEEDOA	:	LEED optics analyzer
LEELS	:	Low-energy energy loss spectroscopy
MDA	:	Magnetic deflection analyzer
MLA	:	Möllenstedt lens analyzer
MTW	:	Minimum trace width
PMA	:	Plane mirror analyzer
RPA	:	Retarding potential analyzer
SAES	:	Scanning Auger electron spectroscopy
SDA	:	Spherical deflector analyzer
SGRPA	:	Spherical grid retarding potential analyzer

SMA	:	Spherical mirror analyzer
TCDS	:	Tandem cylindrical deflector spectrometer
TELS	:	Transmission energy loss spectroscopy
TOFA	:	Time-of-flight analysis
TSDA	:	Tandem spherical deflector spectrometer
UHV	:	Ultrahigh vacuum
UPS	:	UV photoemission spectroscopy
WFA	:	Wien filter analyzer
XPS	:	X-ray photoemission spectroscopy

Acknowledgements

The authors are grateful to A. Adnot for many stimulating discussions, and to Dr. J.P. Hobson (Radio and Electrical Engineering Division, NRC, Ottawa, Canada) who kindly agreed to read the manuscript and made many suggestions to improve it.

References

2.1   J.P. Hobson: Japan. J. Appl. Phys., Suppl. 2, Pt. 1, 317 (1974)
2.2   Y. Ballu: Notes on Instrumentation in Low-Energy Electron Spectrometry (unpublished, 1972)
2.3   C.S. Fadley, R.J. Baird, W. Siekhaus, T. Novakov, S.A.L. Bergström: J. Elect. Spectros. $\underline{4}$, 93 (1974)
2.4   A. Stamatovic, G.J. Schulz: Rev. Sci. Instr. $\underline{39}$, 1752 (1968)
2.5   J.A. Simpson, C.E. Kuyatt, S.R. Mielczarek: Rev. Sci. Instr. $\underline{34}$, 1454 (1963)
2.6   R.Z. Bachrach, F.C. Brown, S.B.M. Hagström: J. Vac. Sci. Technol. $\underline{12}$, 309 (1975)
2.7   D.T. Pierce, T.H. Distefano: Rev. Sci. Instr. $\underline{41}$, 1740 (1970)
2.8   R. Nordberg, J. Hedman, P.F. Heden, C. Nordling, K. Siegbahn: Ark. Fys. $\underline{37}$, 489 (1968)
2.9   U. Gelius, E. Basilier, S. Svensson, T. Bergmark, K. Siegbahn: J. Elect. Spectros. $\underline{2}$, 405 (1974)
2.10 B. Schröder, J. Geiger: Phys. Rev. Lett. $\underline{28}$, 301 (1972)
2.11 J.A. Simpson: In *Methods of Experimental Physics*, Vol. 4A, ed. by V.W. Hughes and H.L. Schultz (Academic, New York, London 1967) pp. 124-135
2.12 C.E. Kuyatt: In *Methods of Experimental Physics*, Vol. 7A, ed. by B. Bederson and W.L. Fite (Academic, New York, London 1968) pp. 1-43
2.13 W. Steckelmacher: J. Phys. E$\underline{6}$, 1061 (1973)
2.14 H.Z. SAR-EL: Rev. Sci. Instr. $\underline{41}$, 561 (1970)
2.15 F.H. Read, J. Comer, R.E. Imhof, J.N.H. Brunt, E. Harting: J. Elect. Spectros. $\underline{4}$, 293 (1974)

2.16 M.E. Rudd: In K.D. Sevier: *Low Energy Electron Spectrometry* (Wiley-Interscience, New York 1972) pp. 17-32
2.17 D. Roy, J.-D. Carette: Can. J. Phys. 49, 2138 (1971)
2.18 D.W.O. Heddle: J. Phys. E4, 589 (1971)
2.19 E.M. Purcell: Phys. Rev. 54, 818 (1938)
2.20 H. Hafner, J.A. Simpson, C.E. Kuyatt: Rev. Sci. Instr. 39, 33 (1968)
2.21 R.W. Hayward: Adv. Electron 5, 97 (1953)
2.22 S. Aksela, M. Karras, M. Pessa, E. Suoninen: Rev. Sci. Instr. 41, 351 (1970)
2.23 K.L. Wang: J. Phys. E5, 1193 (1972)
2.24 W. Schmitz, W. Mehlhorn: J. Phys. E5, 64 (1972)
2.25 M.E. Gellender, A.D. Baker: J. Elect. Spectros. 4, 249 (1974)
2.26 T. Mulvey, M.J. Wallington: Rep. Prog. Phys. 36, 347 (1973)
2.27 K.R. Spangenberg: *Vacuum Tubes* (McGraw-Hill, New York 1948) pp. 329-393
2.28 A.B. El-Kareh, J.C.J. El-Kareh: *Electron Beams, Lenses and Optics*, Vols. 1 and 2 (Academic, London 1970)
2.29 A.W. Blackstock, R.D. Birkhoff, M. Slater: Rev. Sci. Instr. 26, 274 (1955)
2.30 L. Marton, J.A. Simpson: Rev. Sci. Instr. 29, 567 (1958)
2.31 P. Marmet, L. Kerwin: Can. J. Phys. 38, 787 (1960)
2.32 H. Boersch, J. Geiger, W. Stickel: Z. Physik 180, 415 (1964)
2.33 J.A. Simpson: Rev. Sci. Instr. 35, 1698 (1964)
2.34 J.C. Helmer, N.H. Weichert: Appl. Phys. Lett. 13, 266 (1968)
2.35 H. Fellner-Feldegg, U. Gelius, B. Wannberg, A.G. Nilsson, E. Basilier, K. Siegbahn: J. Elect. Spectros. 5, 643 (1974)
2.36 H.G. Nöller, H.D. Polaschegg, H. Schillalies: J. Elect. Spectros. 5, 705 (1974)
2.37 P.C. Kemeny, A.D. McLachlan, F.L. Battye, R.T. Poole, R.C.G. Leckey, J. Liesegang, J.G. Jenkin: Rev. Sci. Instr. 44, 1197 (1973)
2.38 H.D. Polaschegg: Appl. Phys. 4, 63 (1974)
2.39 J.L. Gardner, J.A.R. Samson: J. Elect. Spectros. 6, 53 (1975)
2.40 D. Roy, J.-D. Carette: Rev. Sci. Instr. 42, 1122 (1971)
2.41 D.W.O. Heddle: J. Phys. E4, 981 (1971)
2.42 J.D. Cross, F.H. Read, E.A. Riddle: J. Sci. Instr. 44, 993 (1967)
2.43 R.E. Imhof, F.H. Read: J. Phys. E1, 859 (1968)
2.44 F.H. Read: J. Phys. E3, 127 (1970)
2.45 M.V. Kurepa, M.D. Tasic, J.M. Kurepa: J. Phys. E7, 940 (1974)
2.46 F.R. Paolini, G.C. Theodoridis: Rev. Sci. Instr. 38, 579 (1967)
2.47 G.C. Theodoridis, F.R. Paolini: Rev. Sci. Instr. 39, 326 (1968)
2.48 Y. Ballu: Rev. Phys. Appl. 3, 46 (1968)
2.49 Y. Delage, J.-D. Carette: Can. J. Phys. 49, 2118 (1971)
2.50 P. Allard, J.-D. Carette: Can. J. Phys. 49, 2132 (1971)
2.51 D. Roy, J.-D. Carette: Appl. Phys. Lett. 16, 413 (1970)

2.52 D. Roy, J.-D. Carette: J. Appl. Phys. 42, 3601 (1971)
2.53 D. Roy, J.-D. Carette: Rev. Sci. Instr. 42, 776 (1971)
2.54 D. Roy: Rev. Sci. Instr. 43, 535 (1972)
2.55 D. Roy, P.D. Burrow: J. Phys. E8, 273 (1975)
2.56 D. Roy, A. Delage, J.-D. Carette: J. Phys. E8, 109 (1975)
2.57 C.E. Kuyatt, J.A. Simpson: Rev. Sci. Instr. 38, 103 (1967)
2.58 J.R. Pierce: *Theory and Design of Electron Beam*, 2nd ed. (Van Nostrand, New York 1949) Chap. 9
2.59 G.J. Schulz: Phys. Rev. 125, 229 (1962)
2.60 G. Bosi: Rev. Sci. Instr. 43, 475 (1972)
2.61 G.A. Proca, C. Rüdinger: Rev. Sci. Instr. 44, 1381 (1973)
2.62 G.M. Renfro, H.J. Fischbeck: Rev. Sci. Instr. 46, 620 (1975)
2.63 R.L. Gerlach: J. Vac. Sci. Technol. 10, 122 (1973)
2.64 P. Bryce, R.L. Dalglish, J.C. Kelly: Can. J. Phys. 51, 574 (1973)
2.65 R. Herzog: Z. Physik 97, 596 (1935); Phys. Z. 41, 18 (1940)
2.66 H. Wollnik, H. Ewald: Nucl. Instr. Meth. 36, 93 (1965)
2.67 R.T. Poole, J. Liesegang, J.G. Jenkin, R.C.G. Leckey: Vacuum 22, 499 (1972)
2.68 H. Froitzheim, H. Ibach: Z. Physik 269, 17 (1974)
2.69 E. Bolduc, J.-J. Quemener, P. Marmet: Can. J. Phys. 49, 3095 (1971)
2.70 B. Wannberg, U. Gelius, K. Siegbahn: J. Phys. E7, 149 (1974)
2.71 J.S. Solomon, W.L. Baun: J. Vac. Sci. Technol. 12, 375 (1975)
2.72 J.E. Houston, R.L. Park: Rev. Sci. Instr. 43, 1437 (1972)
2.73 K. Stadler: Vacuum 22, 553 (1972)
2.74 F. Meyer, J.J. Vrakking: Surf. Sci. 33, 271 (1972)
2.75 J.T. Grant, T.W. Haas, J.E. Houston: Surf. Sci. 42, 1 (1974); 44, 617 (1974)
2.76 J.E. Houston: Surf. Sci. 38, 283 (1973)
2.77 D.G. Fedak: Rev. Sci. Instr. 44, 1613 (1973)
2.78 D.J. Pocker: Rev. Sci. Instr. 46, 105 (1975)
2.79 R.J. Hanisch, G.P. Hughes, J.R. Merrill: Rev. Sci. Instr. 46, 1262 (1975)
2.80 P. Staib, J. Kirschner: Appl. Phys. 3, 421 (1974)
2.81 W.M. Mularie, W.T. Peria: Surf. Sci. 26, 125 (1971)
2.82 J.E. Houston: Rev. Sci. Instr. 45, 897 (1974)
2.83 J.T. Grant, M.P. Hooker, T.W. Haas: Surf. Sci. 51, 318 (1975)
2.84 R. Carbonneau, E. Bolduc, P. Marmet: Can. J. Phys. 51, 505 (1973)
2.85 M. De Celles: *Analytical Properties of the Smoothing and Straightening of Ionization Curves* (to be published).
2.86 L. Moore: J. Phys. D1, 237 (1968)
2.87 R.L. Cowperthwaite, H. Myers: J. Chem. Phys. 53, 1077 (1970)
2.88 U. Landman, D.L. Adams: J. Vac. Sci. Technol. 11, 195 (1974)
2.89 A.D. Maclachlan, J.G. Jenkin, J. Liesegang, R.C.G. Leckey: J. Elect. Spectros. 5, 593 (1974)

2.90 G.K. Wertheim: J. Elect. Spectros. $\underline{6}$, 239 (1975); Rev. Sci. Instr. $\underline{46}$, 1414 (1975)

2.91 C.J. Powell: In *Methods of Experimental Physics*, Vol. 7B, ed. by B. Bederson and W.L. Fite (Academic, New York, London 1968) pp. 275-305

2.92 M.E. Rudd: In K.D. Sevier: *Low Energy Electron Spectrometry* (Wiley-Interscience, New York 1972) pp. 32-34

2.93 W.G. Wadey: Rev. Sci. Instr. $\underline{27}$, 910 (1956)

2.94 Y. Petit-Clerc, J.-D. Carette: Vacuum $\underline{18}$, 7 (1968)

2.95 F. Rosebury: *Handbook of Electron Tube and Vacuum Techniques* (Addison-Wesley, Reading 1965) pp. 3-18

2.96 G.J. Schulz: Rev. Mod. Phys. $\underline{45}$, 378 (1973)

2.97 M. Eyb, H. Hofmann: J. Phys. B$\underline{8}$, 1095 (1975)

2.98 H. Froitzheim, H. Ibach, S. Lehwald: Rev. Sci. Instr. $\underline{46}$, 1325 (1975)

2.99 O. Klemperer: Rept. Progr. Phys. $\underline{28}$, 77 (1965)

2.100 L. Kerwin, P. Marmet, J.-D. Carette: In *Case Studies in Atomic Collision Physics*, Vol. 1, ed. by E.W. Mcdaniel and M.R.C. Mcdowell (North-Holland Amsterdam, London 1969) pp. 525-581

2.101 A.L. Hughes, V. Rojansky: Phys. Rev. $\underline{34}$, 284 (1929)

2.102 A.L. Hughes, J.H. Mcmillen: Phys. Rev. $\underline{34}$, 291 (1929)

2.103 E.M. Clarke: Can. J. Phys. $\underline{32}$, 764 (1954)

2.104 M. Cotte: C.R. Acad. Sc. (Paris) B$\underline{265}$, 1388 (1967)

2.105 R. Francois, M. Barat: C.R. Acad. Sc. (Paris) B$\underline{266}$, 1306 (1968)

2.106 J.J. Leventhal, G.R. North: Rev. Sci. Instr. $\underline{42}$, 120 (1971)

2.107 D. Roy, M. De Celles, J.-D. Carette: Rev. Phys. Appl. $\underline{6}$, 51 (1971)

2.108 G.A. Harrower: Phys. Rev. $\underline{102}$, 340 (1956)

2.109 L.A. Harris: J. Appl. Phys. $\underline{39}$, 1419 (1968)

2.110 B. Feuerbacher, B. Fitton: Rev. Sci. Instr. $\underline{42}$, 1172 (1971)

2.111 F.M. Propst, T.C. Piper: J. Vac. Sci. Technol. $\underline{4}$, 53 (1967)

2.112 H. Ibach: J. Vac. Sci. Technol. $\underline{9}$, 713 (1972)

2.113 A. Adnot, Y. Ballu, J.-D. Carette: J. Appl. Phys. $\underline{43}$, 2796 (1972)

2.114 Y. Ballu, J. Lecante: Japan. J. Appl. Phys., Suppl. 2, Pt. 2, 249 (1974)

2.115 A. Adnot, J.-D. Carette: (to be published)

2.116 G. Bosi: J. Appl. Phys. $\underline{46}$, 183 (1975)

2.117 A. Chutjian: J. Chem. Phys. $\underline{61}$, 4279 (1974)

2.118 C.J. Powell, A. Mandl: Phys. Rev. B$\underline{6}$, 4418 (1972)

2.119 R.A. Armstrong: Can. J. Phys. $\underline{44}$, 1753 (1966)

2.120 J.J. Lander: Phys. Rev. $\underline{91}$, 1382 (1953)

2.121 K. Siegbahn: J. Elect. Spectros. $\underline{5}$, 3 (1974)

2.122 R.T. Poole, R.C.G. Leckey, J.G. Jenkin, J. Liesegang: J. Elect. Spectros. $\underline{1}$, 371 (1972/73)

2.123 J.W. Gadzuk, E.W. Plummer: Rev. Mod. Phys. $\underline{45}$, 487 (1973)

2.124 C.E. Kuyatt, E.W. Plummer: Rev. Sci. Instr. 43, 108 (1972)
2.125 T.S. Green, G.A. Proca: Rev. Sci. Instr. 41, 1409 (1970)
2.126 G.A. Harrower: Rev. Sci. Instr. 26, 850 (1955)
2.127 F. Pauty, G. Matula, P.J. Vernier: Rev. Sci. Instr. 45, 1203 (1974)
2.128 F. Edelmann, K. Ulmer: Z. Angew. Phys. 18, 308 (1965)
2.129 S. Aksela: J. Phys. E6, 545 (1973)
2.130 G.A. Proca, T.S. Green: Rev. Sci. Instr. 41, 1778 (1970)
2.131 E. Blauth: Z. Physik 147, 228 (1957)
2.132 W. Mehlhorn: Z. Physik 160, 247 (1960)
2.133 V.V. Zashkvara, M.I. Korsunskii, O.S. Kosmachev: Zh. Techn. Fiz.[English Transl.: Sov. Phys. - Techn. Phys. 11, 96 (1966)] 36, 132 (1966)
2.134 H.Z. Sar-El: Rev. Sci. Instr. 38, 1210 (1967)
2.135 V.V. Zashkvara, V.S. Red'kin: Zh. Techn. Fiz. [English Transl.: Sov. Phys.-Techn. Phys. 14, 1089 (1970)] 39, 1452 (1969)
2.136 J.S. Risley: Rev. Sci. Instr. 43, 95 (1972)
2.137 L.A. Harris: J. Vac. Sci. Technol. 11, 23 (1974)
2.138 H.E. Bishop, J.P. Coad, J.C. Riviere: J. Elect. Spectros. 1, 389 (1972/73)
2.139 S. Aksela: Rev. Sci. Instr. 42, 810 (1971)
2.140 P.W. Palmberg, G.K. Bohn, J.C. Tracy: Appl. Phys. Lett. 15, 254 (1969)
2.141 D.R. Arnott, J.A. Ramsey: Vacuum 22, 355 (1972)
2.142 E. Dietz, K.O. Groeneveld, R. Spohr, R. Staudte: Nucl. Instr. Meth. 105, 467 (1972)
2.143 P.H. Citrin, R.W. Shaw, T.D. Thomas: In *Electron Spectroscopy*, ed. by. D.A. Shirley (North-Holland, Amsterdam 1972) pp. 105-20
2.144 E.B. Bas, U. Bänninger, P. Keller: J. Vac. Sci. Technol. 9, 306 (1972)
2.145 E.N. Sickafus, D.M. Holloway: Surf. Sci. 51, 131 (1975)
2.146 A.S. Shapovalov: Zh. Techn. Fiz. [English Transl.: Sov. Phys. - Techn. Phys. 11, 677 (1966)] 36, 920 (1966)
2.147 S. Aksela: Rev. Sci. Instr. 43, 1350 (1972)
2.148 G.N. Ogurtsov, I.P. Flaks, S.V. Avakyan: Zh. Techn. Fiz. [English Transl.: Sov. Phys. - Techn. Phys. 14, 972 (1970)] 39, 1293 (1969)
2.149 E. Harting: Rev. Sci. Instr. 42, 1151 (1971)
2.150 G. Vassilev, J. Baudon, G. Rahmat, M. Barat: Rev. Sci. Instr. 42, 1222 (1971)
2.151 M. Karras, M. Pessa, S. Aksela: Ann. Acad. Sci. Fennicae A VI, No 289 (1968)
2.152 C.C. Chang: Surf. Sci. 25, 53 (1971)
2.153 N.J. Taylor: In *Techniques of Metals Research*, Vol. 7, ed. by R.F. Bunshah (Interscience, New York 1972) pp. 117-159
2.154 P.W. Palmberg: J. Vac. Sci. Technol. 12, 379 (1975)
2.155 H.Z. Sar-El: Nucl. Instr. Meth. 42, 71 (1966)
2.156 K. Siegbahn: *Alpha - Beta -, and Gamma - Ray Spectroscopy* (North-Holland, Amsterdam 1965) pp. 79-202

2.157 C.S. Fadley, R.N. Healey, J.M. Hollander, C.E. Miner: J. Appl. Phys. 43, 1085 (1972)
2.158 L. Sanche, G.J. Schulz: Phys. Rev. Lett. 26, 943 (1971)
2.159 P.D. Burrow, L. Sanche: Phys. Rev. Lett. 28, 333 (1972)
2.160 E.G. Mcrae, G.H. Wheatley: Surf. Sci. 29, 342 (1972); 25, 491 (1971)
2.161 J.A. Simpson: Rev. Sci. Instr. 32, 1283 (1961)
2.162 P.S.P. Wei, A. Kuppermann: Rev. Sci. Instr. 40, 783 (1969)
2.163 P.S.P. Wei, A.Y. Cho, C.W. Caldwell: Rev. Sci. Instr. 40, 1075 (1969)
2.164 T.H. Distefano, D.T. Pierce: Rev. Sci. Instr. 41, 180 (1970)
2.165 J.A. Simpson, L. Marton: Rev. Sci. Instr. 32, 802 (1961)
2.166 G. Möllenstedt: Optic 5, 499 (1949); 9, 473 (1952)
2.167 P.W. Palmberg: Appl. Phys. Lett. 13, 183 (1968)
2.168 D.K. Skinner, R.F. Willis: Rev. Sci. Instr. 43, 731 (1972)
2.169 N.J. Taylor: Rev. Sci. Instr. 40, 792 (1969)
2.170 J.A. Cross: J. Phys. D6, 622 (1973)
2.171 D.A. Huchital, J.D. Rigden: J. Appl. Phys. 43, 2291 (1972)
2.172 D.A. Huchital, J.D. Rigden: Appl. Phys. Lett. 16, 348 (1970)
2.173 F.M. Harris, P.J. Basset, M. Prutton: J. Phys. E8, 11 (1975)
2.174 P. Staib: Vacuum 22, 481 (1972); J. Phys. E5, 484 (1972)
2.175 J.D. Lee: Rev. Sci. Instr. 44, 893 (1973)
2.176 J.D. Lee: Rev. Sci. Instr. 43, 1291 (1972)
2.177 C.A. Lucchesi, J.E. Lester: J. Chem. Ed. 50, A205, A269 (1973)
2.178 R.L. Gerlach: Surf. Sci. 28, 648 (1971)
2.179 J. Kirschner, P. Staib: Phys. Lett. 42A, 335 (1973)
2.180 J.E. Houston, R.L. Park: Phys. Rev. B5, 3808 (1972)
2.181 A. Kluge: Rev. Sci. Instr. 46, 1179 (1975)
2.182 C.E. Brion, L.A.R. Olsen: J. Phys. B3, 1020 (1970)
2.183 G.J. Schulz: Phys. Rev. 116, 1141 (1959)
2.184 R. Spohr, P.M. Guyon, W.A. Chupka, J. Berkowitz: Rev. Sci. Instr. 42, 1872 (1971)

Recent References Added in Proof

P.E. Best: Apparatus for the measurement of angle-resolved spectra of electrons emerging from single crystals. Rev. Sci. Instrum. 46, 1517 (1975)

E. Harting, F.H. Read: *Electrostatic Lenses* (Elsevier Scientific, Amsterdam 1975)

P. Marchand, P. Veillette: "Straightening through smoothings" seen as a digital filter. Can. J. Phys. 54, 1309 (1976)

J.M. McDavid, S.C. Fain: Leed using a CMA. Surface Science 52, 670 (1975)

A. Sköllermo, B.Wannberg: The fringing fields of a spherical electrostatic spectrometer, and their influence on the aberration coefficients. Nucl. Instr. and Meth. 131, 279 (1975)

R.W. Springer, D.J. Pocker, T.W. Haas: Integral Auger information via tailored modulation techniques. Appl. Phys. Lett. 27, 368 (1975)

3. Electron-Excited Core Level Spectroscopies

J. Kirschner

With 34 Figures

This chapter deals with electron-excited core level spectroscopies for surface analysis. We shall mainly discuss the methods following the pattern "electron in-electron out" but, where appropriate, also those according to "electron in-photon out" or vice versa. The former group comprises Auger Electron Spectroscopy (AES), Ionization Loss Spectroscopy (ILS), Disappearance Potential Spectroscopy (DAPS), Auger Electron Appearance Potential Spectroscopy (AEAPS). The latter group includes Soft X-ray Appearance Potential Spectroscopy (SXAPS), Electron Microprobe Analysis (EMA) and X-ray excited Electron Appearance Potential Spectroscopy (XEAPS). The level of sophistication for these methods is quite different, mainly for historical reasons. It ranges from very recent methods with not yet fully explored potentials like DAPS, to rather old methods which have only recently been revived, like SXAPS. Others, like AES, are based upon well-known effects and are increasingly used in a wide range of applications in research and industry. The degree of theoretical understanding is quite different, too. The emission of characteristic X-rays and of Auger electrons by free atoms can be considered as being relatively well understood. Theoretical work on the influence of solid state effects on these phenomena is in progress. In contrast, for the threshold spectroscopies the detailed theoretical interpretation of the measured spectra is just at the beginning. Nevertheless, all these methods will prove valuable for the elemental and chemical analysis of surfaces.

In the following chapters we first discuss briefly the basic processes of the excitation and deexcitation of free atoms and some of the modifications due to solid state effects. Then we present the various spectroscopies based upon these processes and describe their properties. Finally, some comparisons are made. In anticipating the result here, it seems most advantageous to combine two or more of the methods in one apparatus, which can be very easily done in several cases. This approach yields the maximum information from each of the techniques and compensates for their mutual drawbacks.

3.1 Basic Processes

3.1.1 Free Atoms

First of all, we discuss the excitation of free atoms by primary electrons with energies of about hundred to several thousand eV and the subsequent decay processes. Free atoms are chosen because the presence of other atoms gives rise to several modifications, which are discussed in the next subsection.

The identification of atoms by core level spectroscopies is based upon the characteristic values of the binding energies of the electrons. They are used either directly, when an external electron knocks out a core electron, thereby loosing energy, of which the minimum corresponds to the binding energy. They are used indirectly, when the excited atom rearranges its electron shells, which is accompanied by the emission of photons or electrons with characteristic energies. They correspond roughly to the energy differences of the binding energies of the electrons which take part in the transitions during the rearrangement. In Fig.3.1 we show the electron binding energies for the elements from 1 eV to 10 keV. In the case of gases they are referred to the vacuum level, in the case of solids they are referred to the Fermi level. The latter value is generally accepted as the binding energy to which the work function has to be added when a comparison with gases is to be made [3.1]. The figure is drawn using values of SEVIER [3.2], which have been obtained by a number of different techniques. Depending on the method used, they can be slightly different due to the peculiarities of the process observed. For investigations requiring precise knowledge of the binding energies, the measuring method has to be considered [3.3].

The cross section for electron impact ionization depends on the ratio U of the primary energy E_p to the binding energy E_B: $U = E_p/E_B$. Near $U = 1$ an inner shell excitation must not necessarily be accompanied by ionization. It is sufficient to raise an inner shell electron up to the lowest unoccupied levels of the atom below the vacuum level. As the energy differences involved are of the order of eV, this distinction is of importance only very near to the excitation threshold. Above, the cross section for electron impact excitation can be set equal to that for electron impact ionization. For high values of U, the cross section is proportional to $(\ln U)/U$. This is a result of nonrelativistic quantum mechanical calculations using first or higher Born approximations [3.4]. When the primary energy is very high and/or inner shells of high-Z atoms are considered, relativistic effects have to be taken into account [3.5]. They tend to counterbalance the decrease of the cross section for high U or even convert it into a further increase. In the range of $E_B < 3$ keV and $E_p < 10$ keV which is of interest here, relativistic corrections can be neglected.

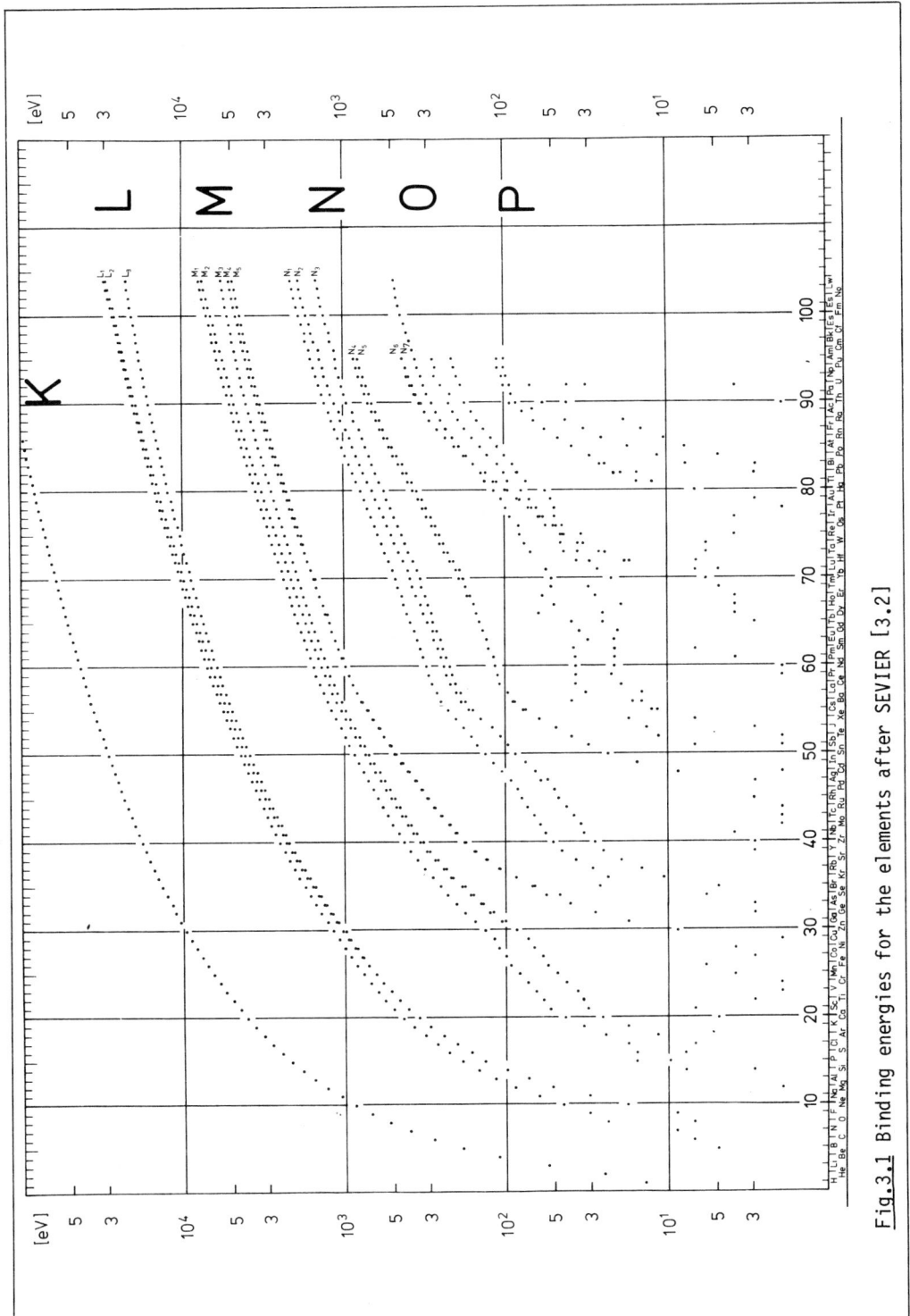

Fig.3.1 Binding energies for the elements after SEVIER [3.2]

The total ionization cross section σ has a maximum at some multiples of the threshold energy (U = 3...4), which is largely independent of the absolute value of the threshold energy and of the symmetry of the electron shells. This range is not well described by the quantum-mechanical calculations, but classical calculations as performed mainly by GRYZINSKI [3.6] do agree quite well with experiments in this range.

At threshold, a general classical theory of excitations is not yet available [3.7]. Quantum mechanical calculations [3.8] have been applied to elastic and inelastic scattering processes involving loosely bound outer electrons, including resonance phenomena between short-lived excited states of metastable atoms and continuum states of the ion (autoionization). When inner shells are excited, additional particle-hole interactions and collective excitations of the electron shells as a whole have to be taken into consideration [3.9]. In this respect, theory is just at the beginning [3.10]. In the case of solids, the band structure and collective excitations of the conduction electrons, too, influence the excitation function near threshold [3.11].

For the analytic description of the cross section per electron over a wide range, with the exception of the region immediately near threshold, a semi-empirical formula found by DRAWIN has proven to be very useful [3.12]

$$\sigma = 2.66 \; \pi a_0^2 (E_B^H/E_B^x)^2 \; f_1(U-1)U^{-2} \; \ln(1.25 \; f_2 U) \quad |cm^2/electron| \tag{3.1}$$

with $a_0 = 8.79 \cdot 10^{-17} cm^2$, E_B^H = the ionization energy of hydrogen (13.6 eV), and E_B^x = the ionization energy of the electron shell in question. The parameters f_1 and f_2 allow fitting in the case of multiply ionized atoms. In general, it can be assumed $f_1 = f_2 = 1$.

For applications of the core level spectroscopies using electronic excitation, the absolute value of the maximum of the cross section near U = 4 is important, because it determines the maximum available intensity and therefore the sensitivity. MEYER and VRAKKING [3.13-14] measured its value for several electron shells relative to that of carbon, which is known from the work of GLUPE and MEHLHORN [3.15]. Fig.3.2 shows some of their results for U = 4. Drawin's formula applied to the K-shell agrees within less than 10% with the measured values over a range of six orders of magnitude. For the $L_{2/3}$-shell the authors find a power law of the form $\sigma_{max} \sim E_B^{-1.56}$, which they assume to approach asymptotically the classical scaling law $\sigma_{max} \sim E_B^{-2}$ at higher binding energies. Therefore, the Drawin formula yields values that are too high at low energies. This could probably be compensated by adjusting the free parameters f_1 and f_2 (for a review of measured ionization cross sections see [3.16]).

Figure 3.2 gives the cross section per electron. When the population of the shells is taken into account, one finds for a given binding energy E_B the maximum cross section to remain approximately constant (within a factor 2 to 3) on passing from one element to another, though it varies over many orders of magnitude within a certain shell. From the point of view of analysis this means that all elements can be detected with similar sensitivity, provided that the measured intensity depends directly on the rate of creation of core holes. As seen below, this is fulfilled to a good approximation for Auger electrons, but not for photons, since the fluorescence yield is generally much smaller than the Auger yield.

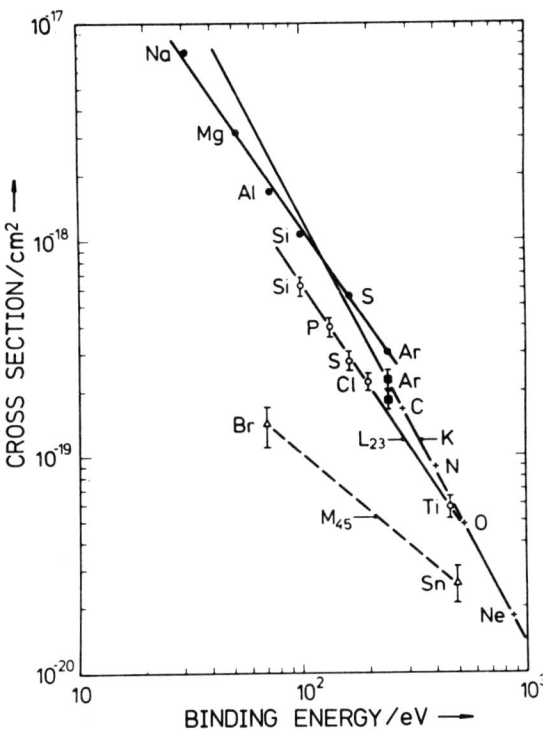

Fig. 3.2 The maximum electron impact ionization cross section per electron (near $U = 4$) versus the binding energy for various electron shells. The K-shell values (+) have been measured by GLUPE & MEHLHORN [3.15]. The L_{23}- and M_{45}-shell values have been measured by VRAKKING & MEYER (O and Δ, respectively). The dots (●) indicate calculations of McGUIRE for the L-shell. The Ar-values have been measured by Cristofzik (after [3.13, 14])

Consider a free ion with a hole in the K-shell, the primary electron and the emitted K-electron both being far away. The lifetime of this highly excited state of the atom is determined by the sum of all possible decay processes. These are either radiative transitions— with transition of an electron from any other shell to the hole—which occur with the probability ω_R, or radiationless transitions with rearrangement of the electron shells and emission of an electron with the probability $1-\omega_R$. The latter are called Auger transitions (ω_A is defined as the Auger yield) if the electron which fills the hole does not belong to the same shell. They are called Coster-Kronig transitions (ω_{CK} is defined as the Coster-Kronig yield) if the hole is filled by an electron of the same shell. The rare cases, where all three electrons belong to the same shell, are called super-Coster-Kronig transitions [3.17]. As no other deexcitation mechanisms are possible, it is

$$\omega_R + \omega_A + \omega_{CK} = 1 \quad . \tag{3.2}$$

According to the uncertainty principle, the natural energy width of the level where the hole is present is $\Gamma = \hbar/\tau$ where τ is the lifetime of the hole [3.18]. As the energy distribution of the emitted electrons or photons is Lorentzian, it is

$$\Gamma = \Gamma_A + \Gamma_R + \Gamma_{CK} \text{ and } 1/\tau = 1/\tau_A + 1/\tau_R + 1/\tau_{CK} \quad . \tag{3.3}$$

From which follows $\omega_R = \Gamma_R/\Gamma$, $\omega_A = \Gamma_A/\Gamma$, $\omega_{CK} = \Gamma_{CK}/\Gamma$.

The lifetime-broadening can be as large as 10 eV for the L-shells and more than 100 eV for the K-shell at high binding energies. The lifetime is often governed by Coster-Kronig transitions which have high transition probability when they are energetically possible. The Coster-Kronig transitions "transform" holes into subshells with lower binding energies, thereby increasing the yield of electrons or photons observed from these shells [3.19]. Recent calculations [3.20] show that Coster-Kronig yields and energies depend quite sensitively on the presence of other holes, even in outer shells which do not take part immediately in the reorganization of the electron shells. These effects can be expected to be even more important in solids. In Fig.3.3 we show the dependence of the fluorescence yield ω_R for some electron shells, where Coster-Kronig transitions do not occur, on the binding energy. It can be realized that below about 2 keV the radiative transitions are highly improbable.

The relatively simplest way of deexcitation is the emission of X-rays. As they are mainly due to dipole radiation, the well-known optical selection rules are valid: $\Delta\ell = 1$, $\Delta j = \pm 0, 1$. The energy of the photons is in rigor equal to the difference of the total energies of the system before and after the electron transition. However, as this many-body problem is not exactly solvable several approximations are used [3.22]. Non-empirical methods use the Hartree-Fock (HF) method to evaluate the electron binding energies and the photon energies from differences of these. Two approaches are in use: the "sudden approximation" or "frozen-orbital approximation" assumes the transition to happen so rapidly that the other "passive" electrons do not adjust to the changing electronic environment. Then the same wave functions may be used to represent the electrons of the system in both the initial and final states. If the "spectator" electrons have enough time to adjust, i.e., if the transitions are slow, an "adiabatic approximation" is required. For relatively large lifetimes as encountered in the X-ray transitions, the latter model is physically more reasonable and its predictions are more accurate [3.23]. For rather rapid processes, such as the Coster-Kronig transitions, the sudden approximation yields good results [3.24].

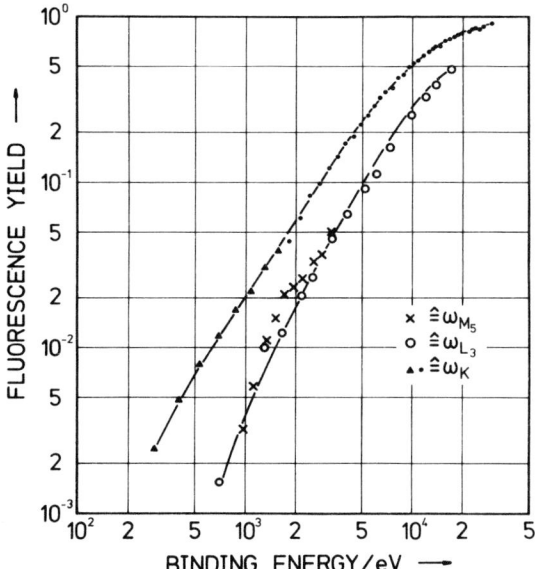

Fig. 3.3 Fluorescence yield for the K-, L_3- and M_5-shells versus the electron binding energy of these shells. The K-shell curve consists of measured (●) and calculated (△) values. The other points are calculated values (after [3.21])

Empirically, the photon energies can be determined from known electron binding energies, measured, for example, by photoelectron spectroscopy:

$$h\nu = E_B^X - E_B^Y \quad .$$

The agreement between empirically or non-empirically calculated values and measured energies is generally of the order of one part in 10^3 or better. For analytical purposes, the X-ray energies can be considered to be sufficiently well known.

More frequently than by dipole radiation (and other radiation) an atom deexcites via nonradiative transitions. They lead from an initial state which can be characterized by a core vacancy and a hole in the continuum, to the final state with two holes in outer shells of the atom. The driving force is the Coulomb interaction of the electrons taking part in the transition in the field of the screened nucleus. The classification of the transitions can be made in terms of jj-coupling if spin--orbit interaction is dominant. In the simplest case of the KLL transitions (initial hole in the K-shell, transition of an L-electron into the hole, followed by the emission of another L-electron) there are 6 different energies of the final state. They are shown in Fig.3.4 on the right-hand side. The other extreme is pure LS-coupling which can be applied when the Coulomb interaction is large compared to the spin-orbit interaction.

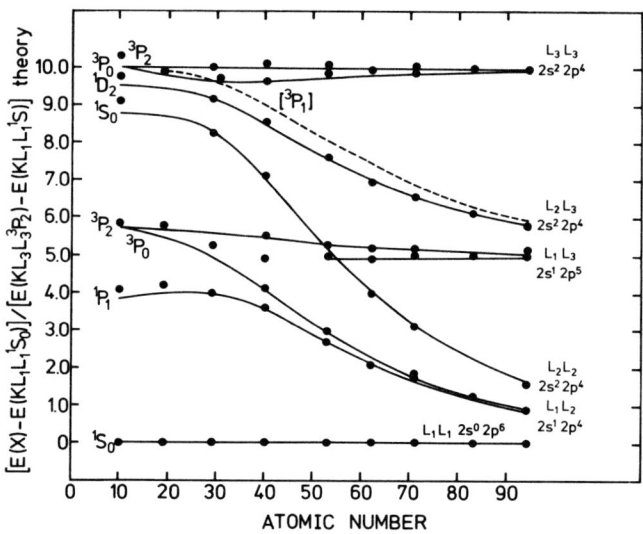

Fig.3.4 Relative positions of the KLL Auger energies versus the atomic number. The figure shows the transition from pure LS coupling (left-hand side) to pure jj-coupling (right-hand side) via intermediate coupling in between. The lines are calculated values, the dots are measured values (after [3.25])

In this case, 5 final states are possible. In the intermediate range a total of 9 lines is found. Accordingly the complete nomenclature describing Auger transitions indicates the shells involved and the final state of the atom, for example KL_3L_3 ($^3P_0$). For low resolution electron spectroscopy, the simplified nomenclature indicating the electron shells in terms of jj-coupling is mostly sufficient. The LXX and MXX Auger spectra are much more complex than the KLL spectra and their identification still raises problems.

The energy of the Auger electrons can in principle be determined in the same way as that of the X-rays: by the difference of the total energies before and after the transition. Exact calculations using the HF model have been made [3.26] but they require considerable computer capacity as the total energies are generally much larger than the Auger energies. Several semi-empirical procedures for determining Auger energies have been developed. Starting from the frozen orbital model, with some simplifying assumptions [3.24], it is possible to derive expressions for the transition energies which include electron binding energy terms. Auger energies are then determined by using experimental values for the binding energies. Whereas relativistic effects can be neglected for low Auger energies, electron correlation effects have to be taken into account. In other words: the deficiencies of the frozen orbital approximation have to be compensated. This is done by considering the one-step Auger transition as a two-step process where the readjustment of the passive orbitals is allowed before the emission of the Auger electron occurs. The correction is called "intra-atomic relaxation" [3.25,27]. This approach seems also well suited to describe chemical and solid state influences on the Auger energies. Today, the best agreement between theory and experiment is obtained for KLL and LMM lines of free atoms with low to medium atomic number.

In spite of good examples [3.28], the state of the theory in general is not yet developed far enough to predict all transition energies and transition intensities for all elements. Therefore, empirical formulas are useful. The one we give here agrees surprisingly well (of the order of 10^{-3}) with measured energies [3.29]:

$$E^{(z)}_{\alpha\beta\gamma} = E^{(z)}_{\alpha} - E^{(z)}_{\beta} - E^{(z)}_{\gamma} - 1/2 \left[E^{(z+1)}_{\gamma} - E^{(z)}_{\gamma} + E^{(z+1)}_{\beta} - E^{(z)}_{\beta} \right]$$

$$= E^{(z)}_{\alpha} - 1/2 \left[E^{(z)}_{\beta} + E^{(z)}_{\gamma} + E^{(z+1)}_{\gamma} + E^{(z+1)}_{\beta} \right] \quad .$$

(3.4)

$E^{(z)}_{\alpha\beta\gamma}$ is the Auger energy of the transition $\alpha\beta\gamma$ of the element Z. The first three terms correspond to the difference of the binding energies of shells α, β, γ of the element Z. The correction term corresponds to the increase in binding energy of the γ-electron when a β-electron is removed. As the transition $\alpha\beta\gamma$ is equivalent to the transition $\alpha\gamma\beta$ and the same reasoning holds for the latter, the average over all four binding energies is a reasonable approximation. Of course, this formula does not take the interaction of final state holes into consideration. Therefore only averaged values for multiplet states can be expected. In spite of this, the formula has proven very useful. Numerical tables derived from (3.4) can be found in [3.30].

Apart from the transition rates, the relative intensities of measured lines may depend strongly on the occurrence of Coster-Kronig transitions. For example, holes in the L_1-shell may be rapidly transformed into $L_{2/3}$-holes by Coster-Kronig transitions of the type $L_1L_{2/3}X$. This is the reason why L_1XX transitions are generally found to have low intensity, whereas $L_{2/3}XX$ transitions are dominant. In addition, those lines are broadened by the short lifetime associated with Coster-Kronig transitions. In measurements where the derivative of the energy spectrum is taken they seem largely suppressed. In quantitative measurements Coster-Kronig transitions are a nuisance because the initial population of holes is in general unknown. This may introduce considerable uncertainty into measurements of the ionization cross section from Auger intensities [3.19].

Every Auger spectrum has a rather complicated shape due to the large number of diagram lines, i.e., lines which can be ascribed to transitions according to the nomenclature. In addition a great number of satellites is to be observed, the origin of which could be explained only in a few cases [3.31]. There may be many reasons: multiple ionization of outer shells, multiple ionization of inner shells and combinations thereof, ionization of an inner shell and simultaneous excitation of the atom as a whole by shake-up processes [3.32] during the ionization. Finally, second-order deexcitation processes also happen to occur: for example the double Auger process [3.33] or the radiative Auger effect [3.34,35].

It has often been observed in experiments that satellite lines are particularly sensitive to changes of the electronic environment. At present, these effects are neither fully explored nor understood. Perhaps they may find analytical applications in the future.

3.1.2 Surface Atoms

When free atoms come together to form a solid, the electron energy levels are broadened and shifted by the mutual interaction of the atoms. If the binding energy is above 100 eV the broadening is negligible compared to the natural linewidth of the radiation following the creation of a core hole. More weakly bound electrons are forming bands with a certain density of states $N(E)$. They are filled by electrons up to the Fermi level E_F. Its energetic position determines the electronic properties of the solid (see Vol.4, page 257 of this series). The band model is valid in rigor only for an infinitely large solid without any imperfections and electronic excitations near the absolute zero of temperature. Though the existence of a surface introduces certain modifications (see Chap.1), it is useful to discuss the results of the various spectroscopies in terms of the band model: electronic excitations and deexcitations are described by one-electron transitions between occupied and unoccupied levels within this scheme, which—to a first order approximation—is assumed to remain unchanged by the transition. This approach is supported by the results of X-ray spectroscopy in emission and absorption, but one has to keep in mind its limitations. For example, many-body effects cannot be properly described by this model. Whereas the width of core levels does not change substantially upon forming a solid, their energetic position as seen by a core level spectroscopy does. Core holes can effectively be screened by valence band electrons, reducing the total energy of the excited state. This is equivalent to a smaller binding energy of core electrons relative to the vacuum level. This "extra-atomic relaxation" [3.36] can lead to the formation of "semilocalized excitations" in metals, due to the small screening radius [3.37]. Similar excitonic states have been postulated in the theory of alloys [3.38] when elements with one or more valence electrons less than the main constituent (corresponding to the positively charged core hole) are alloyed to a metal.

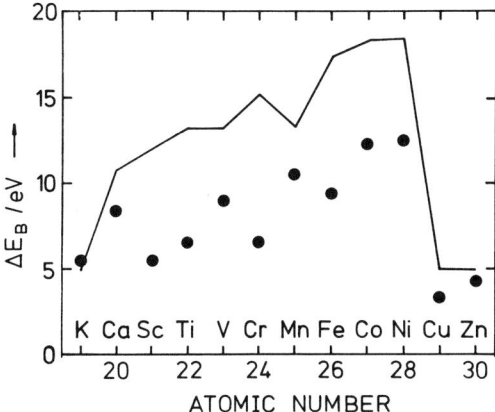

Fig.3.5 Variation of the L-shell binding energies of 3d-elements upon transition from the gaseous to the solid state. The dots are experimental values, the lines are the result of calculations based upon the extra-atomic relaxation concept (after [3.36])

In Fig.3.5 the decrease in binding energy of the L-shell of some 3d-elements upon solidification is shown [3.37]. It is seen to be of the order of 3 to 12 eV. This is somewhat less than the theoretical prediction based on the concept of extra-atomic relaxation but the sharp edge near Ni, which is associated with the filling of the very effectively screening 3d-orbitals, is well reproduced. The energy shift is nearly the same for neighboring shells. Accordingly, X-ray emission spectra show small energy shifts only (0.01 to 0.1 eV) [3.39]. At first glance, one would expect the same behavior for Auger electrons if core levels only are involved. To the contrary, the energy shift is much larger! The reason is that in the final state there are two core holes which are much more efficiently screened by the extra-atomic relaxation in solids than by the intra-atomic relaxation in free atoms, which is the only relaxation process possible in this case. The well-known "chemical shifts" (compare Chap.5) in photoelectron spectroscopy are—at least in part—due to relaxation effects, too.

The symmetry properties of the solid also give rise to several anisotropy effects. In the Auger spectra from single crystal surfaces peaks have been observed which do not correspond to Auger transitions. They appear mainly below 300 eV and are strongly temperature dependent which indicates diffraction effects. They can be explained by the assumption of a slightly different mean free path for electrons of certain energy, emitted into certain directions of the crystal. Their relative importance depends on the degree of perfection of the crystal surface and on the aperture of the analyzer [3.40].

Such an interdependence of energy and angle of emission has also been observed for Auger electrons. Whereas in free atoms anisotropies mostly are averaged out by the statistical orientation (left aside small "memory effects" of the order of percent [3.41]) this is no more the case in solids. As the atoms are rigidly built into the crystal lattice, one should expect to find some influence of the symmetry properties of the wave functions involved in the transitions on the direction of emission of Auger electrons. Simultaneously, diffraction effects like those in LEED may cause anisotropies. Investigations of polycrystalline or strongly disturbed single--crystal surfaces suggest the description of intensity distributions according to $\cos^n \theta$ with $1 \leq n \leq 2$ [3.42]. Similar distributions are found with well-annealed single crystals when high-energy lines are observed. For low-energy lines, however, strong anisotropies have been observed, for example Cu(111), Cu(100) and Fe(100) [3.42,43]. Calculations show [3.44] that by assuming isotropic emission and single scattering only of the outgoing electrons, the measurements can be partly explained. The experiences made in LEED show, however, that multiple scattering must be taken into account. Probably these calculations can be done quite analogously to those in LEED using the same formalism [3.45]. Similar problems await solution in angle--resolved measurements of UPS- and XPS-spectra (see Chap.5).

At present, these anisotropy effects are a nuisance in analysis, especially if cylindrical mirror analyzers or similar equipment with small acceptance angle is used. On the other hand, one can hope to get information about the electronic and crystallographic structure of adsorbate systems from the measurement of angular distributions of Auger electrons, provided the anisotropies of the emission itself are known. A great advantage over LEED would be that ordered structures are not necessary and that the measured signal comes directly from the atoms of interest.

3.2 Threshold Spectroscopies

In the band model, the excitation or ionization of a free atom in inner shells corresponds to an interband transition between a core level and unoccupied states of the valence band. Fig.3.6 shows the case, where a metal sample is bombarded by electrons emitted by a hot filament cathode. According to the Pauli principle, in the ground state all energy levels of the solid are occupied up to the Fermi energy E_F. Therefore, an excitation can only occur if the excitation energy suffices to raise an electron from deep-lying levels into unoccupied states above E_F, the number of which per energy interval is given by N(E), the density of states, which is characteristic for a particular sample. As in free atoms, the excitation is not necessarily accompanied by the loss of an electron from the target though a positively charged core hole has been created. The loss of an electron will occur only if it has enough energy to pass the work function barrier, and if it does not lose too much energy on its way to the surface.

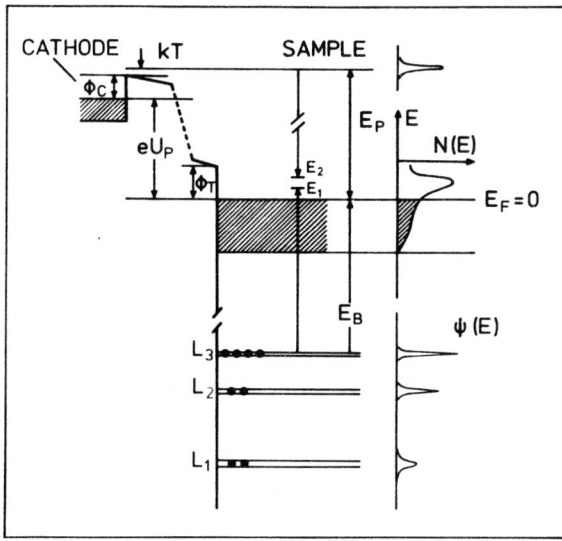

Fig. 3.6 Schematic potential diagram for the transition of a core electron with binding energy E_B into empty states of energy E_1 above E_F which is due to interaction with a kinetic electron emitted from a thermal cathode and accelerated to the energy E_p. The right-hand scale indicates schematically the energy spread of the primary beam, the density of states of the conduction band of the target which is filled up to the Fermi level E_F, and the broadening of the core levels by the creation of a hole

The simplest way to excite the sample by electrons is to bring a hot filament near the surface and to apply the voltage U_p. The energy of the electrons at the sample is then given by the accelerating voltage plus the thermal work function Φ_C of the emitter plus some small correction of the order of kT which takes the energy distribution of the thermally emitted electrons into account, the details of which depend on the beam forming system. The work function of the target Φ_T does not enter. When a field emission gun is used, the primary energy is directly proportional to the applied voltage. As the electron binding energy is referred to the Fermi level, the minimum primary energy sufficient to produce an excitation is equal to the binding energy: $E_{p,min} = E_B$. From that the basic principle of the threshold spectroscopies becomes evident. During scanning of the primary energy some characteristic signal is observed, either in the exciting beam itself or in the secondary radiation following the excitation. As the binding energy of the core levels is known, some change of the characteristic signal at a certain voltage indicates the presence of a particular element. Multiple excitations are also possible, of course, but with small probability.

Not only the excited electron must obey the band structure of the solid, but also the primary electron, which has lost its energy. Therefore, the final state after the excitation is characterized by one core hole and two electrons at or near the Fermilevel. In the one-electron model, the density of states is the same for both electrons. If we assume the transition probabilities to be independent of energy over a small range and the primary electron energy distribution as well as the core levels to be infinitely sharp, the excitation probability will be proportional to the self-convolution of the unoccupied part of the density of states: $W(E_p) \sim N(E) * N(E)$, the asterisk indicating the convolution process.

As, however, the core levels are broadened by the finite lifetime of a core hole (it is indicated in the figure as the function $\psi(E)$) the function above has to be convoluted with the broadening function which can be assumed to be Lorentzian. We arrive at the following result [3.46] for the dependence of the excitation probability on the primary energy near an excitation threshold:

$$W(E_p) = C \int_0^{E_p} \psi(E' - E_B) \left[\int_0^{E_p+E'} N(E_1) N(E_2) \delta(E_1 + E_2 - E' - E_p) dE_1 \right] dE' \tag{3.5}$$

where the δ-function assures the conservation of energy. The energy spread of the primary beam can—at least in principle—be made arbitrarily small and is neglected here. In most cases there is a large background signal which can be suppressed by differentiating electronically the measured signal. Thus we obtain for the recorded intensity

$$I(E_p) \sim dW(E_p)/dE_p \quad . \tag{3.6}$$

We will illustrate the rather complex operations in two typical examples in Fig.3.7, which shows schematically the density of states of two 3d-metals. The peak results from the 3d-orbitals which are partly (left) or completely filled (right). The cross-hatched area indicates that part of the conduction band, which is occupied by electrons. The curve named N(E)*N(E) corresponds to the self-convolution of the unoccupied part of N(E). The left-hand figure is representative for Ni, showing an asymmetric peak. The right-hand figure stands for Cu with its filled 3d-band, showing a step of relatively small intensity. In any case, the onset of the features is at the Fermi level, thus allowing a determination of the binding energy and of the element from that. When the density of states has much structure above E_F, the experiment can yield quite complex spectra which are not easy to interpret.

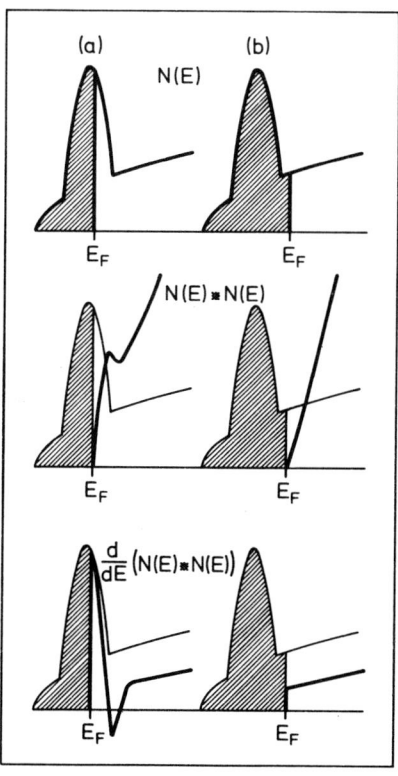

Fig.3.7 Schematic representation of the density of states of 3d-metals and its influence on the excitation function near threshold. Case (a) with a partially filled d-band is realized for example in Ni, case (b) is typical for Cu. N(E)*N(E) is the self-convolution of the empty part of the density of states. The measured quantity is (if broadening effects are absent) proportional to the derivative of the self-convolution. In case (a) it exhibits a large peak near E_F, whereas it is only a small step in case (b). This dependence of shape and strength of the signal on the unoccupied density of states is the main reason for the large variations in sensitivity for different elements found in the threshold spectroscopies (after [3.47])

They might also be explained as characteristic losses of the primary electrons. In the two-step model [3.48], core level excitation and characteristic losses (plasmon excitation, interband transitions) are independent of each other (extrinsic excitation). Therefore, at the high-energy side of an excitation feature, it may be reproduced with lower intensity and broadened by the convolution with the line shape of the loss feature. The intensity ratio should approximately equal that of characteristic loss spectra. When intrinsic excitations occur [3.49], i.e., core level excitation and plasmon generation happen simultaneously, the loss structure will show similar intensity as the no-loss peak. But such an interpretation requires some knowledge of the density of states. It should be emphasized that the density of states used here is not necessarily identical to that obtained from band structure calculations. A transition of the type we consider here involves a strongly localized inner orbital. Therefore $N(E)$ should be considered as a "local density of states", i.e., the density of states weighted by the probability density of the wave function at the location of the atom (compare to Vol.4, Chap.1 of Topics in Applied Physics). In composite materials, the local density of states is influenced by neighbor atoms, but the excitation process still resembles, at least partly, an intra-atomic process.

In summary: simple signal shapes can only be expected for a simple local density of states and absence of intrinsic loss mechanisms. The signal strength depends sensitively on the details of the electronic structure of the atom and its environment, in particular on the density of states near E_F. This feature is common to all spectroscopies where threshold excitations are observed.

We come back to the question of by which signal we can observe the threshold excitation. In principle there are two ways: either we observe the secondary particles emitted during the deexcitation, that means the Auger electrons or characteristic X-rays, or we analyze the exciting beam and observe its variation near threshold. Both alternatives have been followed, giving rise to the methods "Auger Electron Appearance Potential Spectroscopy" (AEAPS) and "Soft X-ray Appearance Potential Spectroscopy" (SXAPS) on the one hand, and "Dis-Appearance Potential Spectroscopy" (DAPS) on the other hand. The results of the former ones depend on the deexcitation mechanism, whereas those of the latter do not.

3.2.1 Observing the Excitation: *Disappearance Potential Spectroscopy (DAPS)*

In analogy to the FRANCK-HERTZ experiment [3.50] the primary energy is varied and the quasielastically reflected electrons are observed. When the primary energy is sufficient to produce a new excitation, those electrons which created a core hole lose their energy, remain in the solid, and "disappear" from the reflected beam [3.51]. Consequently, the coefficient of quasielastic reflection $R(E)$ decreases at an excitation threshold, being a measure of the excitation probability. As no secondary process is involved, this method offers the unique possibility to observe the excitation directly, without any interference with decay processes. The variation of R(E) is rather small. In the case of the L_3-shell of vanadium the decrease has been observed to be of the order of 1 in 10^3 [3.52]. Therefore, electronic differentiation of the reflected current with respect to the primary energy, which suppresses the continuous background and enhances the threshold features, has proven advantageous.

The experimental requirements of the method are: a simple electron gun which produces a beam with variable energy at constant intensity and shape, and a simple analyzer which separates out the quasielastic part of the electron spectrum. In principle every energy analyzer can be used, provided its pass energy can be varied simultaneously with the primary energy. In particular a cylindrical mirror analyzer or similar equipment can be used. A certain disadvantage of dispersive analyzers in this respect is their constant relative resolution which makes the transmission and with it the measured intensity vary over the energy range. It is also advisable to use an analyzer with large acceptance angle in order to maximize the signal and to minimize diffraction effects of the electrons at the surface. A particularly simple device using a retarding field analyzer is shown schematically in Fig.3.8. In the switch position "DAPS" the accelerating potential applied to the filament is varied between 50 V and 2000 V. A special emission control keeps the beam current constant to within 20%. The electrons coming from the target are decelerated

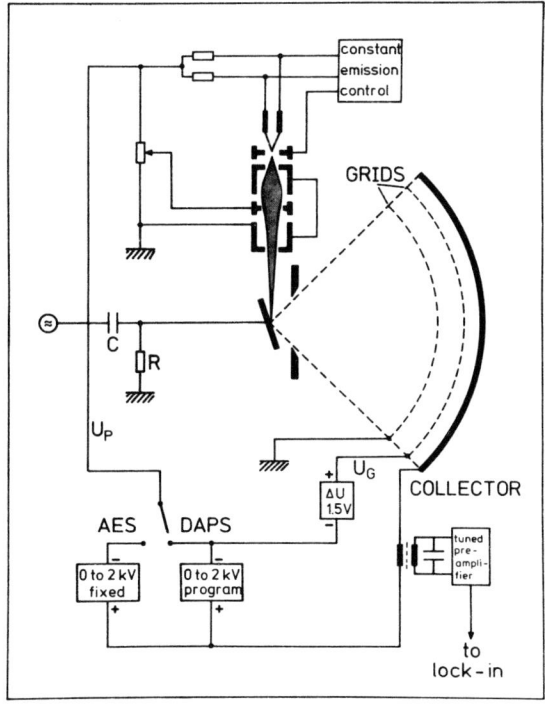

Fig. 3.8 Schematic experimental setup for surface analysis in a simple LEED system using DAPS and AES [3.52]

between two concentric spherical grids. The voltage of the second grid is at the potential $U_p - \Delta U$ with ΔU adjusted so that only the quasielastically reflected electrons can reach the collector (ΔU is of the order of 1 V). In order to enhance the threshold structures, the following modulation technique is applied: the target potential is modulated by 0.1 to 1 V_{rms} around ground potential via an RC coupler. The impact energy of the electrons at the target is modulated by the same value, but not their energy with respect to the analyzer. The potential difference between target and first grid is so small that no loss of resolution is to be observed. The AC component of the collector current is fed into the lock-in amplifier via a tuned preamplifier. The output of the lock-in is (for small modulation voltages) proportional to the derivative of the reflection coefficient R(E) with respect to the primary energy. By tuning to the first harmonic of the modulation frequency, the second derivative can be obtained. If the primary energy is kept constant by a second supply (switch position "AES"), Auger spectra can be obtained by the same experimental setup.

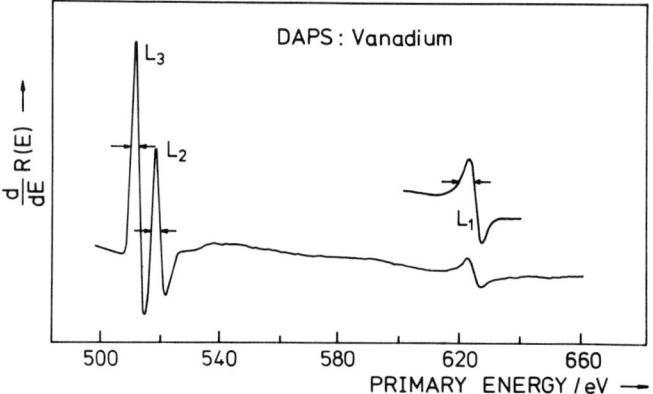

Fig.3.9 Typical DAP spectrum for L-shell excitation of a clean polycrystalline vanadium surface. The negative derivative of R(E), the coefficient of quasielastic reflection, is shown as a function of the primary energy (uncorrected for the work function of the cathode). R(E) decreases at the threshold. Note the increasing width from L_3 to L_1 which is due to the increasing lifetime-broadening of the core levels

As a typical example for DAPS, Fig.3.9 shows the results from a clean vanadium surface in the region of the L-shells. The y-axis indicates the quantity dR(E)/dE which is the differential decrease of the elastically reflected intensity for constant primary beam current. The spectrum shows the typical shape of the L-shell excitation of 3d-metals. The L_3-excitation is the most intense and sharpest one. The L_2-excitation is smaller, partly due to the smaller electron population (2 instead of 4 in the L_3-shell) partly due to the broadening by the reduced lifetime of the L_2-hole (Coster-Kronig transitions are energetically possible). The L_1-intensity seems to be smaller than that of L_2. This is in part due to the even smaller lifetime and increased linewidth, in part also to the smaller excitation cross section for the shells with higher binding energy (approximately proportional to E_B^{-2}; compare with Fig.3.2).

It is interesting to see how the results obtained with DAPS compare with other spectroscopies which observe a very similar electronic transition. In Fig.3.10 we show the integrated DAP spectrum of the $L_{2/3}$-excitation together with the ionization-loss spectrum and the X-ray obserption spectrum. Ionization-loss spectroscopy (ILS), which is explained in more detail below, observes the electronic excitation by means of the loss spectrum of the inelastically scattered electrons. In the schematic of Fig.3.6 the energy loss is small, i.e., E_2 is large and the line shape of the spectrum is determined directly by the density of states above E_F (not by its self-convolution), which is convoluted with the core level function $\psi(E)$. In X-ray absorption, the photon gives away its energy completely, but the final state is the same: one core hole and one electron near E_F. As the selection rules are valid in X-ray absorption, the small s-like part of the density of states is suppressed and

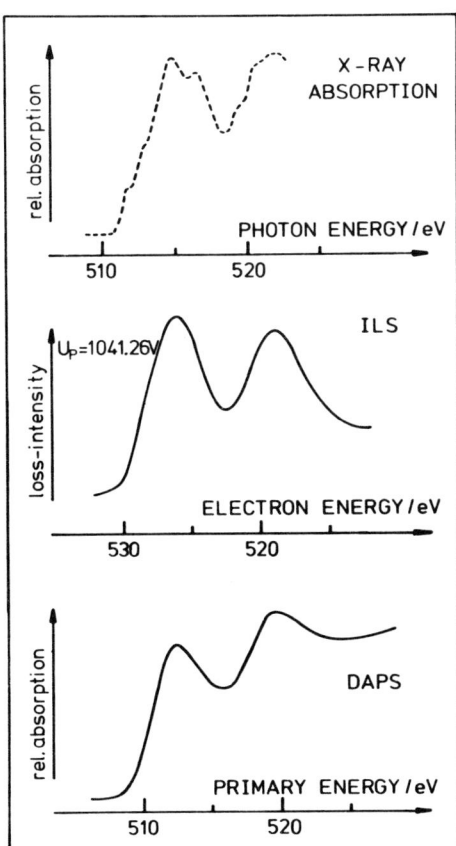

Fig. 3.10 Comparison of the reintegrated DAP spectrum (bottom) of V with an ionization loss spectrum and an X-ray absorption spectrum (top) of the same subshells. The fine structure found in the latter is not resolved in ILS due to the lack of sufficient energy resolution. In DAPS the fine structure is smeared out by the self-convolution process. All spectra agree in that the density of states above E_F is mainly determined by a large single peak

the spectra are slightly different in principle [3.53]. The high-energy resolution of X-ray absorption is not obtained in ILS, partly due to the primary beam width, but mainly due to signal-to-noise problems which require a relatively large modulation amplitude which in turn deteriorates the energy resolution. Taking this into account, the spectra look quite similar. DAPS has, in principle, better experimental energy resolution than ILS. The broadening due to the energy spread of the primary beam and to the modulation is about 0.6 to 0.8 eV. Therefore the apparent broadening in DAPS is mainly a direct consequence of the self-convolution procedure. Also the fine structure is smeared out. From this comparison of different methods, applied to the same material, we can conclude that the band model is useful in the interpretation of spectra and that DAPS is indeed sampling the empty density of states which in this case consists mainly of one peak. The width of this peak or the core level width can be estimated if one of them is known.

For purposes of analysis, the simplicity of the apparatus and the relatively small number of lines for each element is attractive. The latter point is a definite advantage over methods which observe the deexcitation products like Auger analysis and X-ray analysis. This feature is found in photoelectron spectroscopy also, but at the price of much more sophisticated instrumentation.

The principal disadvantage of DAPS (and all other threshold spectroscopies) is its strong variation in sensitivity for different elements. As already shown in Fig.3.7 the intensity and the shape of the observed lines depend largely on the density of states available above E_F. Elements with a peak structure at or near E_F are detected with high sensitivity, elements without it, for example copper, are barely detectable. Besides a low density of states, small transition probabilities near threshold (which we assumed being constant up to now) can also inhibit the detection of elements. It is known from X-ray absorption measurements that some noble metals have very small absorption at the threshold ("delayed transitions") which is explained by the centrifugal barrier which core electrons have to surmount in order to get into orbitals of high angular momentum [3.54,71]. A similar effect seems to exist for the electron excited threshold spectroscopies. Up to now, no useful spectra of Au and Pd have been published, though it is known that these materials do have a high density of states above E_F.

One possibility to remove this obstacle partially is to re-integrate the spectra, to subtract the background, and to measure the slope of the spectrum well above the threshold. In the case of the 3d-metals, this is equivalent to measuring the s-like part of the density of states neglecting the d-like peaks. It was reported that in this way the large variations in sensitivity of SXAPS could be removed [3.55] and that about equal sensitivity was found for Ti through Cu. However, this method relies on some assumptions about the background and further tests with other materials seem necessary.

A general advantage of the threshold spectroscopies is the good energy resolution which normally can only be obtained by the use of high-quality analyzers. This property allows one to get chemical information by simple means which may well be comparable to those from photoelectron spectroscopy. It is well known that upon chemical bonding of an element the binding energies of the core levels are shifted. As demonstrated in Fig.3.11 for the oxidation of Ti, this can be observed also in DAPS. The surface has been treated with a certain dose of oxygen, metallic Ti and Ti-oxide being present simultaneously. The chemical shift of the Ti $L_{2/3}$-levels amounts to 1.9 eV to higher binding energy as determined from the splitting of the lines (1st harmonic detection). This value agree quite well with the 1.9 eV found by X-ray absorption in Ti_2O_3 [3.53].

Shifts can also be produced by modification of the unoccupied part of the band structure, for example if a band gap opens upon oxidation. Therefore, one should not expect to observe the same shifts as in photoelectron-spectroscopy which samples the core levels only [3.56]. Line splitting as in the present case can be observed only if the chemical shift is large and/or the line is sufficiently sharp.

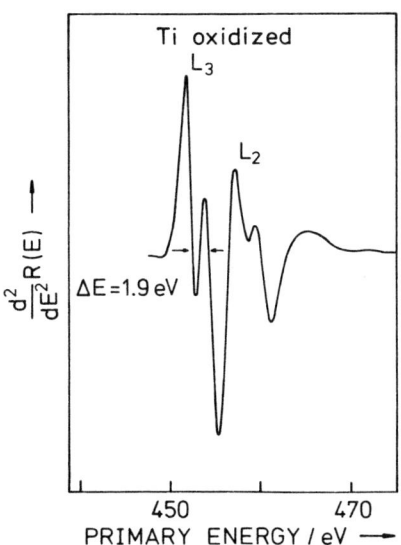

Fig. 3.11 Line splitting of Ti($L_{2/3}$) observed in DAPS attributable to the chemical shift of the core levels during oxidation (second derivative representation), which indicates the simultaneous presence of metallic and oxidized Ti at the surface. Experimental conditions: 100 μA primary current, 0.4 V_{rms} modulation, 0.3 s time-constant

If this is not so, line broadening only is to be observed, which allows less detailed conclusions to be drawn.

A special property of DAPS is its surface sensitivity. As the quasielastically backscattered electrons only are measured, their penetration depth determines the number of atomic layers which contribute to the signal. Let us assume for simplicity normal incidence and normal exit and the exponential law to be valid. The probability that an electron travels down to the depth x and back to the surface without energy loss is given by $\exp(-(2x/\lambda))$ where λ is the mean free path at the primary energy (compare Chap.1). In other spectroscopies like Auger spectroscopy, photoelectron spectroscopy, ultraviolet photoelectron spectroscopy, SXAPS, and AEAPS, the contribution from deeper layers is determined by the mean free path either of the incident or of the outgoing electron. Assuming similar geometry, we see that in DAPS for a certain depth x the no-loss probability is smaller by the factor $\exp(x/\lambda)$. From that a particularly high sensitivity of DAPS for the first one to three atomic layers results, depending on the material and the electron energy. In this respect it is comparable only to ILS where similar conditions exist. The high surface sensitivity of DAPS is illustrated in Section 3.6 by a direct comparison with AEAPS.

3.2.2 Observing the Deexcitation

A core hole state decays after a time of the order of 10^{-15} s, accompanied by the emission of Auger electrons or X-rays. Therefore, the decrease of the quasielastically reflected electron current at an excitation threshold corresponds to the simultaneous increase of the true secondary electron current and the total X-ray intensity. The decay must not necessarily occur in a single transition of a conduction electron into the hole, but may be split into several steps, including Coster-Kronig transitions, each of them being capable of creating one or more "hot" secondary electrons. Thus a cascade of electrons with different energies may result from one single excitation. Similarly, the decay via photon emission can lead to the creation of several photons with different energies. Combined processes are possible, too. The relative importance of the different decay mechanisms may vary from one shell to the other. Thus it is clear, that the "appearance" of secondary particles must be preceded by the creation of a hole as a necessary condition, but it is not necessarily true that the secondary particle flux is exactly proportional to the excitation probability. In Section 3.6 we show an example where this becomes evident.

Auger Electron Appearance Potential Spectroscopy (AEAPS)

The simplest way to do AEAPS is shown in Fig.3.12. A gun irradiates the target with electrons. The electron energy is modulated by adding a sinusoidal voltage to the accelerating potential. The target current, which is the difference between primary current and secondary current leaving the probe, is fed into the lock-in amplifier. A certain difficulty is encountered by the fact that the energy modulation may induce an intensity modulation which disturbs the measurement. This is removed by applying a part of the modulation voltage to the focusing electrodes. The measured current contains also the quasielastically reflected electrons, the intensity of which decreases at thresholds, counterbalancing the increase of secondary electrons. However, this influence is negligible since the reflection coeffecient is very small (10^{-4} to 10^{-3}). Instead of measuring the target current, a simple collector, biased slightly positive, can be placed near the target which collects the slow secondary electrons only [3.52]. A third alternative is to suppress the secondaries by a broad-band energy filter built by retarding fields [3.58] and to make use of the high-energy Auger electrons only. Though all three methods yield very similar spectra, there are differences in the signal-to-noise ratio. For the first and second method it is about equally good, whereas for the third one it is about one to two orders of magnitude worse. AEAPS using secondary electrons is the most sensitive of all threshold spectroscopies.

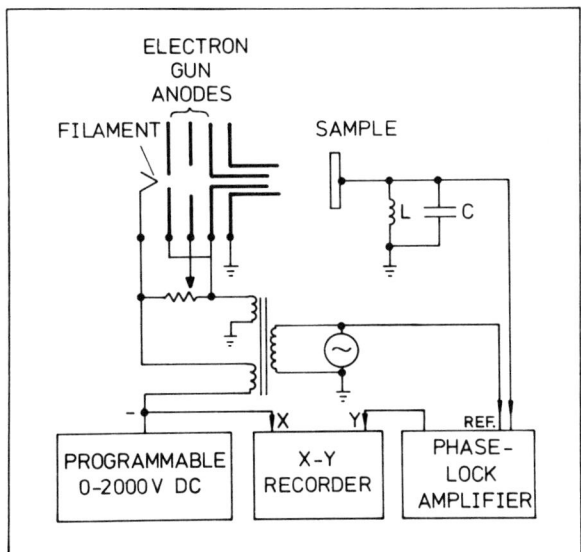

Fig.3.12 Schematic representation of the method AEAPS. The potential of the electron gun is oscillated by the isolation transformer and swept by the programmable supply. The AC current from the target, which is proportional to the derivative of the secondary electron yield $\gamma(E)$, is detected by the tuned tank circuit and phase-lock amplifier (after [3.57])

Referred to the same primary current and the same signal-to-noise ratio, it is about 100 times more sensitive than DAPS and 10^4 to 10^5 times more sensitive than SXAPS in its conventional form. This enormous gain over SXAPS, which is mainly due to the absence of the high bremsstrahlung background, is neutralized partly by the presence of a "not well behaved background" [3.58]. With single crystals, below about 100 eV strong diffraction effects completely cover the relatively small variations in the secondary electron yield. At higher energies and/or for polycrystalline samples the method is well applicable. As an example, Fig.3.13 shows results for polycrystalline Ba in 1st and 2nd derivative representation. All ten excitation thresholds from M_1 to N_5 are discernible with good signal-to-noise ratio even at low primary current.

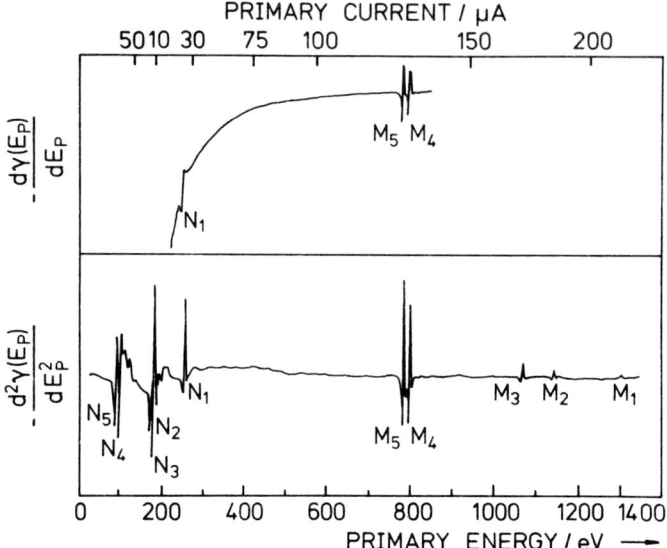

Fig 3.13 First and second derivatives of the secondary electron yield with respect to the primary energy E_p as a function of E_p for amorphous Ba. A signal-to-noise ratio of 100 : 1 is obtained for the double differentiated M_5- and M_4- peaks with 1 V_{rms} demodulation, 1 s integration time and the indicated primary current I_p (after [3.57])

Soft X-Ray Appearance Potential Spectroscopy (SXAPS)

Since the revival of SXAPS by PARK et al. [3.59], this method has been widely used and several recent reviews are available [3.47,60]. Therefore we will discuss briefly the principle, some problems of the method, and some recent developments.

The total X-ray yield from a probe irradiated by electrons results mainly from continuous bremsstrahlung. Its relative increase at an excitation threshold is therefore quite small and electronic differentiation has to be employed to make the structures clearly visible. The experimental set up is shown schematically in Fig.3.14. From the filament, which is slightly positive with respect to ground, electrons are accelerated onto the sample by a varying voltage which is modulated sinusoidally. The photons emitted from the sample pass through the grid or a thin metal foil, which rejects the electrons from the filament, and produce photoelectrons at the wall of the photocathode. These are collected by a thin wire and fed into the lock-in amplifier via a tuned LC circuit. Derivatives of first or higher order are obtained by tuning to the corresponding multiples of the modulating frequency. A typical result is shown in Fig.3.15 for a clean Cr film.

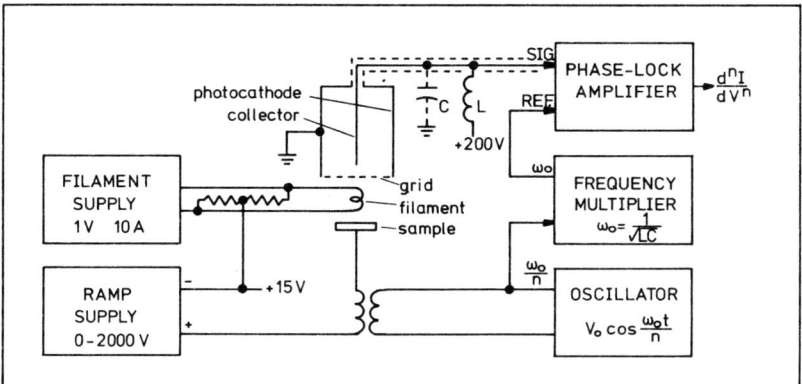

Fig.3.14 Schematic diagram of an SXAPS instrument using a simple photocathode (mostly Au) as a photodetector. To obtain the nth derivative of the collector current the target is modulated by the frequency ω_0/n where ω_0 is the resonant frequency of the LC circuit (after [3.47])

The total collector current rises approximately linearly with the electron energy. The increase near the Cr $L_{2/3}$-thresholds (exaggerated in the drawing) is below 1% and becomes clearly visible only in the derivative (curve B). The noise results from the bremsstrahlung and agrees quite well with the value calculated from the total current. The complete spectrometer can be mounted on one single flange of 35 mm inner diameter [3.56,62]. Probably it is this attractive feature which led to the widespread use of SXAPS, in spite of the considerable problems with the signal-to-noise ratio. The principal reason for these is the small X-ray yield for light elements and low binding energies which is of the order of only 10^{-2}. A further problem consists in the photocathode. If it is made of a material which itself shows SXAP structure, "ghost peaks" may appear which have nothing to do with the elemental composition of the sample [3.61]. They are produced by absorption of the high energy cutoff of the bremsstrahlung from the sample (which corresponds to an X-ray induced electron appearance potential). This problem can be avoided by using materials without sharp excitation thresholds like Au, but in any case the detection efficiency, i.e., the probability that a photon creates at least one electron, is very small [3.63]. A great improvement, at the expense of simplicity, however, has been achieved by the replacement of the metal cathode by a cooled silicon diode, which yields a detection efficiency near 1. When the low energy bremsstrahlung background is removed by a thin Al foil in front of the diode, the sensitivity is about an order of magnitude higher, though the solid angle is somewhat smaller than in the conventional detector [3.63]. A proportional counter has also been used successfully [3.64]. Elemental analysis with SXAPS is possible and has been performed for many of the elements of the periodic table [3.65]. Due to the unfavorable noise conditions rather high primary currents have to be used. The thermal load of the target may therefore be considerable and special precautions for cooling have to be observed. As to the variations in sensitivity, SXAPS suffers from the same drawbacks as all other threshold spectroscopies namely the strong influence of the density of states on the signal shape and intensity. In addition, the fluorescence yield is not only small, but also varies with the binding energy and the electron shell. (It depends on the chemical environment also [3.94].) An advantage of SXAPS is its relatively well-behaved background at low energies (though not every structure in SXAP spectra can be explained in terms of excitation thresholds [3.61]).

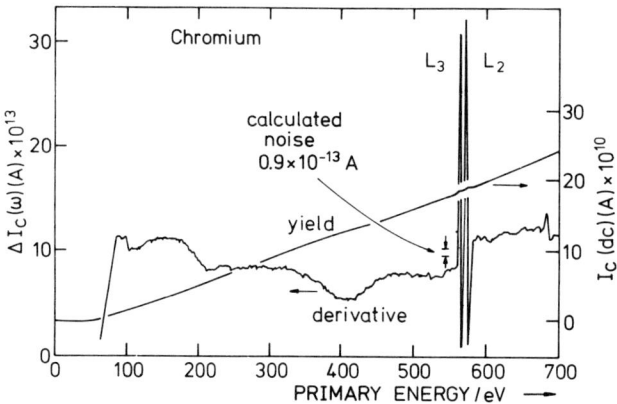

Fig. 3.15 A typical SXAP spectrum for a clean Cr film. Target current 2 mA, modulation 1 V_{rms}, bandwidth 0.25 Hz. ΔI_C (ω) is the relative collected current at the modulation frequency (after [3.61])

Chemical effects in SXAPS have been investigated quite extensively, namely in alloys [3.66], oxides [3.67] and other simple chemical compounds [3.68]. The observed phenomena like peak splitting, broadening, and chemical shift correspond to those found in AEAPS and DAPS. In general it must be said that chemical effects are not yet well understood in detail. At present their role resembles that of "fingerprints", which can be useful, too. As an example we show the SXAP spectra of O and Ni in various stages of oxidation (Fig.3.16), which have been obtained with the improved technique using a Si diode. During the chemisorption phase the oxygen signal is characterized by a single peak whereas the Ni peaks remain sharp. When oxidation occurs, the Ni lines split and the O spectrum becomes strongly structured. Thus it becomes evident that much information on the electronic structure and its variation is contained in the spectra.

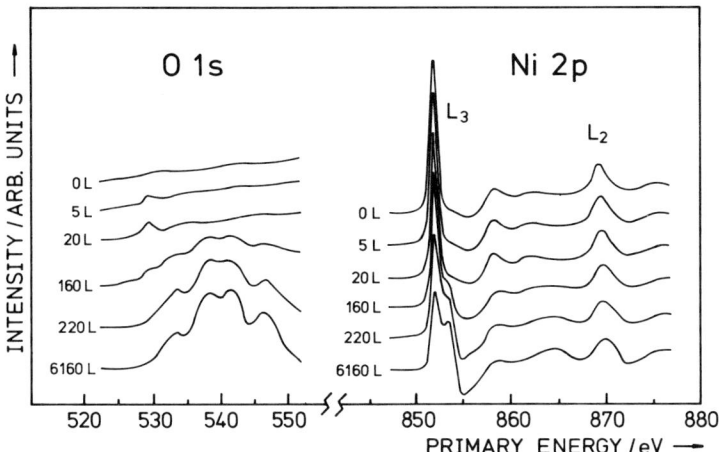

Fig 3.16 SXAPS results for the oxidation of Ni. Experimental data: 150 µA primary current, 0.5 V_{rms} modulation for oxygen (0.2 V_{rms} for Ni), 10 s time constant (after [3.69])

Though the one electron band model is able to explain many of the features of the threshold spectroscopies, there are cases where its failure is obvious. Responsible is the neglect of the interaction of the core hole with the outer electrons in the final state, which is able to modify the empty density of states. Such an example is shown for the M_5-shell of Ba in Fig.3.17. SXAPS yields a peak near 790 eV with a shoulder near 785 eV. Also shown is the spectrum obtained with X-ray--induced Electron Appearance Potential Spectroscopy (XEAPS), a method which is si-

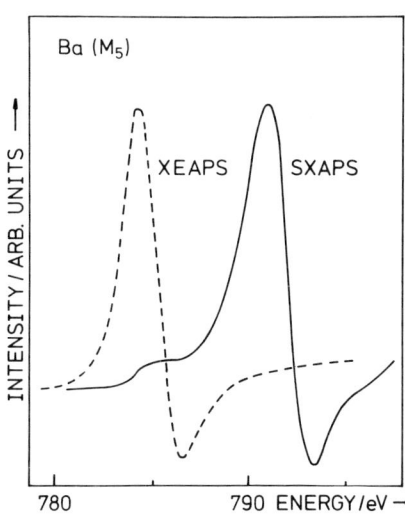

Fig. 3.17 Comparison of X-ray-induced electron appearance potential spectroscopy (XEAPS) and soft X-ray appearance potential spectroscopy (SXAPS). The difference of the spectra may result from the different final state after the excitation: one electron near E_F in the case of XEAPS and two interacting electrons in the case of SXAPS (after [3.70]

milar to the photo-yield curves obtained by scanning the energy of monochromatic synchrotron radiation [3.70-72]. The electron transition being induced by a photon, the main difference to SXAPS is that only one electron resides near the Fermi level in the final state. The peak of the SXAP spectrum is placed at the position of the shoulder in SXAPS. An explanation for this phenomenon has been given with a model which employs two different densities of states, depending on the excitation energy [3.73]. Though this explanation seems plausible, another phenomenon may also play a role.

One of the assumptions underlying SXAPS is that the measured photon yield is directly proportional to the number of core hole excitations. In the light of the following observations, however, this seems questionable. In investigations of the characteristic X-ray radiation from the lanthanides, line-like resonances of the bremsstrahlung were found near excitation thresholds [3.74]. Fig.3.18 displays the X-ray spectra from La at electron energies around the threshold energy for M_α-radiation which follows the transition of M_5-electrons to the Fermi level. Even if the primary energy is *below* the threshold, a line-like radiation appears which is energy dependent and runs through a maximum near the threshold. Compared to this radiation, the M_α-emission has low intensity. As an explanation, it has been proposed [3.74,75] that below the threshold a virtual localized state is excited by the primary electron, which is characterized by a 3d-core hole and two electrons in 4f-orbitals above the Fermi level. This resonant state decays via the transition of an electron to the core hole and emission of a bremsstrahlung photon. Such resonant emission processes could have large influence on the SXAP spectra. It depends on the fluorescence yield, to what extent they mask the characteristic radiation. This example shows that the detailed interpretation of SXAP spectra should

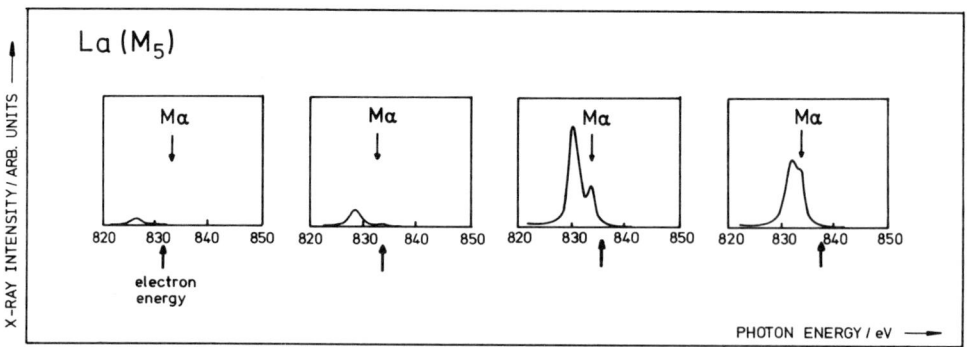

Fig.3.18 Occurrence of line-like resonances in the bremsstrahlung spectra of La when the energy of the primary electrons approaches the threshold for characteristic X-ray excitation. The bremsstrahlung resonance can be much more intense than the characteristic radiation (after [3.74])

be done very carefully [3.76,77]. There is no doubt that a comparison with other threshold spectroscopies like DAPS, which samples the excitation mechanism, would be very useful.

3.3 Ionization Loss Spectroscopy (ILS)

Contrarily to photons, electrons can transfer their energy to bound electrons in arbitrary amounts, the minimum of which is given by the lowest excitation energy of the system. Consequently, the energy loss spectra from solids show an edge at some minimum energy, followed by a broad shoulder extending down to zero energy. Considering the one-electron model of Fig.3.6, and assuming the primary energy to be very high, we see the minimum energy loss to be determined by the transition of a core electron to the lowest unoccupied states, i.e., to the Fermi level. When the relative energy loss is small, the energetic position of the primary electron after the excitation is still far above E_F. The density of states in this region is given by that of free electrons (proportional to $\sqrt{E}$) which can be assumed constant for small energy changes. Assuming constant transition probability, the shape of the loss function near threshold is proportional to the density of unoccupied states above E_F, *not* to its self-convolution as in the threshold spectroscopies. As far as nomenclature is concerned, one could think of including ILS in the threshold spectroscopies as the onset of a characteristic loss is observed. But this method is based upon an analysis of the energy spectrum, the results depending on the resolving power of the analyzer, whereas the threshold spectroscopies measure some integral yield, the resolution depending on the primary beam only. If the core level broadening is not negligible, one obtains the convolution of the function $\psi(E)$ with the density of unoccupied states. Additional characteristic energy losses of the incident or outgoing electrons can produce a weak replica of the main spectrum, folded with the line shape of the losses.

The applicability of ILS to elemental analysis is evident. As the core level excitation process is the same as in the threshold spectroscopies, a sensitivity of ILS to chemical effects can be expected either by modification of the density of states or by shifts of the core levels. ILS resembles DAPS in that it samples the excitation itself, independent of the subsequent decay. That means that Auger or X-ray yields play no role. When characteristic losses are excluded from IL spectra, the surface sensitivity should be high, because the condition of no other loss than that for core level excitation is imposed. In this respect it resembles DAPS, too, with the difference that the mean free path for entering and leaving electrons is not the same.

ILS can be carried out by means of simple analyzers, like the retarding field analyzer, utilizing electronic differentiation [3.78] (the IL spectrum of Fig.3.10 has been obtained with a analyzer of this type). In general however, the signal-

-to-noise ratio is too unfavorable. In this respect, dispersive analyzers are superior if the acceptance angle is sufficiently large. This is the case in cylindrical mirror analyzers, which are often used for this purpose. The energy resolution required in ILS is rather high: it should be well below 1 eV to obtain high-quality spectra. The easiest way to obtain this resolution is to apply pre-retardation to the electrons before entering the analyzer. An example for a system of this type is shown in Fig.3.19. On the axis of the cylindrical mirror an electron gun is mounted, which has to produce a beam of variable energy, high intensity and narrow cross section (for example 1 mA at 1000 eV into a spot of 0.25 mm diameter [3.80]). Two concentric hemispherical grids surround the target, which is thus placed in a field free region. The electrons leaving the target are retarded between the grids down to the fixed pass energy of the analyzer (100 to several hundred eV). The spectrum is scanned by increasing the accelerating potential at the gun. The collector current changes, when the loss structure, which is referred to the primary energy, passes by the fixed energy window of the CMA. As usual, the structures are enhanced by using the lock-in technique, mostly in the form of the second derivative of the energy spectrum. The advantage of this technique of obtaining IL spectra is the constancy of the absolute energy resolution throughout the entire spectrum. This is gained at the expense of varying transmission as the primary beam has a finite diameter.

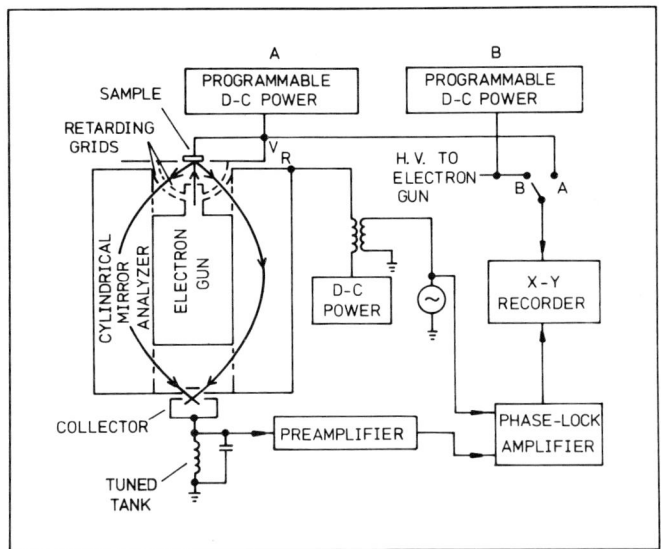

Fig.3.19 Simplified schematic of a high-resolution spectrometer with pre-retardation used for ILS and AES. Loss spectra are obtained with power supply B (= primary energy) programmed, supply A fixed, switch in position B. Auger spectra are obtained with A programmed, B fixed and switch in position A. The pass energy of the cylindrical mirror analyzer is fixed by the dc supply ($\sim$ 100 eV) (after [3.79])

The same apparatus can be used for Auger spectroscopy. Then the primary energy and the pass energy are kept constant and the retarding potential is varied. In this way a better energy resolution at high Auger energies can be obtained than in the common CMA.

Using this type of analyzer, the scattering angle is fixed to about 135° with an angular resolution of about 20° [3.81]. The analyzed electrons have mostly suffered single scattering only. Consequently, the measured intensity is not determined by the total ionization cross section, but by the differential cross section, integrated over the angular resolution. This fact opens the possibility to measure the differential cross sections for electron impact ionization of inner shells as a function of primary energy.

Some examples for K- and L-shells are shown in Fig.3.20. The measured values are generally higher than those calculated from the Burhop theory. The scaling factors indicated result from magnifying the theoretical curves to be tangent to the experimental ones. The measured cross sections are decreasing steadily, a maximum

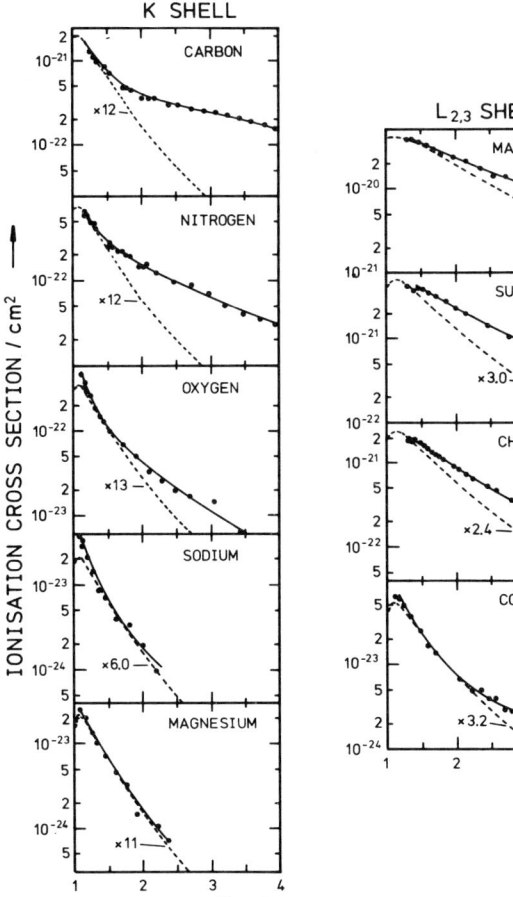

Fig. 3.20 Differential cross section for inelastic scattering by 128° to 143° with ionization of the K- or L-shell of some elements adsorbed on W (100) versus the reduced energy U (after [3.81,82])

near U = 1.1 to U = 1.4 can hardly be recognized. The decrease is in general slower than predicted and the rate of decrease is different for the elements.

For the use of ILS in elemental analysis, some important conclusions follow from these considerations. First, the sensitivity is expected to depend on the density of unoccupied states. A sharp ionization loss peak will be observed only if the density of states near E_F is high and concentrated in a narrow band. This is the same problem as encountered in the threshold spectroscopies. Secondly, in order to obtain high intensity, it is advisable to adjust the primary energy close to the binding energy, i.e., $U \approx 1$. That means, the ionizing electron leaves the target with low energy. However, it must not be chosen too low, as it may be buried in the large secondary electron peak, causing unfavorable signal-to-noise ratio. Thus, the minimum pass energy of the described apparatus will be between 50 and 100 eV. Thirdly, as the rate of decrease of the cross section for U > 1 is not the same for different shells and different elements, the relative sensitivity for two elements depends sensitively on the absolute primary energy, even if the binding energies are similar. This is illustrated in Fig.3.21 for Be (K-shell) and Si (L-shell) with 111 eV and 100 eV binding energy, respectively. The ratio of the cross sections (not only their absolute values) decreases with increasing primary energy.

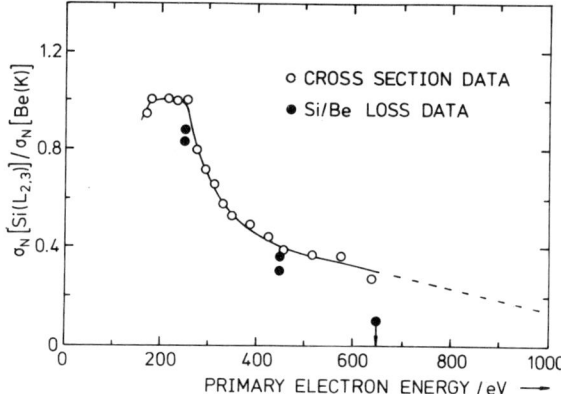

Fig. 3.21 Ratio of the differential ionization cross section of Si (L_{23}) and Be (K) calculated from the data of Fig. 3.20 for C(K) and Mg (L_{23}) versus the primary energy (o). Dots: experimental loss amplitude ratio Si/Be for a Be surface contaminated by Si. The signal strength in ILS depends on the ratio of primary energy to binding energy, but is different for different electron shells with the same binding energy. The strong decrease of the signal indicates the segregation of Si to the surface (after [3.83])

If this effect is not taken into account, the presence of some elements may be overlooked in the analysis. Since the surface sensitivity decreases with increasing primary energy the deviation of the measured Si/Be data from the predicted curve can be interpreted as evidence for the segregation of Si to the surface of

Be. These particular properties of ILS are an outcome of the fact that differential cross sections rather than total cross sections play a role. Therefore, the observed intensities depend also on the type of analyzer and the solid angle it subtends. The total cross section has a maximum near U = 3 to 4 which is approximately the same for all electron shells. Therefore effects like the one described here are not to be expected in spectroscopies sampling the total cross section, like Auger or X-ray analysis. At U >> 1 the total cross section is mainly determined by inelastic small angle scattering ($0°$ to $40°$) [3.81]. At U ≈ 1 the total cross section is small and is determined by large angle backscattering, i.e., those events which are observed in ILS.

The main advantage of ILS for elemental analysis lies in the simplicity of the spectra. As in the threshold spectroscopies, the number of lines is relatively small, the lines are sharp, and overlap is rare. An example for a spectrum is shown in Fig.3.34 in a comparison with Auger spectroscopy. It shall be mentioned that ILS is also successfully applied in transmission electron microscopes at high primary energy. In this case, no use of its surface sensitivity is made, the bulk composition only being sampled [3.84].

As mentioned above, ILS is sensitive to chemical effects. In Fig.3.22 we show a typical example for carbon on tungsten. After bakeout of the vacuum system, the W(100) is covered with graphitized carbon. Spectrum (a) displays the large doublet peak (in second derivative representation); the tungsten is barely visible. After heating to about 1500 K, the tungsten lines are clearly visible and the carbon peak becomes a singlet. Though a detailed explanation cannot be given at present, the result show the sensitivity of ILS to the chemical state of carbon. ILS can be ap-

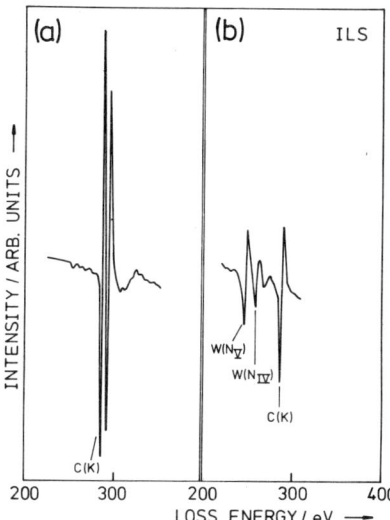

Fig. 3.22 Ionization loss spectra for two chemical states of C on W(100). (a) After bake out, the sample is covered with partially graphitized carbon, producing the doublet peak structure. (b) Upon heating to about 1500 K, the carbon peak becomes a singlet (after [3.85])

plied to single crystals, too, without particular problems. A special "chemical effect" has been observed for clean Si single crystals [3.86] (compare with Chap. 6). The IL spectra for Si(100) and Si(111) at the L-shell excitation threshold are very similar, showing multiplet peaks around 100 eV. When the surface is bombarded with Ar^+ ions, the multiplet peaks vanish and a large singlet peak appears which is shifted by about 1 eV from the main line of the single crystal spectra. Thus the distinction between crystalline and amorphous Si becomes possible. Due to the high surface sensitivity of ILS it has also been possible to detect surface states present at the clean single crystal surface.

Thus, ILS seems to be a promising technique, in spite of the rather high cost of the equipment, which is comparable to that of photoelectron spectroscopy. It seems to be of particular value in basic surface studies, where it can complement other techniques.

3.4 Auger Electron Spectroscopy (AES)

Of all spectroscopies discussed here, AES has reached the highest level of sophistication and has found the most widespread use. Up to now, more than 1500 papers have been published dealing with Auger spectroscopy (see the bibliography compiled by HAWKINS [3.87]). The method itself can be considered well known; several reviews are available [3.88-91]. In particular we wish to mention the paper of BAUER in Volume 4, Chapter 6 of Topics in Applied Physics. In our context, we will discuss the influence of the chemical environment on Auger lines, some of the more recent application-oriented developments in AES, and the relationship of AES to the other core level spectroscopies, including X-ray analysis.

3.4.1 Influence of the Atomic Environment

If one compares Auger lines of the same element in gaseous and in solid state, they are found to be shifted in energy as discussed above, but also to be considerable broader than in the gaseous state. There are two main reasons: lifetime broadening of the initial state [3.92] (i.e., of the core hole), and lifetime broadening of the final state [3.93]. The lifetime of the initial state contributes most to the width of Auger lines from free atoms. The variation of core level lifetimes upon solidification cannot be excluded, but it seems improbable that these should increase by several eV. There remains the lifetime broadening of the final state. After an Auger transition in gases, a further deexcitation of the highly ionized atom is often not possible for energetic reasons. Therefore the final state is long-lived. In the solid, there are many electrons available from the valence band which may take part in further rearrangements of the electron shell. Therefore the final state may be short-lived. It was shown in several examples [3.93] that measured solid state spectra can be obtained from gaseous state spectra by broadening with a Lorentz function and convolution with the energy loss spectrum of inelastically scattered electrons. This seems to indicate that the lifetime broadening of the final state is an important factor in solid state Auger emission. Another reason might be the overlapping of Auger lines from surface atoms with those from bulk atoms which might have different energies due to different extra-atomic relaxation energies. However, those effects are difficult to separate out. In composite materials, low-energy Auger lines might be broadened by lifetime broadening of the final state due to cross transitions with neighbor atoms of different kind [3.94]. Whereas these are improbable for high-energy transitions, the transition rates for inter-atomic transitions below 50 eV may become comparable to those for intra-atomic transitions or even exceed them [3.95].

If transitions involving the conduction band are analyzed, one should expect the band structure to be reflected in the line shape of the Auger lines. In particular, in transitions of the type XVV, the self-convolution of the occupied part of the conduction band should be observed. In investigations of the nearly free electron metals Al and Li [3.96,97] broad lines were found which could be correlated to peaks in the band structure, but the agreement is quantitatively not satisfying. On the other hand, for Ag [3.98] and for the LVV-transitions of Cu and Zn [3.99], spectra have been observed which are similar to those of the free atoms. No influence of the band structure has been found. In view of the broadening effects mentioned above, the decision seems difficult, whether a measured spectrum is a broadened quasi-atomic line spectrum or some complicated convolution of the band structure with itself, or parts of it.

This question touches the problem of the validity of the one-electron band model for the description of electron transitions in solids. A recent work [3.100] indicates that, at least for low energy transitions, it seems not to be wrong. In oxidized Mg, the 34 eV line can be interpreted as an $L_{2/3}VV$ transition involving a common valence band. This agrees with the calculated high inter-atomic transition rates for MgO at low electron energies [3.95].

The latter example is a special case of "chemical shifts" in Auger spectra. Those shifts have been observed in purely atomic transitions also (see, for example, Ref.14-37 in [3.101] and [3.102,103]. A modification of the transition energy can be caused by an alteration of the charge distribution around the nucleus upon chemical bonding. A chemical shift then may result from a variation of the extra-atomic relaxation energy, i.e., of the screening of the two core holes by the outer "passive" electrons. In [3.104] a comparison is made between chemical shifts of the core levels, as determined by photoelectron spectroscopy, and those of Auger energies, mostly determined from Auger emission following the photoionization. Contrary to earlier considerations, based upon the shifts of the core levels alone, the Auger line shifts are often much larger than XPS shifts. They do appear even in cases where the XPS shifts are close to zero. Values around 5 eV, which are an exception in XPS, are easily found in AES. Therefore, chemical shifts in AES should be well measurable, provided the natural line width is sufficiently small and the experimental resolution of the analyzer approaches that of XPS instrumentation. Then, AES would be capable of yielding chemical information which is up to now only accessible in photoelectron spectroscopy. In the widely used low resolution Auger analysis, which trades off resolution for signal-to-noise ratio, chemical effects might be rather a nuisance than an advantage. If a sharp Auger peak lies near a broad one and is not well resolved experimentally, a chemical shift of the sharp peak may appear as "vanishing" in the broad peak. This may change substantially the observed peak shape, especially if the derivative representation is used. In this way, intensity modifications and "chemical shifts" can be observed which bear little or no relationship to the real ones. An illustrative example is given in Fig.3.23, showing AES spectra of clean and oxidized vanadium in $N(E)$ and $dN(E)/dE$ representation.

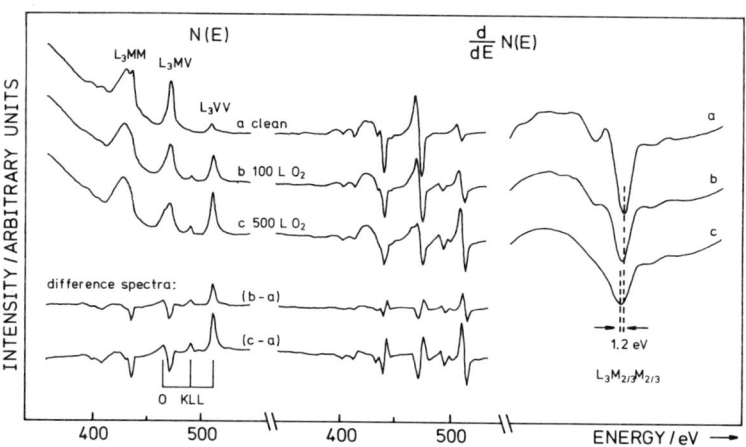

Fig. 3.23 Auger spectra of clean and oxidized vanadium in $N(E)$ and $dN(E)/dE$ representation together with their difference spectra in two stages of oxidation. The "vanishing" of the small sharp peak in the $LM_{23}M_{23}$ group, which is due to chemical shift of this line and does not reduce noticeably the over-all intensity in the energy spectrum $N(E)$, causes strong modification of the $dN(E)/dE$ spectra. The negative excursion of the main $LM_{23}M_{23}$ peak is reduced and the position of the maximum is shifted by about 1 eV. This "chemical shift" is only very indirectly related to the true alterations of the Auger spectra and allows no quantitative conclusions

The reduction in intensity and the shift of the line as observed in the derivative spectra are not real as proven in the $N(E)$ representation. Such effects clearly are an obstacle to quantitative analysis in Auger spectroscopy. It will probably best be overcome by using $N(E)$ data which can be obtained either by measuring directly the energy spectrum $N(E)$ or by re-integrating the derivative data $dN(E)/dE$.

3.4.2 Quantitative Auger Analysis

The problem of quantitative Auger analysis can be divided into two parts:
1) Determination of the relevant physical parameters entering the calculation, and
2) data recording and manipulation to extract the Auger intensities from the measured curves.

For the intensity of an Auger line, or better a group of lines, relative to the primary intensity, one may write

$$I_A/I_0 = A \cdot \sigma(E_B,E_P) \cdot \omega_A(E_B) \cdot r(E_B,E_P, \text{matrix}) \cdot \lambda(E_A) \cdot n_A \quad . \tag{3.7}$$

In the constant A we summarized the experimental factors, like the analyzer effects, angle of incidence of the primary beam, surface roughness of the sample. In principle, for quantitative work the retarding field analyzers offer some advantages over the cylindrical mirror analyzers though the signal-to-noise ratio is inferior. They are much less sensitive to anisotropy effects in the Auger emission due to the large solid angle subtended. The energy resolution can be made rather high by simple means and the transmission remains constant. In dispersive analyzers the solid angle is small and the transmission varies as a function of energy. If a multiplier is used as a current preamplifier, the dependence of gain on the electron energy and its ageing have to be taken into account. The angle of incidence of the primary beam has considerable influence on the creation of Auger electrons. If the primary energy is large and the mean free path λ_A of the Auger electron is small compared to the penetration depth of the primary beam, one would expect a dependence of the intensity proportional to the inverse cosine of the angle of incidence. In Fig.3.24 this function corresponds to the dotted line. The other two curves have been measured for an Al single crystal using a retarding field analyzer with 2π solid angle. Curve a represents the total quasielastically backscattered intensity whereas curve b shows the Auger current of the Al $L_{2/3}VV$ line. Crystalline effects are seen in the large variations of the backscattered intensity, and, less intense, in the Auger current. It should be emphasized that they are not due to the anisotropies of the Auger emission in this case, but only to different conditions for propagation of the primary electrons if the angle of incidence is changed [3.106]. By means of glancing incidence of the primary beam the signal intensity can therefore be increased. However, the quantitative work becomes more difficult in general, as at very large angles the backscattering of the primary electrons and the surface roughness effects become more important. To give an order of magnitude illustration for the latter point, a recent investigation [3.107] shall be cited. For evaporated Au layers with a mean roughness of only 0,28 μ, at 70° angle of incidence a reduction of the Auger signal by about 40% was observed as compared to a polished flat surface.

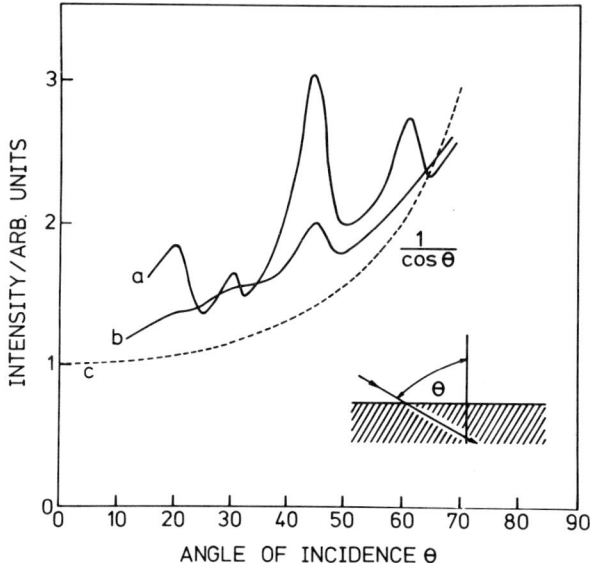

Fig.3.24 Influence of the angle of incidence of the primary beam on the Auger intensity (b) and the quasielastically reflected intensity (a). Sample: Al(001) single crystal, 68 eV Al-Auger line, primary energy 1500 eV. Curve c shows the expected increase in Auger intensity with increasing angle of incidence when only the length of the path within the surface region determines the Auger yield (after [3.105])

The total ionization cross section σ is relatively well known for inner shells, but for outer shells more data are needed. Another major problem, is the evaluation of the number of core holes which contribute to the Auger line when Coster-Kronig transitions are possible. For example, in L_3XX Auger lines there is a certain contribution from L_1- and L_2-holes which are transformed into L_3-holes via Coster-Kronig transitions. Measurements of CK yields being very scarce, one has to rely on theoretical values to estimate their contribution. When CK transitions are taken into account in this manner, the Auger yield ω_A can be set equal to 1 without large errors, the fluorescence yield being negligible for low binding energies.

The backscattering factor r is of considerable importance for quantitative work as well as for Auger microanalysis [3.108]. It describes the enhancement of an Auger line due to backscattered primary electrons and energetic secondary electrons. It depends on the binding energy of the electron shell from which the Auger electrons come, relative to the primary energy, on the angle of incidence [3.109], on the energy distribution of the backscattered electrons [3.110] and, of course, on the backscattering properties of the matrix [3.111]. For the experimental determination of the backscattering factor several methods have been developed, which since recently yield rather similar values. As an example of the importance of the effect, Fig.3.25 shows measured backscattering factors for various elements as a function of the reduced energy $U = E_p/E_B$ at primary energies up to about 2.5 keV. For higher primary energies (10 to 20 keV) Monte Carlo calculations as well as measurements exist, aiming mainly at the quantification of X-ray microanalysis [3.113,114]. The extrapolation of these results to lower energies yields r-values as a function of U which are higher than those measured at low primary energy. It is to be expected that the backscattering factor may easily reach the value r = 2 at high primary energies.

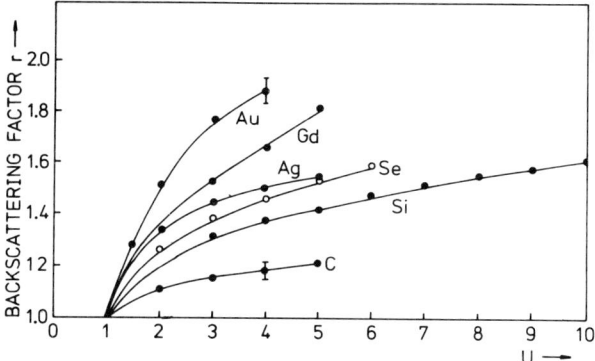

Fig.3.25 Measured backscattering factors r versus the reduced primary energy for light and heavy materials. Absolute primary energy below 2 keV (after [3.112])

As in X-ray analysis, the concentration n_A determined from the Auger current is an average value only, averaged over the information depth with some weight factor given by the free path of the Auger electrons (assuming the primary energy to be sufficiently high that no attenuation of the primary beam occurs). As seen in the compilation in Chapter 1, the scatter in the measurements is quite large, an uncertainty by a factor of 2 being not exceptional. The mean free path is at present one of the most uncertain parameters, both from the point of view of its questionable concept, as well as from the available experimental data.

After all, it seems rather difficult to determine Auger intensities from "first principles" with an accuracy better than a factor of two, though its possibility has been demonstrated in some cases [3.115]. The other approach is to compare the measured intensity from an unknown specimen with that from a standard with known composition under equal conditions. This method works quite well as long as matrix effects can be neglected, i.e., as long as pure elements can be used as standards. The use of composite materials as standards brings about serious problems due to the instability of a surface composition exposed to an environment other than extreme UHV. Even when this is assured, and a clean surface has been obtained, in general the surface composition does not equal that of the bulk [3.116]. At present, i.e., at the beginning of quantitative analysis, one mostly compares measured derivative spectra from the probe with those from the pure elements taken at similar primary energy. In spite of the many uncertainties of this method, an accuracy of 10 to 30% can be obtained [3.117].

As the line shape of the measured derivative spectra depends sensitively on many physical and experimental parameters, it seems most desirable to make use of the direct spectra, or even better, of the area under the Auger peaks as a quantitative information. This requires some procedure to separate the Auger lines from the large background. Two such methods shall be discussed in the following. They both approximate the background by some polynomial function and consider the difference between this function and the measured curve as being the true Auger spectrum. The first method, called "dynamic background subtraction" (DBS) [3.118] is based on the assumption that a line superimposed onto an arbitrary background can be separated from it by n-fold differentiation followed by n-fold integration, provided that the line is sufficiently sharp, that the background is sufficiently smooth, and that the order n is sufficiently high. This procedure is equivalent to the interpolation of the background by a polynomial of the degree (n-1). By the lock-in technique the spectrum is usually obtained differentiated to 2nd order. (In principle higher orders can be obtained also, but with increasing electronic problems). Accordingly, the peak areas are determined by parabolic interpolation (n=3). The method requires a decision by the operator, which part of the spectrum belongs to the background and which does not. Therefore, it does not work automatically. On the other hand, the technique can easily be performed by means of an analog integrator or a small multichannel analyzer. The second method [3.115] works fully automatically, requiring a mini-computer at least. The spectrum is recorded in derivative representation, is corrected for the analyzer properties (transmission, sensitivity variations, and so on) and then re-integrated. Assuming the background to decrease continuously, a number of points corresponding to the background are selected in an automatic procedure applying a tangent criterion. These points are connected by polynomials of 3rd order (spline polynomials) and the resulting curve is subtracted from the measured one. Both procedures are compared in Fig.3.26 using test functions. A gaussian function and a superposition of two gaussians, and a Lorentz curve in the case of DBS are added to an exponential background. The resulting curve is treated by the retrieving procedures.

DBS yields satisfying results in a small range around the line only with 10th-order correction or higher. The automatic procedure retrieves the lines over the complete range with good accuracy. The lines reduced by a factor of 50, which are no more visible, are retrieved also (with somewhat larger error). Real Auger lines are difficult to handle due to their multiplet structure and due to the characteristic loss features mixed up with the lines. Consequently, in both procedures a certain arbitrariness as to the low-energy cutoff of the lines is present. In DBS it is introduced by the choice of the minimum order to which the correction is performed. The spline technique tends to identify loss features as belonging to the Auger lines. An improvement would be to unfold the measured spectra with the characteristic loss spectrum taken at the same energy as the Auger lines. The treatment of the background is a crucial point in every data manipulation procedure in AES, which affects directly the precision. Even the depth of information is concerned because the inclusion of Auger electrons with energy losses extends the information depth of the analysis.

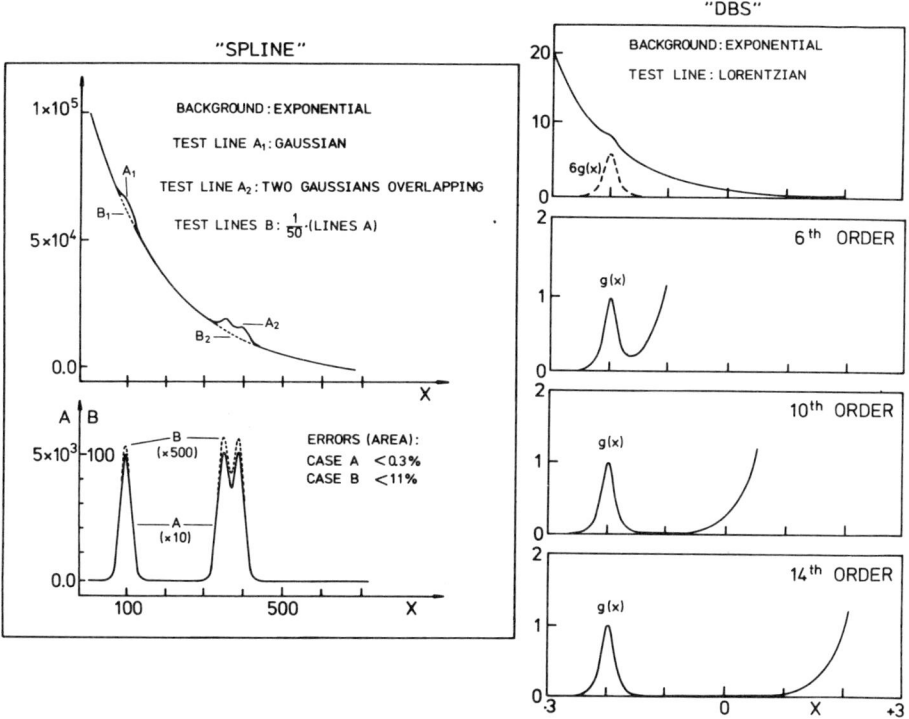

Fig.3.26 Comparison of two methods for background subtraction in AES ("dynamical background subtraction" (DBS) and the "spline" method) applied to small lines on an exponential background. In DBS interpolation by polynomials of high order is needed to obtain satisfactory results. The spline method uses polynomials of third order and retrieves very small lines also (test lines B) (after [3.115,18])

3.4.3 Auger Microanalysis

In principle, any of the core level spectroscopies using electron excitation is capable of microanalysis (which means a resolution of 1μm or less) provided a sufficiently intense beam can be focused into a small spot. Due to space charge effects this becomes the more difficult the lower the primary energy is and the higher the beam current is. In AES, microanalysis has become possible by the use of high primary energies, though a slight loss of sensitivity is caused by the smaller ionization cross section at high energy. (In practical use are beam voltages up to 10 keV). Since in AES the signal-to-noise ratio is the decisive factor, one has to use relatively high currents which in turn limit the obtainable resolution. Therefore, a compromise must be found between resolution and measuring time, which depends on the brightness of the electron gun. With the tungsten hairpin cathode, which is the most frequently used system, about 1 μA into 1 μm diameter is possible. With the LaB_6-cathode, which is more delicate in practice, an order of magnitude can be gained [3.119]. The field emission gun offers high brightness, but the total emission current is limited to some μA for stability reasons. As only a small part of it can be focused into the beam, a high resolution is obtainable only at low current levels. In addition, noise problems due to fluctuations of the surface coverage of the emitter tip are particularly unpleasant in AES [3.120]. Up to now, beam currents of some 10^{-8} A can be focused into some 10 to 100 nanometers [3.121].

If the noise in the complete analysis system is due to the shot noise of the primary beam only, the following approximate formula for the signal-to-noise ratio (S/N) of a cylindrical mirror of high transmission (therefore moderate resolution) assuming intense Auger lines is valid [3.122]:

$$S/N \sim u \cdot (\tau \cdot I_p)^{1/2} \tag{3.8}$$

where I_p is the beam current in μA, τ is the time constant of the lock-in in ms, and u is the modulating voltage applied to the analyzer in volts (peak to peak). The linear relation with u is true only for u < 10 V. If we take S/N = 5 as a tolerable value, for u = 5 V and I_p = 1 μA results a time constant τ = 1 ms. Thus, for the analysis of an area with 100 × 100 points a time of about 100 s is needed. Reducing the current to 10^{-8} A means an increase of the measuring time by a factor of 100, which is hardly tolerable for routine investigations. Thus, an extremely good resolution can be expected only for point analyses where time is no problem. A resolution of 0.1 μ in Auger images will represent a good performance. In any case, the measuring times are prolonged. Present commercial apparatus with a resolution better than 3 μm (Physical Electronics Industries, Varian Associates) take an Auger image of 200 lines within 1 to 5 min. Fig.3.27 shows some examples (using a 5 μ diameter beam) from the analysis of a Si photodiode. The impression of plasticity is produced by the simultaneous modulation of intensity and y-deflection. In the Al image a cross-shaped marker is visible which corresponds to the dark shadow in the O image. With silicon, one can take advantage from the chemical shift of the low energy lines upon oxidation. Therefore, different images of Si in elemental and in oxidized state can be obtained. The Auger image of oxidized Si agrees with that of oxygen, whereas that of the elemental Si shows the areas which were protected during the oxidation. For example note the edges of the Al cross. Though it is not always possible to make chemical effects visible in Auger images, this capability is a very useful feature of Auger microanalysis.

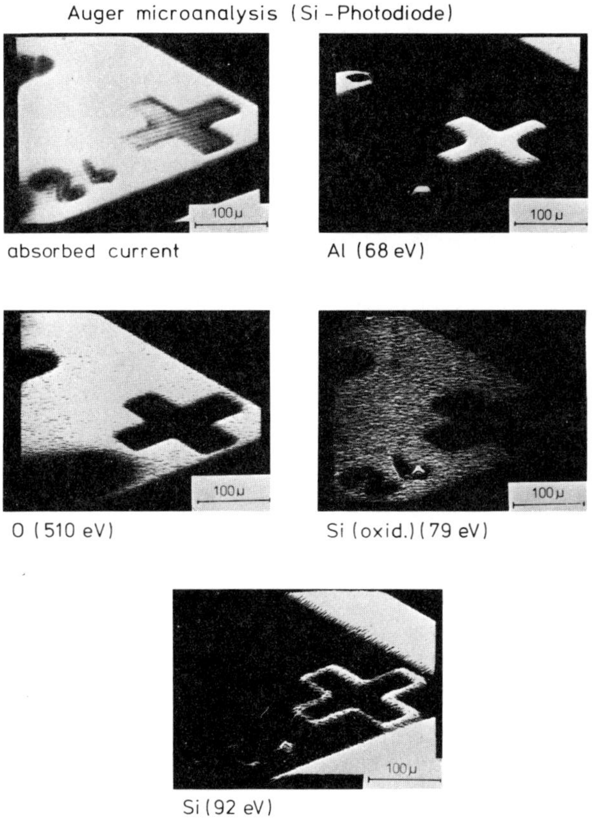

Auger microanalysis (Si-Photodiode)

absorbed current

Al (68 eV)

O (510 eV)

Si (oxid.)(79 eV)

Si (92 eV)

Fig.3.27 Auger images of a semiconductor. The chemical effects in AES are used to detect different distributions for elemental Si (92 eV) and oxidized Si (79 eV)

As shown above, backscattering effects become particularly important at high primary energies. Due to their influence, the Auger image contains information not only from the surface but also from bulk material, which may not be visible directly in the Auger spectrum. In the example of Fig.3.28 a Si chip had been covered with Au strips (20 nm thick) and a continuous film of Al (10 nm thick). The contrast in the Auger images is mainly produced by the backscattering properties of the substrate, not by variations of the surface concentration. The Au strips backscatter primary electrons and secondary electrons to a larger extent than the Si. The backward current enhances the measured Al intensity above the Au and produces the contrast though no trace of Au is visible in the Auger spectrum. The effect is largest for high-Z materials in a low-Z matrix. For the interpretation of Auger images this means that a high Auger intensity does not necessarily indicate the presence of a high concentration. Therefore, some further knowledge of the sample may be necessary in order to interpret Auger images correctly.

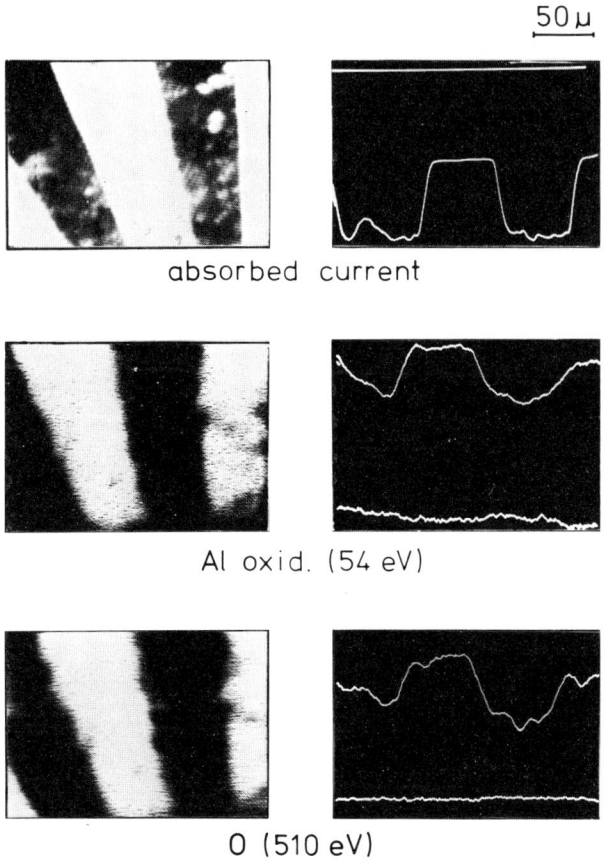

absorbed current

Al oxid. (54 eV)

O (510 eV)

8 KeV; 100Å Al / 200Å Au / Si substrate

<u>Fig.3.28</u> Influence of backscattered electrons on Auger images from a sandwich structure. The signals from the Al film are enhanced by electrons backscattered from thin Au strips. The line scans along a horizontal line at the center of the images are indicated together with the zero lines. The Au itself is not visible in the Auger spectrum

3.4.4 Combined Auger/X-Ray Microanalysis

As indicated in Fig.3.3 the fluorescence yield may rise above 20% for binding energies above 5 keV. The primary energy used in Auger microanalysis is therefore sufficient to produce also characteristic X-rays in measurable quantity. The characteristic radiation is inevitably accompanied by continuous bremsstrahlung. Therefore, a signal-to-noise problem exists, but it is by far less important than in AES. For this reason, analyzers can be used which have a small entrance aperture, and single photon counting is appropriate. Two analyzer system are in use: crystal spectrometers, so-called wavelength dispersive systems, and Si(Li) detectors, called energy-dispersive systems. The former have excellent resolution, but they are expensive and not well compatible with UHV requirements. Cooled Si(Li) detectors yield an energy resolution of 150 to 160 eV only, but they are easy to use and can be adapted to UHV systems.

If a thin Be window (8 μ) is placed in front of the detector, the low-energy cutoff is near 1 keV due to the absorption in the window, which limits the range of elements detected to $Z > 10$. If the window is suppressed, the cutoff is near 200 eV, caused by the electronic noise of the pulse amplifier. However, the detection of light elements is difficult due to the low fluorescence yield.

The combination of an Auger analyzer with an X-ray detector offers the possibility to apply two different microanalysis techniques to the same specimen, which complement each other in more than one respect. They are complementary in the physical principle, they have different sensitivities for heavy and light elements, and they offer two different depths of information. In AES it is determined mainly by the mean free path of the secondary electrons, which is of the order of 1 nm. In X-ray analysis it is determined by the penetration depth of the primary electrons, as the absorption of the X-rays can be neglected. The primary beam is broadened by scattering from the ion cores and from free and bound electrons in the target, leading to a pear-shaped distribution of energetic electrons below the surface, which corresponds approximately to the distribution of the points in which X-rays are produced. The center of gravity of this distribution is near $\rho \cdot z = 0.2$ to 0.4 mg/cm^2 (at 10 keV) below the surface, shifting to higher values with increasing primary energy. The lateral extension ranges from some 100 nm to several 1000 nm, depending on the atomic number of the target. The depth of information is of the same order, being larger by a factor of 100 at least than that of Auger analysis. Thus a simultaneous analysis of the surface composition of a specimen and its bulk composition at the same spot becomes feasible. This is desirable in a number of cases some of which shall be mentioned:

- The problems in the interpretation of Auger images can be solved as the depth of the X-ray generation correlates with the penetration depth of the backscattered electrons. Therefore the elements falsifying the Auger signals but remaining undetected in AES can be easily identified.

- The surface composition of unknown compounds or alloys may differ significantly from that of the bulk, for example when one of the constituents segregates to the surface. The combination of both methods permits such effects to be detected.

- In depth-profiling measurements using AES and sputter-etching, the interesting layers are often covered by contamination and are therefore inaccessible to AES. X-ray analysis can give nondestructively the average composition of the sample and —by varying the primary energy— rough information, which parts of the depth profile have to be observed at higher resolution.

3.5 Comparisons

Far from claiming completeness, we wish to give some comparisons of the methods described which will further elucidate some of their characteristic properties. Examples are given for the influence of the different decay mechanisms on the spectra of threshold spectroscopies and of ILS, for the influence of the density of states on the spectra, for the different surface sensitivities, and for the capabilities for elemental analysis.

3.5.1 Threshold Spectroscopies Inter Se

As explained above, DAPS only is capable of observing the threshold excitation directly. The spectra of AEAPS and SXAPS depend also on the decay mechanisms, i.e., on the radiative and radiationless transition yields. Thus the sensitivity of SXAPS suffers from a small fluorescence yield for low binding energies, whereas AEAPS has no problems in this respect. Apart from the general reduction in intensity, the shape of the spectra may depend on variations of the fluorescence yield from one subshell to the other. Fig.3.29 demonstrates this effect in a comparison of the $L_{2/3}$-spectra of clean chromium in SXAPS and AEAPS. Whereas the AEAPS spectrum reflects approximately the ratio of the populations of the L_3-shell and the L_2-shell (4 : 2 = 2) these thresholds are of nearly equal intensity in SXAPS. From that is concluded that the excitation probability is proportional to the electron population

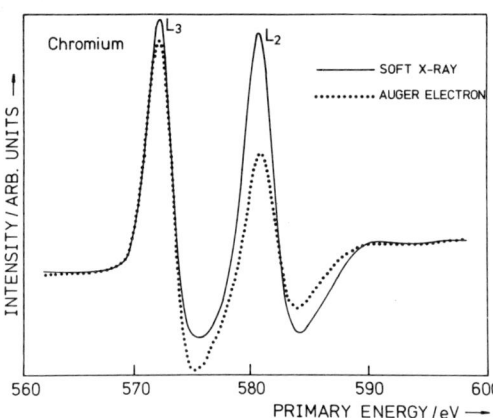

Fig. 3.29 Comparison of the L_{23}-spectra of clean chromium in SXAPS and AEAPS. The anomalous intensity ratio of the L_3- and L_2-lines in SXAPS indicates a particularly high fluorescence yield for the L_2-shell (after [3.58])

of the subshells, whereas the radiative decay of an L_2-hole is twice as probable as that of an L_3-hole. In view of the complexity of the deexcitation processes which may also occur in several consecutive steps, this point seems to need further investigation. The inverse anomaly of the relative intensities should be observed in a comparison of AEAPS with DAPS. As, however, the nonradiative transitions are highly probable, the relative change is very small and both spectra agree within the experimental uncertainty.

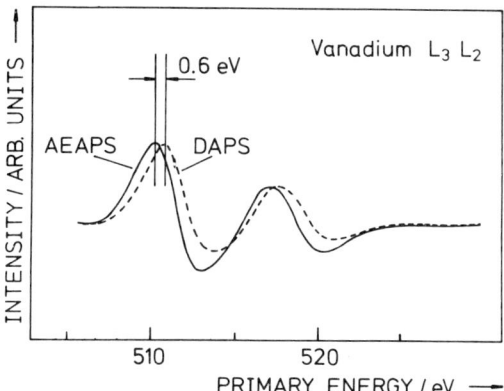

Fig. 3.30 Comparison of the L_{23}-spectra of clean vanadium in AEAPS and DAPS. (Energy scale not corrected for the work function of the cathode)

The surface sensitivity of SXAPS and AEAPS is expected to be nearly the same as it is determined by the mean free path of the incident electrons. The absorption of X-rays is weak and the mean free path of the secondary electrons which determine the signal in AEAPS is large. To the contrary, the surface sensitivity of DAPS is high as the electrons backscattered quasielastically travel at least twice the path corresponding to the penetration depth. Therefore, alterations in the electronic environment of surface atoms leading to shifts of the core level binding energy (or of the density of states) should become visible in a comparison of DAPS with AEAPS. The result of a comparison using clean polycrystalline vanadium is shown in Fig.3.30. The DAPS spectrum is shifted slightly to higher binding energy, both for the L_3- and the L_2-excitation, and the low-energy shoulder increases less rapidly than in the AEAPS spectrum.

The same features are observed in the re-integrated form of the spectra. An explanation could be given in terms of "surface chemical shifts". If we assume the extra-atomic relaxation to be less effective at the surface due to the reduced mean electron density, we expect the binding energy to approach that of the free atoms, i.e., to be higher than in the bulk. The broad low-energy shoulder in the DAPS spectrum can be interpreted as the contribution from the bulk atoms which are sampled with less sensitivity [3.123]. This explanation contradicts the conclusions drawn from the comparison of SXAPS with XEAPS [3.72], namely that the binding energy of surface atoms is lower than that of bulk atoms. It must be emphasized however, that the excitation mechanisms are not the same in SXAPS and XEAPS which can result in different spectra (compare Subsec.3.3.2, Fig.3.17). It should be mentioned also that—at least in principle—some anomalies of the total backscattering cross section could modify the DAP spectra, though up to now no evidence for this could be found in DAPS or the other threshold spectroscopies.

A further experiment illustrates the difference in surface sensitivities of DAPS and AEAPS. The chemical shift due to oxidation of V results initially in a broadening of the lines as shifted and unshifted core levels coexist. (After full oxidation the lines become smaller again.) This behavior is the same for DAPS and AEAPS lines. In Fig.3.31 the width of the vanadium L_3-line is plotted versus the oxygen exposure of the initially clean surface [3.52]. The broadening due to oxidation begins at lower doses and rises more steeply than in AEAPS. This is consistent with a high sensitivity of DAPS to the first layers, whereas AEAPS averages into larger depth. To give a numerical example, we assume normal incidence and a mean free path of 0.9 nm for electrons of 500 eV. Then the first layer contributes by about 45% to the DAPS signal and only by 25% to the AEAPS signal.

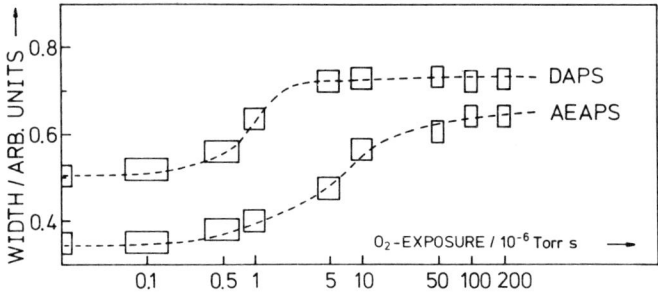

Fig. 3.31 Comparison of the line broadening of the vanadium L_{23}-lines due to oxidation versus the oxygen exposure as observed by DAPS and AEAPS. DAPS, which is particularly surface sensitive, indicates the oxidation process at lower exposures than AEAPS, which has a larger depth of information

The most sensitive of all threshold spectroscopies is AEAPS when the secondary electron yield is employed. Its sensitivity is four to five orders of magnitude higher than that of conventional SXAPS [3.58], and one to two orders of magnitude higher than that of DAPS. Besides the small background, this results mainly from the fact that no analyzer is needed at all. Though the problems with diffraction effects at low energies are a nuisance, at high energies this method should be well applicable.

3.5.2 Threshold Spectroscopies Versus ILS

In the one-electron model, both types of spectroscopies observe the transition of a bound electron to the Fermi level. Therefore, the loss energy in ILS and the excitation energy in the threshold spectroscopies are the same and equal to the binding energy of a particular electron shell. Therefore the spectra look similar, notably the relative positions of the lines are the same. The line shapes are similar, too, but not the same. The reason is the self-convolution procedure in the threshold spectroscopies, which is absent in ILS.

There is often considerable fine structure in the spectra, especially for the light elements (compare with the oxygen spectra in Fig.3.16) and the question arises whether these are due to the density of states or to extrinsic or intrinsic loss mechanisms. An energy loss is called intrinsic when it occurs simultaneously to the core level excitation, it is called extrinsic when these processes are not coupled. In particular, the rich fine structure of the carbon K-excitation in the threshold spectroscopies has attracted much attention [3.124]. Intrinsic losses are expected to be independent of the angle of incidence, whereas extrinsic losses depend on the path length of the exciting electron in the surface region, which depends on the angle of incidence. Intrinsic losses are expected to have high intensity [3.125], comparable to that of the no-loss peak, whereas for extrinsic losses the relative intensity should be similar to that in electron-loss spectroscopy. As shown in Fig.3.32a, the satellite peaks in SXAPS are similar in intensity to the main peak or even higher.

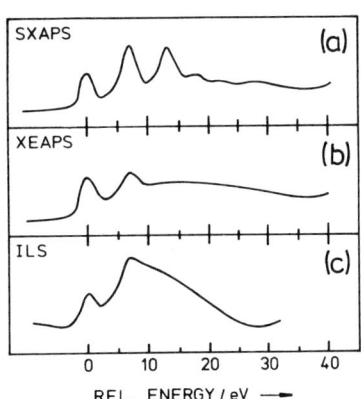

Fig. 3.32 Comparison of the results of SXAPS, XEAPS, and ILS for the K-shell excitation of graphitized carbon. ILS indicates a density of states with two peaks

Angle-resolved measurements [3.64] show some weak dependence of the relative intensities on the angle of incidence, but intrinsic plasmon excitation or density-of-states effects could not be ruled out. On the other hand, a spectrum like that of Fig.3.32a with three main peaks could also be obtained by the self-convolution of a density of states with two peaks. In DAPS and AEAPS the same features have been observed. Thus, these spectroscopies cannot help to decide the question.

In ILS, however, the density of states enters directly. The result is shown in Fig.3.32c, obtained with a retarding field analyzer. It exhibits two main peaks (in agreement with the derivative spectrum of Fig.3.22) suggesting a density of states with two peaks. In addition, in curve b the result of XEAPS is shown, also with two peaks. In this method intrinsic plasmon excitation should not occur [3.49, 125] but due to the optical selection rules one might suspect this method to suppress the third peak. Numerical calculations [3.126] using the derivative of the self-convolution of the ILS spectrum showed that the SXAPS spectrum can be reproduced quite well. Arguments based on the band structure [3.68] support the conclusions drawn from the ILS measurements, namely that the results of the threshold spectroscopies can be explained on the basis of a one-electron band model, at least in this case. Similar comparisons of ILS with the threshold spectroscopies would probably contribute much to the understanding of the complex effects, which occur upon chemical bonding.

3.5.3 Elemental Analysis

The main advantage of the threshold spectroscopies and ILS for elemental analysis is the relatively small number of lines for each element and their sharpness which makes line overlap a rare event [3.47]. In AES the number of lines is large and they are not very sharp in general. As an illustration, Fig.3.33 shows the analysis of stainless steel as seen by DAPS in first and second derivative representation, and by AES as obtained by a cylindrical mirror analyzer, for the same energy range. In AES, many elements overlap others and it is not easy to find a line which can unambiguously be attributed to one element. For example oxygen, vanadium, and chromium overlap strongly and it is difficult to decide whether manganese is present or not. DAPS has no problems in this respect: there are only three lines per element at most and these are relatively sharp (broadened by the modulation in the second derivative spectrum). The presence of vanadium and manganese can be decided easily.

The drawback of the threshold spectroscopies, their dependence on the density of states, finds its expression in the spectra, too. The sample is strongly oxidized in both cases, but whereas the oxygen line dominates the Auger spectrum, just the opposite is found in the DAPS spectrum. It is this large variation in sensitivity which make the threshold spectroscopies somewhat inconvenient in practical use [3.127]. In order to evaluate the sensitivity of DAPS in relation to AES, a direct comparison was made for a contaminated Ti surface [3.52]. A retarding field analyzer was employed for both, according to the configuration of Fig.3.8.

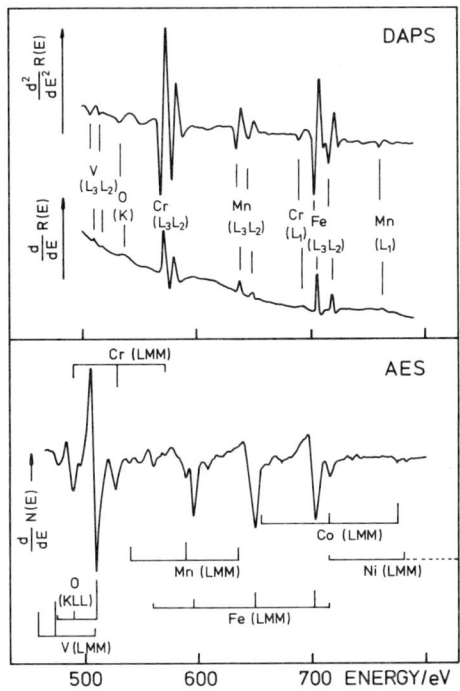

Fig. 3.33 Comparison of the results of DAPS and AES for stainless steel. The energy scale indicates the primary energy for DAPS and the secondary electron energy in AES. Note the overlap of V, Cr, O, Mn, and Fe in AES

The first harmonic of the modulation voltage was detected and the peak-to-peak amplitude was evaluated. Because in AES generally a high primary energy is preferred, the signal was normalized to the same power applied to the specimen. It was normalized also to the same modulation amplitude, which is equivalent to the same resolution (as far as the broadening due to the modulation is concerned) as the effect of analyzer resolution was negligible. Referred to the same signal-to-noise ratio, comparable sensitivities were found in both methods, but with differences in detail: DAPS is more sensitive to Ti than AES, which in turn is more sensitive to O and C. However, as the natural linewidth of Auger lines is generally large, high modulation amplitudes with corresponding gain in sensitivity can be applied without degrading the resolution too much. High modulation amplitudes in DAPS do mostly not result in increased sensitivity but in decreased resolution. In practice, AES is therefore mostly superior as far as sensitivity is concerned. If comparison is made with AES performed by a dispersive analyzer such as the cylindrical mirror, the competition is undoubtedly won by AES. Only AEAPS might perhaps compete. On the other hand, one should not forget to include the complexity and cost of the equipment in a comparison.

ILS is also capable of elemental analysis, with the same advantages as the threshold spectroscopies and also their disadvantages in part. At first glance the method seems somewhat simpler as the problems arising from the self-convolution of

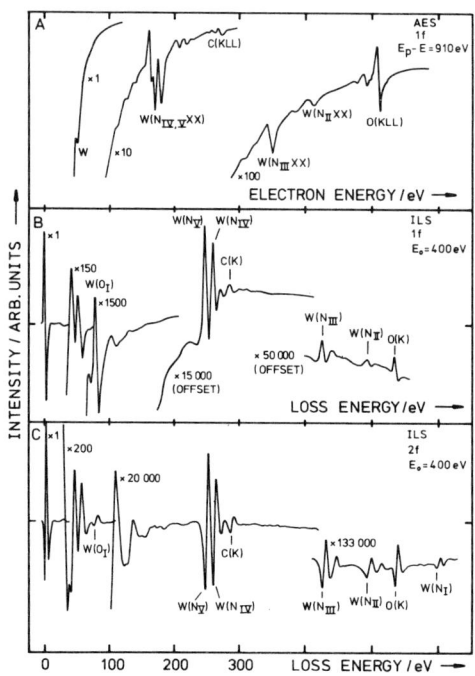

Fig. 3.34 Comparison of Auger and ILS spectra from W obtained with the instrumentation of Fig. 3.19. Pass energy 400 eV for ILS. Note the sensitivity factors (after [3.85])

the density of unoccupied states do not exist and the line shapes consequently are less complex. On the other hand, the intensity of the loss features depends on the primary energy and the type of analyzer as the ionization cross section is a function of both the ratio of binding energy to primary energy and of the scattering angle. In addition, the dependence of the intensity on the incident energy for a particular analyzer is not necessarily the same for all elements. In order to estimate the relative sensitivity of ILS and AES, we cite the analysis of a contaminated W surface which was performed with the analyzer system described in Fig.3.19. Fig.3.34 shows the Auger lines and the ionization loss features in first and second derivative representation. In all spectra the sensitivity has to be changed several times, with the larger range in ILS. If one compares the peak-to-peak amplitudes as a measure of sensitivity, ILS is found to be one to two orders of magnitude less sensitive than AES. Therefore it will find its main application as a tool for elemental analysis in conjunction with AES to remove ambiguities resulting from line overlap. The prospects for quantitative analysis with ILS are not too good in view of its complexity, which is equal to or slightly greater than that of the threshold spectroscopies.

Quantitative analysis by means of the threshold spectroscopies and of ILS is hampered by the sensitivity of the line shapes to chemical effects. This is mainly due to the inherently good energy resolution of the method (or the indispensability of a good analyzer in ILS) which makes small changes in the line shape clearly visible. One should not forget, however, that AES suffers from the same problems as soon as high resolution is obtained. Due to chemical shifts of the Auger lines the line shape of the Auger spectra is subject to changes of the chemical environment, too, and measurements of the peak-to-peak amplitudes in AES are as doubtful as in the other methods. Hence, for quantitative analysis, re-integration of the measured spectra seems to be a necessity in all techniques.

Acknowledgements

The author wishes to express his gratitude to H. Ibach for discussing the manuscript and for his valuable suggestions. Special thanks are due to H.P. Bonzel and H. Froitzheim for reading and correcting the manuscript. The able assistance of Mrs. G. Ehser and Mrs. U. Müller is gratefully acknowledged.

References

3.1   H.D. Hagstrum: Surf. Sci. 54 197 (1976)
3.2   K.D. Sevier: *Low Energy Electron Spectrometry* (Wiley Interscience, New York, London, Sydney, Toronto 1972) p. 356
3.3   K.D. Sevier: *Low Energy Electron Spectrometry* (Wiley Interscience, New York, London, Sydney, Toronto 1972) pp. 242-270
3.4   D.H. Madison, E. Merzbacher: *Atomic Inner-Shell Processes*, Vol. 1 (Academic Press, New York, San Francisco, London 1975) pp. 2-23
3.5   C.B.O. Mohr: *Adv. Atomic and Molec. Phys.*, Vol. 4 (Academic Press, New York, London 1968) pp. 221-230
3.6   M. Gryzinski: Phys. Rev. 138 A, 336 (1965) and references therein
3.7   A. Burgess, I.C. Percival: *Adv. Atomic and Molec. Phys.*, Vol. 4 (Academic Press, New York, London 1968) pp. 126-127
3.8   H.S.W. Massey, E.H.S. Burhop, H.B. Gilbody: *Electronic and Ionic Impact Phenomena*, 2nd ed., Vol. 1 (Oxford Univ. Press, Oxford 1969) pp. 594-664
      M.R.H. Rudge, M.J. Seaton: Proc. Roy. Soc. A 238, 262 (1965)
3.9   G. Wendin: *Vacuum Ultraviolett Rad. Phys.* (Pergamon, Vieweg, Braunschweig 1974) pp. 225-239
3.10 S. Lundquist, G. Wendin: J. Electr. Spectr. Rel. Phen. 5, 513 (1974)
      M.Ya. Amusia: *Vacuum Ultraviolett Rad. Phys.* (Pergamon, Vieweg, Braunschweig 1974) pp. 205-224
3.11 R.L. Gerlach, A.R. DuCharme: Japan. J. Appl. Phys., Suppl. 2, Pt 2, 675 (1974)
3.12 H.W. Drawin: Z. Phys. 164, 513 (1961)
3.13 J.J. Vrakking, F. Meyer: Phys. Rev. A 9, 1932 (1974)
3.14 J.J. Vrakking, F. Meyer: Surf. Sci. 47, 50 (1975)
3.15 G. Glupe, W. Mehlhorn: Phys. Lett. 25 A, 274 (1967)
3.16 C.J. Powell: Rev. Mod. Phys. 48, 33 (1976)

3.17 E.J. McGuire: Phys. Rev. A $\underline{5}$, 1043 (1972)
3.18 V. Weisskopf, E. Wigner: Z. Phys. $\underline{63}$, 54 (1930)
    A. Hoyt: Phys. Rev. $\underline{36}$, 860 (1930)
    A. Hoyt: Phys. Rev. $\underline{40}$, 477 (1932)
3.19 J.C. Tracy: Surf. Sci. $\underline{38}$, 265 (1973)
    A.R. DuCharme, R.L. Gerlach: J. Vac. Sci. Tech. $\underline{11}$, 281 (1974)
3.20 F.P. Larkins: J. Phys. B $\underline{7}$, 37 (1974)
3.21 B. Crasemann, R.W. Fink, H.-U. Freund, H. Mark, C.D. Swift, R.E. Price, P. Venugopala Rao: Rev. Mod. Phys. $\underline{44}$, 716 (1972)
3.22 H.P. Kelly: *Atomic Inner-Shell Processes*, Vol. 1 (Academic Press, New York San Francisco, London 1975) pp. 331-350
3.23 P.E. Best: *X-Ray Spectroscopy* (McGraw-Hill, New York, Toronto, London, 1974) pp. 1-20
3.24 E.J. McGuire: *Atomic Inner-Shell Processes*, Vol. 1 (Academic Press, New York, San Francisco, London 1975) pp. 311-315
3.25 D.A. Shirley: Phys. Rev. A $\underline{7}$, 1520 (1973)
3.26 C.A. Nicolaides, D.R. Beck: Chem. Phys. Lett. $\underline{27}$, 269 (1974)
    C.A. Nicolaides: Phys. Rev. A $\underline{6}$, 2078 (1972)
3.27 J.A.D. Matthew: Surf. Sci. $\underline{40}$, 451 (1973)
3.28 P. Weightman, E.D. Roberts, C.E. Johnson:
    J. Phys. C $\underline{8}$, 550 (1975)
    H. Aksela, S. Aksela: J. Phys. B $\underline{7}$, 1262 (1974)
    M.O. Krause: Phys. Rev. Lett. $\underline{34}$, 633 (1975)
3.29 M.F. Chung, L.H. Jenkins: Surf. Sci. $\underline{22}$, 479 (1970)
3.30 W.A. Coghlan, R.E. Clausing: USAEC Rep. ORNL-TM-3576, Oak Ridge National Lab. (1971)
    W.A. Coghlan, R.E. Clausing: Surf. Sci. $\underline{33}$, 411 (1972)
3.31 W. Mehlhorn: Z. Phys. $\underline{187}$, 21 (1965)
3.32 see Chap.5 of this volume
3.33 W. Mehlhorn, W. Schmitz, D. Stalherm: Z. Phys. $\underline{252}$, 399 (1972)
    T.A. Carlson, M.O. Krause: Phys. Rev. Lett. $\underline{17}$, 1079 (1966)
3.34 A. Servomaa, O. Keski-Rahkonen: J. Phys. C $\underline{8}$, 4124 (1975)
    O. Keski-Rahkonen, J. Utriainen: J. Phys. B $\underline{7}$, 55 (1974)
3.35 T. Åberg: *Atomic Inner-Shell Processes*, Vol. 1 (Academic Press, New York, San Francisco, London 1975) Chapt.9
3.36 S.P. Kowalczyk, L. Ley, F.R. McFeely, R.A. Pollak, D.A. Shirley: Phys. Rev. B $\underline{9}$, 381 (1974)
    S.P. Kowalczyk, R.A. Pollak, F.R. McFeely, L. Ley, D.A. Shirley: Phys. Rev. B $\underline{8}$, 2387 (1973)
3.37 L. Ley, S.P. Kowalczyk, F.R. McFeely, R.A. Pollak, D.A. Shirley: Phys. Rev. B $\underline{8}$, 2392 (1973)

3.38 J. Friedel: Adv. Phys. 3, 446 (1954)
3.39 D.J. Nagel, W.L. Baun: *X-Ray Spectroscopy* (McGraw-Hill, New York, Toronto, London 1974) pp. 445-517
3.40 L. McDonnell, B.D. Powell, D.P. Woodruff: Surf. Sci. 40, 669 (1973)
      G.E. Becker, H.D. Hagstrum: J. Vac. Sci. Tech. 11, 284 (1974)
3.41 B. Cleff, W. Mehlhorn: J. Phys. B 7, 593 (1974)
      B. Cleff, W. Mehlhorn: J. Phys. B 7, 605 (1974)
3.42 T. Matsudaira, R. Watanabe, M. Onchi: Japan J. Appl. Phys. Suppl. 2, 181 (1974)
3.43 L. McDonnell, D.P. Woodruff, B.W. Holland: Surf. Sci. 51, 249 (1975)
3.44 D.P. Woodruff: Surf. Sci. 53, 538 (1975)
3.45 J.B. Pendry: J. Phys. C 8, 2413 (1975)
3.46 B. Dev, H. Brinkmann: Ned. Tijdschr. Vacuumtechn. 8, 176 (1970)
3.47 R.L. Park: Surf. Sci. 48, 80 (1975)
3.48 C.J. Davisson, B. Germer: Nature 119, 558 (1927)
3.49 D.C. Langreth: Phys. Rev. B 1, 471 (1970)
      D.C. Langreth: Phys. Rev. Lett. 26, 1229 (1971)
      J.J. Chang, D.C. Langreth: Phys. Rev. B 5, 3512 (1972)
3.50 J. Franck, G. Hertz: Verh. Deutsch. Phys. Ges. 16, 12 (1911)
3.51 J. Kirschner, P. Staib: Phys. Lett. 42 A, 335 (1973)
3.52 J. Kirschner, P. Staib: Appl. Phys. 6, 99 (1975)
3.53 D.W. Fischer, W.L. Baun: J. Appl. Phys. 39, 4757 (1968)
3.54 D.M. Smith, T.E. Gallon, J.A.D. Matthew: J. Phys. B 7, 1255 (1974)
3.55 J.T. Grant, M.P. Hooker: Surf. Sci. 51, 433 (1975)
3.56 C. Webb, P.M. Williams: Phys. Rev. Lett. 33, 824 (1974)
      K. Wandelt: Thesis, Ludwig-Maximilian-Univ. Munich (1975)
3.57 R.L. Gerlach: Surf. Sci. 28, 648 (1971)
3.58 J.E. Housten, R.L. Park: Phys. Rev. B 5, 3808 (1972)
3.59 R.L. Park, J.E. Houston, B.G. Schreiner: Rev. Sci. Instr. 41, 1810 (1970)
3.60 A.M. Bradshaw: *Surface and Defect Properties of Solids*, Vol. 3 (Chemical Soc., Burlington House, London 1974)
      S. Kato: Oyo Buturi 43, 36 (1974)
3.61 J.C. Tracy: J. Appl. Phys. 43, 4164 (1972)
3.62 R.G. Musket, S.W. Taatjes: J. Vac. Sci. Tech. 9, 1041 (1972)
3.63 S. Andersson, H. Hammarquist, C. Nyberg: Rev. Sci. Instr. 45, 877 (1974)
3.64 J. Verhoeven, J. Kistemaker: Surf. Sci. 50, 388 (1975)
3.65 R.L. Park, J.E. Houston: J. Vac. Sci. Tech. 11, 1 (1974)
3.66 J.E. Houston, R.L. Park: J. Vac. Sci. Tech. 9, 579 (1972)
      G. Ertl, K. Wandelt: Phys. Rev. Lett. 29, 218 (1972)
3.67 G. Ertl, K. Wandelt: Z. Naturforsch. 29a, 768 (1974)
      G. Ertl, K. Wandelt: Surf. Sci. 50, 479 (1975)
3.68 C. Webb, P.M. Williams: Surf. Sci. 53, 110 (1975)
      A.M. Bradshaw, U. Krause: Ber. Bunsen-Ges. 79, 1095 (1975)

3.69 S. Andersson, C. Nyberg: Surf. Sci. 52, 489 (1975)
3.70 J. Kanski, P.O. Nilsson: Physica Scripta 12, 103 (1975)
3.71 R. Haensel, G. Keitel, P. Schreiber: Phys. Rev. 188, 1375 (1969)
3.72 J.E. Houston, R.L. Park, G.E. Laramore: Phys. Rev. Lett. 30, 846 (1973)
3.73 G. Wendin: *Vacuum Ultraviolett Radiation Physics* (Pergamon, Vieweg, Braunschweig 1974) pp. 252-254
3.74 R.J. Liefeld, A.F. Burr, M.B. Chamberlain: Phys. Rev. A 9, 316 (1974)
3.75 M.B. Chamberlain, A.F. Burr, R.J. Liefeld: Phys. Rev. A 9, 663 (1974)
3.76 M.B. Chamberlain, W.L. Baun: J. Vac. Sci. Tech. 11, 441 (1974)
3.77 M.B. Chamberlain, W.L. Baun: J. Vac. Sci. Tech. 12, 1047 (1975)
3.78 R.L. Gerlach, J.E. Houston, R.L. Park: Appl. Phys. Lett. 16, 179 (1970)
3.79 R.L. Gerlach, A.R. DuCharme: Surf. Sci. 29, 317 (1972)
3.80 R.L. Gerlach: J. Vac. Sci. Tech. 8, 599 (1972)
3.81 R.L. Gerlach, A.R. DuCharme: Phys. Rev. A 6 1892 (1972)
3.82 DuCharme, R.L. Gerlach: Phys. Rev. A 9, 197 (1974)
      R.L. Gerlach, A.R. DuCharme: Phys. Rev. Lett. 27, 290 (1971)
3.83 R.G. Musket: Surf. Sci. 44, 629 (1974)
3.84 A.V. Crewe, J. Wall: Optic 30, 461 (1970)
3.85 R.L. Gerlach: *Electron Spectroscopy*, ed. by D.A. Shirley (North-Holland Publishing Comp., Amsterdam, London 1972) pp. 885-893
3.86 A. Koma, R.R. Ludeke: Phys. Rev. Lett. 35, 107 (1975)
3.87 D.T. Hawkins: *Auger Electron Spectroscopy 1967-1975* (Physical Electronics Industries, Libraries and Information Systems Center, Eden Prairie 1975)
3.88 E.N. Sickafus: J. Vac. Sci. Tech. 11, 299 (1974)
3.89 G. Ertl, J. Küppers: *Low Energy Electrons and Surface Chemistry* (Verlag Chemie, Weinheim 1974)
3.90 J.C. Tracy: *Electron Emission Spectroscopy* (D. Reidel, Dordrecht, Boston 1973) pp. 295-340
3.91 K. Müller: *Springer Tracts in Modern Physics*, Vol. 77 (Springer, Berlin, Heidelberg, New York 1975) pp. 97-125
3.92 R.W. Shaw Jr., T.D. Thomas: Phys. Rev. Lett. 29, 689 (1972)
3.93 T.E. Gallon, J.D. Nuttall: Surf. Sci. 53, 698 (1975)
3.94 P.H. Citrin: Phys. Rev. Lett. 31, 1164 (1973)
      P.H. Citrin: J. Elec. Spectr. Rel. Phen., 5, 273 (1974)
3.95 J.A.D. Matthew, Y. Kominos: Surf. Sci. 53, 716 (1975)
3.96 J.E. Houston: J. Vac. Sci. Tech. 12, 255 (1975)
      J.W. Gadzuk: Phys. Rev. B 9, 1978 (1974)
3.97 A.J. Jackson, C. Tate, T.E. Gallon, P.J. Basset, J.A.D. Matthew: J. Phys. F 5, 363 (1975)
3.98 P.J. Bassett, T.E. Gallon, J.A.D. Matthew, M. Prutton: Surf. Sci. 35, 63 (1973)
3.99 L. Yin, J. Adler, T. Tsang, M.H. Chen, B. Crasemann: Phys. Lett. 46 A, 113 (1973)

S. Aksela, J. Väyrynen, H. Aksela: Phys. Rev. Lett. 33, 999 (1974)
3.100 M. Salmerón, A.M. Baró, J. Rojo: Surf. Sci. 53, 689 (1975)
3.101 G. Schön: J. Electr. Spectr. Rel. Phen. 2, 75 (1973)
3.102 C.D. Wagner, P. Biloen: Surf. Sci. 35, 82 (1973)
3.103 M.P. Hooker, J.T. Grant: J. Vac. Sci. Tech. 12, 325 (1975)
H. Nozoye, Y. Matsumoto, T. Onishi, T. Kondow, K. Tamaru: J. Phys. C 8, 4131 (1975)
3.104 C.D. Wagner: Analytical Chem. 47, 1203 (1975)
3.105 G. Allié, E. Blanc, D. Dufayard: Surf. Sci. 46, 188 (1974)
3.106 M. Baines, A. Howie: Surf. Sci. 53, 546 (1975)
3.107 P.H. Holloway: J. Electr. Spectr. Rel. Phen. 7, 215 (1975)
3.108 H.E. Bishop, J.C. Rivière: J. Appl. Phys. 40, 1740 (1969)
3.109 J.J. Vrakking, F. Meyer: Surf. Sci. 35, 34 (1973)
P.W. Palmberg: Appl. Phys. Lett. 13, 183 (1968)
3.110 J.H. Neave, C.T. Foxon, B.A. Joyce: Surf. Sci. 29, 411 (1972)
T.E. Gallon: J. Phys. D 5, 822 (1972)
3.111 K. Goto, K. Ishikawa: Surf. Sci. 47, 477 (1975)
K. Goto, T. Koshikawa, R. Shimizu, K. Ishikawa: Japan. J. Appl. Phys. Suppl. 2, Pt 2, 633 (1974)
3.112 D.M. Smith, T.E. Gallon: J. Phys. D 7, 151 (1974)
3.113 H.E. Bishop: Br. J. Appl. Phys. 18, 703 (1967)
3.114 K. Murata: J. Appl. Phys. 45, 4110 (1974)
3.115 P. Staib, J. Kirschner: Appl. Phys. 3, 421 (1974)
3.116 J.J. Burton, E. Hyman, D.G. Fedak: J. Catalysis 37, 106 (1975)
H.H. Brongersma, T.M. Buck: Surf. Sci. 53, 649 (1975)
M. Lagües, J.L. Domange: Surf. Sci. 47, 77 (1975)
3.117 J.M. Morabito: *Scanning Electron Microscopy 1976*, ed. by O. Johari (IIT Res. Inst., Chicago 1976) p.221
3.118 J.E. Houston: Rev. Sci. Instr. 45, 897 (1974)
3.119 A.N. Broers, M. Hatzakis: Scientific Amer. 227, 34 (1972)
3.120 J.A.R. Cleaver: Int. J. Electronics 38, 513 (1975)
3.121 B.D. Powell, D.P. Woodruff, B.W. Griffiths: J. Phys. E 8, 548 (1975)
3.122 M.P. Seah, C. Lea: *Scanning Electron Microscopy 1973* (Inst. of Phys., London 1973)
3.123 J. Kirschner: Thesis, Technical Univ. Munich 1974
3.124 A.M. Bradshaw, D. Menzel: phys. stat. sol. (b) 56, 135 (1973)
J.E. Houston: Solid St. Comm. 17, 1165 (1975)
3.125 A.M. Bradshaw, S.L. Cederbaum, W. Domcke, U. Krause: J. Phys. C 7, 4503 (1974)
A.M. Bradshaw, W. Wyrobisch: J. Electr. Spectr. Rel. Phen. 7, 45 (1975)
3.126 U. Krause: Diplomarbeit, Techn. Univ. München, 1974
3.127 J.C. Tracy: Appl. Phys. Lett. 19, 353 (1971)

4. Electron Diffraction and Surface Defect Structure

M. Henzler

With 17 Figures

Although single crystal surfaces frequently are assumed to be atomically flat and strictly periodic along the surface, in reality surface defects of various kinds and various amounts are always present. In nearly all surface physics laboratories, equipment for electron diffraction is available. It is therefore desirable to check which information on surface structure besides periodicity may be derived from diffraction experiments.

The diffraction pattern of a perfect crystal with a perfect instrument is obvious: infinitely sharp spots with zero intensity between. Finite instrumental limitations yield a finite width of the spots and possibly some background. Deviations from periodicity like defects in structure or varying composition may alter spot shape and background dramatically. Additionally inelastic scattering, if this contribution is not separated by the detector, may change the pattern. Therefore the interpretation of a diffraction pattern may yield information on the defect structure. The defects under consideration are those which are possible on a single crystal substrate. Therefore polycrystalline or amorphous bulk samples are excluded. This restriction gives always a long-range correlation due to the strict periodicity of the substrate.

Surface defects may be zero-, one- or two-dimensional. The zero-dimensional defects are point defects due to single contamination atoms, missing substrate atoms, or atoms which are displaced temporarily or permanently out of their regular lattice positions. The one-dimensional defects are formed by atomic steps or superstructure domain boundaries. For two-dimensional defects the surface is completely altered as in facets or with amorphous top layers. The defects may be present just in the top layer or may affect several layers. A classification in this respect has not been tried.

To detect defects, direct imaging would be the simplest technique. Unfortunately so far only field ion microscopy is surface sensitive, of atomical resolution, and capable of handling clean surfaces. Its applicability, hovewer, is restricted to sharp points and to the presence of high electric fields. Therefore it is only used in special cases. The technique used most widely is electron diffraction. To get sufficient surface sensitivity high energy electrons (5-30 keV) have to have grazing incidence. In most cases low-energy electron diffraction (LEED) with 10-500 eV is used; this yields, even with normal incidence, a penetration depth of only a few atomic layers (see Preface). Therefore defects of the top layer are easily found in the diffraction pattern. In principle the diffraction pattern is just the Fourier transform of the crystal, that is in this case of the top layer(s). Therefore by using the X-ray evaluation techniques [4.1-3], the surface structure determination should be possible within the same framework as bulk structure determination. Additional difficulties, however, give so far additional limitations.

a) Most electron sources are not sufficiently monochromatic and parallel in orientation, so that the coherently diffracted area on the surface is relatively small (5-30 nm diameter).

b) The low penetration depth is due to a strong interaction, therefore multiple scattering and inelastic processes are very important in low-energy electron diffraction. Nevertheless LEED provides much useful information on defect structure within kinematical approximations. Problems which are solved only by dynamical computations are not treated in this chapter.

The purpose of this chapter is to discuss several principal influences on the diffraction pattern so that a qualitative interpretation of the pattern is possible (Sec. 4.1). Point defects are discussed in Section 4.2, especially with respect to model calculations. Steps, superstructure domains, and facets are treated in Sections 4.3 and 4.4. The last section is devoted to the interpretation of the LEED pattern with emphasis on a distinction of the different possibilities.

Frequently LEED is thought to be very insensitive to defects. It is hoped that the present chapter will clearly demonstrate the various possibilities of LEED to detect and evaluate surface defects qualitatively and quantitatively.

4.1 Principles of Defect Detection by Electron Diffraction

4.1.1 Validity of the Kinematical Approximation

The amplitude of a scattered wave is described by summation of all waves coming from the atoms of the crystal. If the sample is described by repetition of identical units, the absolute square of the amplitude A is given $A^2 = F \cdot G$, with F being the absolute square amplitude of a single unit (atom of group of atoms) and being called "atom factor" (for single atom) or "structure factor" or "brick factor" (for group of atoms). The factor F contains all information on the scattering properties a single atom. Since it describes the outgoing wave, it also includes multiple scattering, that is those contributions to the outgoing wave due to scattering of the primary wave on the considered atom and also those due to multiple scattering on different atoms. The second factor G is called interference factor or lattice factor and is just due to the repetition of identical units. It does not contain any information about the scattering properties of the atoms or group of atoms, it reproduces only the arrangement of identical, otherwise arbitrary scatterers. This factor therefore safely is calculated within kinematical models. For periodic or nearly periodic lattices the lattice factor has appreciable values only for definite angles and their immediate neigborhood (spot pattern). If the F factor varies slowly with angle compared to G, the existence and shape of a spot are completely explained by G. Therefore the unit mesh in size and orientation is derived from the existence of diffraction spots completely within kinematical approximations (for extensive descriptions see textbooks [4.1-4]). Additionally the spot shape like splitting or broadening (if not due to instrumental artifacts) is under the same restrictions explained by the approximate lattice factor G. On the other hand the spot intensity may not be interpreted without the F factor, that is without dynamical calculations. There is extensive literature available on LEED intensities, that is on the form factor with the mostly implicit assumption that the surface is strictly periodic and therefore the lattice factor that of the ideal lattice. There is, however, nearly no literature available on the lattice factor under nonideal conditions. It turns out that the F factor varies in most investigated cases sufficiently slowly. In the few remaining cases the angle dependence of the F factor is recognized by comparing measurements at varying angles. By averaging, this effect is approximately eliminated. Therefore the spot shape may be evaluated with kinematical aprroximations. Of course, the content of the unit cell (like atomic distances and displacements within the unit cell) is not available. The arrangement of units especially with respect to defects however is easily that is, without large computers evaluated.

4.1.2 Construction and Calculation of the Ideal LEED-Pattern

The diffraction pattern is not an image of the surface. The lattice factor of the intensity is described by a Fourier transform of the surface. Since only the intensity of the diffration pattern is observed without its phase structure, the absolute square of the complex Fourier transform has to be taken for a comparison with experimentally obtained patterns or, mathematically completely equivalent, the Fourier transform has to be taken not from the surface atom arrangement itself, rather from the autocorrelation of this arrangement (as treated in detail in text books on X-ray structure analysis [4.1.2]). The autocorrelation function is a function of real space (contrary to reciprocal space). If the atoms of the crystal are represented by points, the autocorrelation ϕ $(\underline{x})$ describes the number of pairs with the distance of the vector $\underline{x}$. The intensity function $I(k)$ as Fourier transform of the autocorrelation $I(k) = F\{\phi(\underline{x})\}$ is a function of coordinates in reciprocal space. Whereas there is a unique way from atom positions in real space via autocorrelation function or scattering amplitude to the intensity function (lattice factor of intensity), the reverse way is difficult due to the phase problem in going from intensity to amplitude or from autocorrelation function to atom positions (see [4.1]). In this chapter some possible interpretations of observed patterns and the influence of possible surface defects on the diffraction pattern are discussed.

The LEED pattern to be expected is easily derived from Ewald's construction as used in X-ray analysis. For an ideal lattice, that means strictly periodic in three dimensions, the reciprocal lattice again is a point lattice. Since the penetration of the electron beam is only a few layers, the periodicity normal to the surface is lost and all points form rods normal to surface. Nevertheless the points of the three-dimensional reciprocal lattice still have a special meaning: no matter which surface is formed, those points are always part of the reciprocal lattice. This is also true for nonideal surfaces like rough or curved surfaces, with or without missing atoms, as long as all atoms are on regular bulk atom positions. This fact will be a big help in interpretation of patterns of random stepped surfaces. Fig. 4.4 shows the construction of the LEED pattern of an ideal surfaces with $\underline{k}_0$ and $\underline{k}$ the wave factor of the incident and scattered wave, respectively. The difference $\underline{K} = \underline{k} - \underline{k}_0$ is called scattering vector. Therefore from LEED patterns the intensity function in reciprocal space is first derived. By comparison with LEED patterns derived from assumed surface structures an evaluation including several quantitative parameters is possible, as shown below.

4.1.3 Instrumental Limitations

For an ideal crystal and an ideal instrument the diffraction spots should be infinitely sharp. In reality the incident electron beam is not an ideal plane wave showing variations in wave length mainly due to thermal energy spread from hot cathode emission and variations in the direction of propagation due to finite cathode size and due to lack of ideal focusing of the electron optics. In practice this means each LEED spot has a finite width on the screen, which may also be expressed as a width in reciprocal space via the Ewald construction. Mathematically the experimentally observed intensity function $I_{exp}(\underline{K})$ is produced by a convolution of the intensity function of the lattice $I_{lattice}(\underline{K})$ and the instrument response function $T(\underline{K})$: $I_{exp}(\underline{K}) = I_{lattice}(\underline{K}) * T(\underline{K})$ [4.5-7]. The asterisk * means convolution operation. Taking the Fourier transform the lattice term yields the Patterson function (that is the autocorrelation function of the infinite lattice) $P(\underline{x})$, and $T(\underline{K})$ yields the transfer function $t(\underline{x})$. With the help of the convolution theorem the following equation is obtained:

$$F\{I_{lat}(\underline{K}) * T(\underline{K})\} = F\{I_{lat}(\underline{K})\} \cdot F\{T(\underline{K})\} = P(\underline{x}) \cdot t(\underline{x}) \quad . \tag{4.1}$$

If $T(\underline{K})$ is approximated by a gaussian function, $t(\underline{x})$ is also a gaussian function, which is negligible for x larger than a certain limit. The effect is therefore to suppress any correlation or interference for distances larger than a given instrumental limit. This distance is called coherence width. Its meaning may be described by assuming an ideal instrument and reducing the crystal to that small size giving the same width of the diffraction spot: now the crystal has just the diameter of the coherence width. A careful discussion of instrumental parameters has been given by PARK et al. [4.7]. The LEED systems commercially available today provide half width down to 0.5 mm at about 20-50 eV which gives a maximum coherence width of 30 nm. More typical is 10 nm which decreases for lower and higher voltages. Due to the low coherence width compared with the actual beam diameter of about 0.5 mm, the observed LEED pattern is therefore not just the diffraction pattern of the crystal surface, it is the average of a very high number of diffraction patterns (that is incoherent addition of intensities) each of which is obtained from a fraction of the surface in the size of the coherence width. This will be important in interpretation of patterns of nonperfect surfaces.

4.1.4 Diffraction Pattern of Simple Defect Structures

To get a feel for which parameters are important for defect detection from diffraction, some simple one-dimensional examples are described. For this purpose several atom arrangements and their diffraction patterns are shown in Fig. 4.1. The diffraction pattern may be obtained by straightforward calculation or by optical diffraction of a slide taken from the atom arrangement shown [4.8]. A single atom yields no intensity modulation, since within the approximation the atom is assumed to be an ideal point scatterer (Fig. 4.1a). Two atoms give a sinusiodal modulation as known from the optical diffraction pattern of two narrow slits (Fig. 4.1b). The half width of the spots is half of the spot separation. For a regular array of N atoms the spot position remains the same, the sharpness increases proportional to the number of atoms (Fig. 4.1c). All reflexes have the same shape. If several identical groups of N atoms each are regularly arranged with a connecting distance other than the atom distance within a group (e.g., a/2), then some of the spots are split with the splitting depending on the number of atoms within a single group and the sharpness (identical for all spots) depending on the total width of the object (Fig. 4.1d). If the number of atoms within a group is varied, but the arrangement from group to group retained then some spots are unchaged and some are split or broadened or both depending on the distribution of sizes of the groups (Fig. 4.1e). If N atoms are randomly distributed over 2 N regularly spaced sites and the other sites are kept empty, then spot position and sharpness are the same as if all 2 N sites were occupied. The intensity, however, in the spots is lower and the missing intensity is found in the background between the spots (Fig. 4.1f). If finally N atoms are distributed at random over a range of N·a (with the probability of a gaussian function) then only one central peak is found (Fig. 4.1g). Its half width is determined by the size of the range. Due to lack of periodicity no other peaks are found.

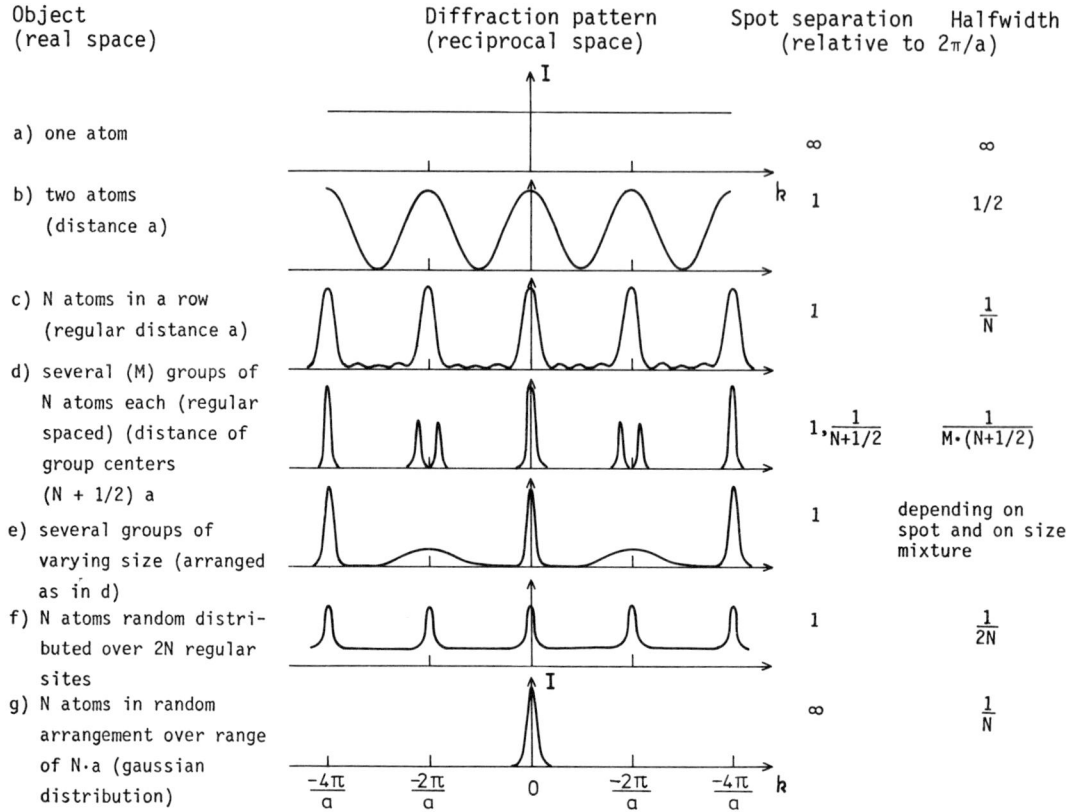

Fig. 4.1.a-g Schematic representation of atom arrangements and the corresponding diffraction patterns together with numbers on spot separation and half width in reciprocal space

Besides straightforward calculations the diffraction patterns of Fig. 4.1 are also easily produced by optical diffraction [4.8]. For this purpose some of the atom arrangements of Fig. 4.1 have been drawn as gratings. The diffraction pattern of the one-dimensional arrangements as obtained with a small laser is shown in Fig. 4.2. The first three patterns show the increase in sharpness if the number of regularly spaced atoms is increased. For the last three patterns each group of atoms has the same regular distance from atom to atom. The distance from group to group is always half the distance from atom to atom. As discussed in Subsection 4.4.2, each spot with an even index corresponds to in-phase scattering, those with odd index to out-of-phase scattering. The odd spots therefore reflect by splitting (Fig. 4.2d,e) or by broadening (Fig. 4.2f) the arrangement of groups. These models may be used both for step and domain arrangements as discussed in Sections 4.3 and 4.4 and in [4.8].

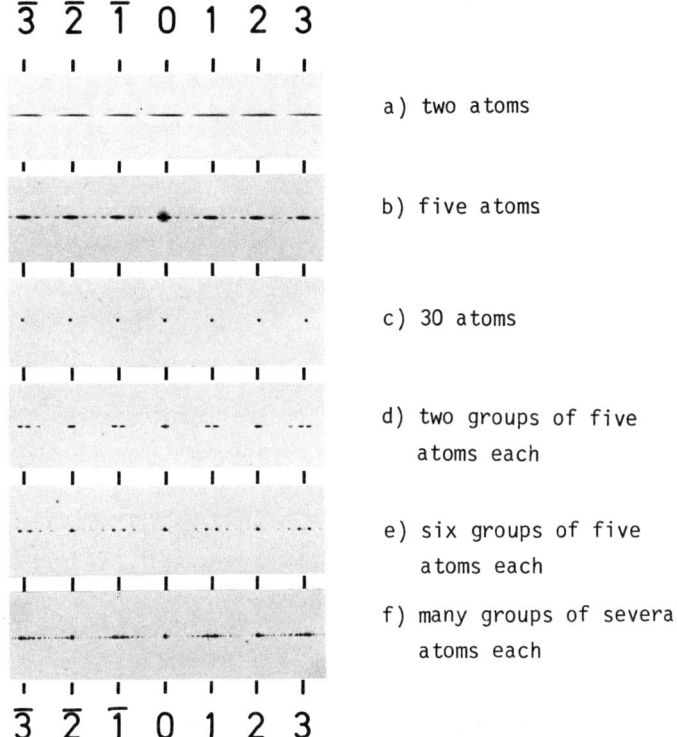

Fig. 4.2a-f Diffraction patterns on some atom arrangements produced by optical diffraction of a grating with a slit arrangement representing the indicated atom arrangement

4.1.5 The Kind of Information in the Diffraction Pattern

In principle for a complete interpretation the whole reciprocal space, that is the intensities of all spots and all sites between spots at all electron energies, should be known and used. Within the kinematical approximation (see Subsec. 4.1.1) only the spot existence and its shape additionally background intensity are used, that is elastically scattered intensity between the areas of the spots. As seen from the examples in the preceding section and Figs. 4.1 and 4.2, a spot separation is always reciprocal to a frequently occurring distance. This may be the regular atomic distance due to substrate structure or to superstructure or it may be the average distance between groups of atoms (see Fig. 4.1d). The second useful information lies in the spot shape. The reciprocal half width reflects the range of coherent or in-phase scattering. If this range is smaller than the coherent range due to instrumental limitations or smaller than sample size as in Fig. 4.1c, then this length indicates the average size of regular atomic positions (for example domain size of superstructures or terrace width of step structure). The third information lies in the background. In principle it contains the number and correlations of point defects. In reality also the ideal lattice shows some intensity due to inelastic processes with very low energy change (like phonon interactions) so that the instrument is not able to separate clearly those electrons from elastically scattered electrons. Therefore the background information is not easily evaluated with the present available commercial instruments. Its information is so far used only qualitatively.

4.2 Point Defects

With point defects all those deviations from periodicity which refer to single sites and are randomly or nearly randomly distributed are described. For example missing or displaced atoms and atoms with a different scattering factor are point defects. Also lattice vibrations are included, although here the random displacement is only approximately given.

Within the kinematical approximation, as described in Subsection 4.1.1, the scattered intensity I of a group of N atoms with the real form factors f_n is given by

$$I = |\sum_n^N f_n \exp(i\underline{K}\underline{r}_n)|^2 = \sum_{mn}^N f_m \cdot f_n \exp[i\underline{K}(\underline{r}_n - \underline{r}_m)] \quad . \tag{4.2}$$

Here $\underline{r}_n$ is the position of the nth atom (with so far no restriction) and $\underline{K}$ is the scattering vector, $\underline{K} = \underline{k} - \underline{k}_0$ ($\underline{k}_0$ and $\underline{k}$ are wave vector of the incident and scattered plane wave, respectively).

4.2.1 Variation of Scattering Factor

First the effect of varying form factor is discussed. If the distribution of the f_n is at random, that means any two f_n are not correlated, the factor $f_m \cdot f_n$ may be replaced for $m \neq n$ by its average $<f_m \cdot f_n> = <f_m><f_n> = <f>^2$ where the brackets $<>$ indicate the average over all atoms. For $m = n$ the average is $<f_m \cdot f_n> = <f^2>$. In general therefore $<f_m \cdot f_n> = <f>^2 + \delta_{mn}(<f^2> - <f>^2)$. The intensity shows two contributions.

$$I = N (<f^2> - <f>^2) + <f>^2 \sum_{mn} \exp[i\underline{K}(\underline{r}_m - \underline{r}_n)] \quad . \tag{4.3}$$

The first term is due to the N terms with $\delta_{nm} = 1$ and does not depend on the scattering factor $\underline{K}$; it yields a homogeneous background of uniform intensity. In neutron scattering it is called the incoherent term. It vanishes if all f_n are identical. The second term describes the usual diffraction pattern for the case that all atoms scatter with the average $<f>$ which is called coherent term in neutron diffraction. The effect of varying f is just to add some uniform background by diminishing the intensity in the spot pattern.

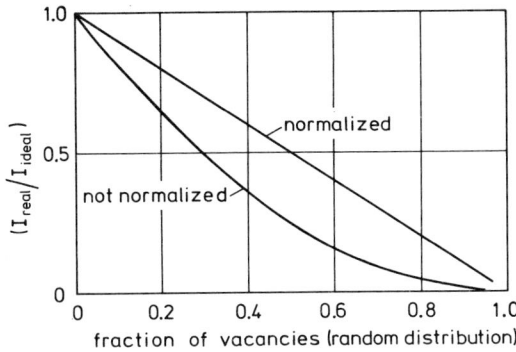

Fig. 4.3 Calculated intensity in the spot pattern for an increasing number of empty sites (kinematical approximation, (4.3). The intensity missing in the spots is found in a homogeneous background

For illustration purposes a two-dimensional regular array of atoms is considered, where a random fraction of the sites is unoccupied, that is their scattering factor is zero. In Fig.4.3 the intensity of the spot pattern is plotted versus the fraction of missing atoms. It decreases with the square of the number of remaining atoms. Since the total scattered intensity decreases proportional to the number of remaining atoms, the plot called "normalized" gives the fraction of the total intensity found in the spot pattern. The rest forms the uniform background. The effect of varying form factor may be estimated in the following way. If half of the sites show $f = 1$, the other half $f = 2$, then only 10% of the intensity is found in the background; if $f = 1$, and $f = 1{,}5$ only 4% of the intensity forms the background.

To compare with real situations: only very drastic changes in the distribution of f_n give measurable results. Since more than a few percent of empty sites and more than a factor of two in varying f_n seems not very likely, in most cases the contributions to background due to this effect should be negligible compared to those described in Subsection 4.2.2.

Finally some comments on the assumption of random distribution should be given: if half of a monolayer is adsorbed on a perfectly flat surface, the distribution may be at random. Then a uniform background is formed. If the atoms attract themselves, larger or smaller islands are formed, a severe deviation from random distribution. This case is treated in Section 4.3. No uniform background is found: depending on wavelength additional intensity is found only in the vicinity of the spots. If on the other hand a repulsion exists between single atoms in the top layer an increase of intensity in the center between regular spots (Sec.4.4) is found. If the degree of order increases, a superstructure with sharp spots is formed. It is therefore clear that the condition of random distribution is decisive with respect to homogeneity of the intensity distribution within background. The LEED pattern easily indicates such deviations from random distribution. Nevertheless a quantitative evaluation is so far difficult.

4.2.2 Variation of Atom Position

In the preceding subsection no restriction has been made with respect to atomic positions. The second term of (4.3) which is dependent both on scattering vector and atom positions, is therefore discussed now. Since a perfect single crystal is always assumed as substrate the atom positions $\underline{r}_n$ may be described by their regular positions $\underline{r}_n^o$ and the deviations $\underline{u}_n$ thereof: $\underline{r}_n = \underline{r}_n^o + \underline{u}_n$. In the following for the sake of simplicity all scattering factors f are taken the same, so that now only the second term in (4.3) remains.

$$I = f^2 \sum_{m,n} \exp[i,\underline{K}(\underline{r}_m - \underline{r}_n)] = f^2 \sum_{mn} \exp[i\underline{K}(\underline{r}_m^o - \underline{r}_n^o)] \exp[i\underline{K}(\underline{u}_m - \underline{u}_n)] \quad . \quad (4.4)$$

As discussed in Subsection 4.1.3 the observed LEED pattern is produced by addition of the intensities of a very high number of small coherently illuminated areas. In Fig. 4.4 therefore the average has to be taken. Since the first exponential term on the right side of (4.4) is identical for all coherent areas, the average has to be taken only over the second exponential term.

$$<I> = f^2 \sum_{mn} \exp[i\underline{K}(\underline{r}_m^o - \underline{r}_n^o)] < \exp[i\underline{K}(\underline{u}_m - \underline{u}_n)] > \quad . \tag{4.5}$$

As shown in textbooks [4.2,9] the second exponential term may be approximated by

Reciprocal space

hk=3̄0 2̄0 1̄0 00 10 20 30

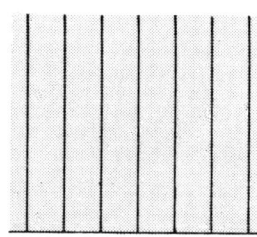
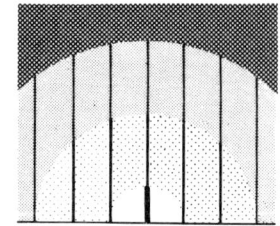

a) for ideal lattice   b) with vacancies   c) with random displacements

Fig. 4.4a-e Schematic representation of reciprocal space a) for an ideal, defect free surface, b) a flat surface with random distribution of vacancies, c) for a complete surface with random displacements (vertical and horizontal). The decrease in spot intensity (thickness of rods) and increase of background with increasing scattering vector should be continuous

$$\exp\left\{-\frac{1}{2} < [\underline{K} \cdot (\underline{u}_m - \underline{u}_n)]^2>\right\}$$

since the odd terms of the Taylor expansion average to zero. If all deviations are completely independent of each other, that means true random displacements, the exponent for $m \neq n$ is approximated by $2M = \underline{K}^2 \cdot <u_{\underline{K}}^2>$ with $\underline{u}_{\underline{K}}$ the component of $\underline{u}$ in the direction of $\underline{K}$. $\exp(-2M)$ is called the Debye-Waller factor. In the summation again the N terms with $m = n$ have to be treated separately. Then the following equation is obtained:

$$<f> = f^2 \left| N[1-\exp(-2M)] + \exp(-2M) \sum_{mn} \exp[iK(r_m^o - r_n^o)] \right| . \tag{4.6}$$

The sum just represents the diffraction pattern of the ideal lattice. The intensity of the spots is reduced by the Debye-Waller factor. The missing intensity is found in the background with similar $\underline{K}$. Here the background intensity and the spot intensity are dependent on the scattering vector $\underline{K}$. For illustration purposes Fig. 4.4 shows the reciprocal space with Ewald construction for the ideal surface (Fig. 4.4a), for a surface with random point defects as discussed in Subsection 4.2.1 (Fig. 4.4b) and with random displacements, where vertical and horizontal displacements are taken as having the same magnitude (Fig. 4.4c). Whereas the vacancies produce homogeneous background, here the intensities within the spot pattern are decreased with increasing scattering vector. To give numbers, Fig. 4.5 shows the decrease of the intensity in the 00-beam for normal incidence with increasing vertical random displacement of the surface atoms. It is calculated for gaussian distribution with a mean square displacement identical to that of a homogeneous distribution of the indicated width. A distribution homogeneous in a definite width yields only for low displacements approximately the same plot. The decrease of Fig. 4.5 may be accomplished either by increasing the electron energy (that is going up the 00-rod in Fig. 4.4c) or by increasing the disorder at constant energy (e.g., increasing temperature or bombarding

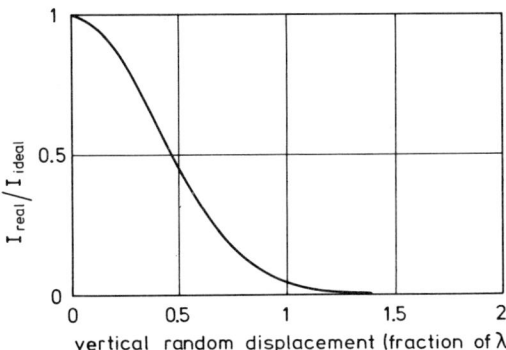

Fig. 4.5 Decrease in spot intensity with increasing vertical atom displacements for the 00-spot at normal incidence

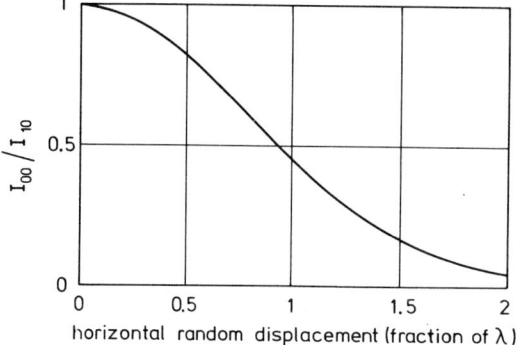

Fig. 4.6 Decrease of the ratio of 10 and 00 intensities with increasing horizontal displacement at normal incidence. The 00-beam is not affected at all

damage). Fig.4.6 shows the corresponding results, if only horizontal displacements are present. Here the 00-beam is not affected at all. With increasing spot index the intensity goes down due to the increasing component of the K-vector parallel to the surface. It corresponds to a horizontal movement of the end point of vector K in Fig. 4.4c starting at the 00-rod. A numerical example is given in Fig. 4.7. Here both vertical ($1.2 \times \lambda$) and horizontal ($0.6 \times \lambda$) random displacements are assumed. Several basic features are seen from the plot: 1) The sharpness of the spots is not affected by the displacements. It is completely given by the coherence width due to the instrumental limitations. The width has been assumed to 20 atomic distances.
2) The intensity of the higher order spots goes down due to horizontal displacements.
3) A homogeneous background is formed. In the computation only 80 zones of 20×20 atoms each have been averaged. With the number of 10^{10} zones according to experimental

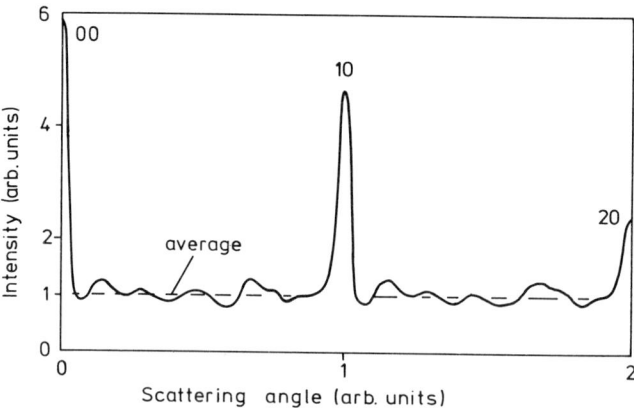

Fig. 4.7 Calculated diffraction profile between 00 and 20 spot of a square atom array with random vertical ($1.2 \times \lambda$) and horizontal ($0.6 \times \lambda$) displacements. The coherence width (with gaussian shape) has been taken as 20 atomic distances. The plot is an average of 80 independent areas

situation a completely homogeneous background is obtained. As seen from Fig. 4.5 in this example more than 98% of the total intensity is already in the background. Therefore the increase in background with scattering angle is not visible in this case. The indicated average corresponds to the expected pattern without any diffraction spots.

In a more rigorous and more general treatment the atomic displacements $\underline{u}_n$ are described by their Fourier expansion,

$$\underline{u}_n = \sum_q V_q \exp(i\underline{q}\underline{r}_n^o) \quad .$$

Any displacement is described, if all vectors $\underline{q}$ are taken as given by atom arrangement and crystal dimension (see any textbook on lattice vibrations). For static displacements (like ion bombardment damage) the expansion is used as it is. For lattice vibrations a factor $\exp(i\omega_q t)$ is added with ω_q the frequency of the lattice vibration with wave vector $\underline{q}$ according to the respective dispersion relation. Following textbooks (Ref.4.9, Chap.19), scattering with scattering vector $\underline{K}$ occurs if one of several conditions is met: 1) The scattering vector may be equal to a vector $\underline{G}$ of the reciprocal lattice. Here the diffraction spots of the ideal lattice are produced, although reduced in intensity by the Debye-Waller factor. The wave length of the electron is not changed. The remaining intensity forms the background via the following processes. 2) The scattering vector $\underline{K}$ may equal $\underline{K} = \underline{G} + \underline{q}$, with $\underline{q}$ the wave vector in the Fourier expansion of $\underline{u}_n$. In lattice dynamics the case $\underline{G} + \underline{q}$ is described as phonon emission, since one phonon of vector $\underline{q}$ and frequency ω_q is produced during scattering. Consequently the energy of the electron is reduced by $\hbar\omega_q$. Similarly $\underline{G} - \underline{q}$ corresponds to phonon absorption. 3) Finally multiphonon processes may be included by allowing the change of $\underline{K}$ with several phonons.

If all lattice distortions are due to lattice vibrations the above description with phonon absorption and emission is complete. Static distortions are not included. This case, however, may be treated exactly the same by setting all frequencies of lattice oscillations equal to zero. Therefore scattering occurs with $\underline{K} = \underline{G}$ or $\underline{K} = \underline{G} + \underline{q}$, etc., with energy conserved (strictly elastic scattering). Since the Debye-Waller factor is not affected by phonon frequencies, the intensities of the spots are exactly the same for static and dynamic distortions as long as the appropriate average of the distortions is the same. It is not necessary to take a time average since in LEED experiments an average over the high number of coherent areas is always taken. In background a difference occurs due to the energy loss or gain with phonon scattering. Since the commercial LEED systems cannot discriminate phonon energies, static and dynamic disorders are not distinguished; they produce exactly the same LEED pattern. The results obtained with lattice vibrations may also be used therefore to interpret the pattern of static disorder.

The intensity for a scattering with vector $\underline{K}$ is roughly determined by all those Fourier components which fulfill the condition $\underline{K} = \underline{G} + \underline{q}$. (In a rigorous treatment transition probabilities have to be included.) If the displacements $\underline{u}_n$ are at random, the Fourier coefficients are all the same and a uniform background is formed (as discussed at the beginning of this section). In thermal vibrations and most likely also in static disorder neighboring displacements are dependent on each other. Therefore deviations from uniform background are expected and have been found in thermal vibrations (for X-rays see [4.2]); a LEED experiment of this kind is reported by DENNIS and WEBB [4.10].

Since the effect of lattice vibrations on the LEED pattern has been discussed extensively elsewhere [4.11,12] it should only be pointed out here that static disorder may be included in much the same way.

4.3 Atomic Steps

Whereas so far the surface of the crystal has always been in one atomic plane, with atomic steps the surface varies over several planes.

Since LEED is sensitive to only a few atomic layers, interference occurs between neighboring terraces due to horizontal and vertical shift. Depending on wave length and angles of incident and outgoing waves this shift may cause constructive or destructive interference. In case of constructive interference, the spot shape is not changed compared to the ideal crystal. With destructive interference a splitting or broadening of the spot occurs, as shown in Figs. 4.1d,e and 4.2d,e,f. It is therefore necessary to study the spot shapes and their variations with experimental conditions (like wave length) to derive the number and distribution of atomic steps.

4.3.1 Regular Step Arrays on Primitive Lattices

The derivation of the diffraction pattern is very much simplified if all terraces have the same width and are arranged in a monotonic regular step array. Additionally it is assumed that all atoms are in regular atomic sites as given by the three-dimensional periodicity of the crystal. Then the surface is described by a periodic arrangement of identical terraces. Whereas the scattering pattern of a single terrace is described only approximately by a kinematical calculation, the effect of the periodic repetition of identical terraces is exactly reproduced within such a calculation (as discussed in Subsec. 4.1.1 [4,13-15]). In Fig. 4.8 the pattern is constructed first by considering the diffraction pattern of a single terrace (upper portion of Fig. 4.8). In principle is should reproduce the intensity which comes from a single terrace (including all underlying atoms), when the whole sample is illuminated by the primary beam. In the kinematical approximation used only the single scattering of the surface atoms of one terrace is considered. Therefore the

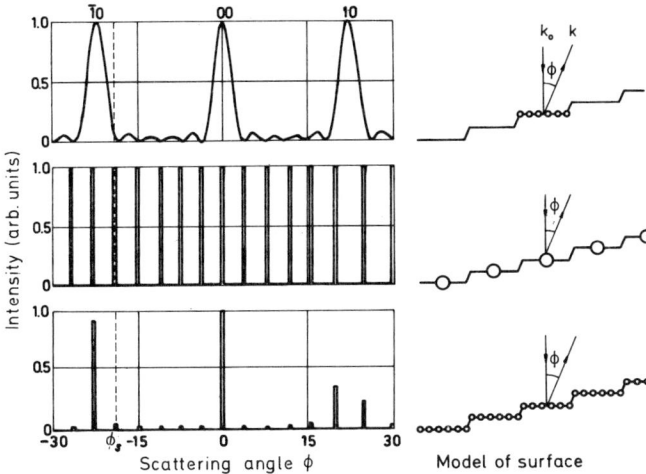

Fig. 4.8 Demonstration of spot splitting due to regular step arrays. Upper portion: atom arrangement and diffraction pattern of a single terrace, center protion: regular array of identical (arbitrary) scatterers, lower portion: atom arrangement is produced by folding of the upper two-diffraction pattern by multiplication. ϕ_S is the angle of specular reflection at the inclined face (two times the angel of inclination)

diffraction pattern is identical to that of an optical grating with (in this case) six slits. The width of the beams is due to the finite terrace width. With increasing electron energy (that is decreasing electron wave length) the spots move towards the not moving 00-beam, as usual with the LEED pattern of a flat surface. The center portion of Fig. 4.8 describes exactly the effect due to the infinite periodic array of identical scatterers with a periodic array of δ functions. Due to the macroscopic inclination towards the terrace plane, the zeroth order beam is at the doubled inclination angle (= angle of specular reflection ϕ_S). With increasing electron energy the spots therefore move towards the angle of specular reflection. Since the crystal may be described by a convolution of a single terrace (upper part of Fig. 4.8) with its periodic arrangement (center part), the diffraction pattern of the stepped surface according to the convolution theorem is given by the product of the two intensity patterns (lower part of Fig. 4.8). The pattern consists of single and double spots depending on the relative situation of the broad maxima and δ functions. Since this relative situation changes with electron energy, a periodic change of each diffraction spot between appearance as a single and double spot is expected. Small satellite peaks may be observed depending on the exact shape of the diffraction pattern of a single terrace. Especially a "single spot" may show weak satellites on both sides, which has been observed experimentally.

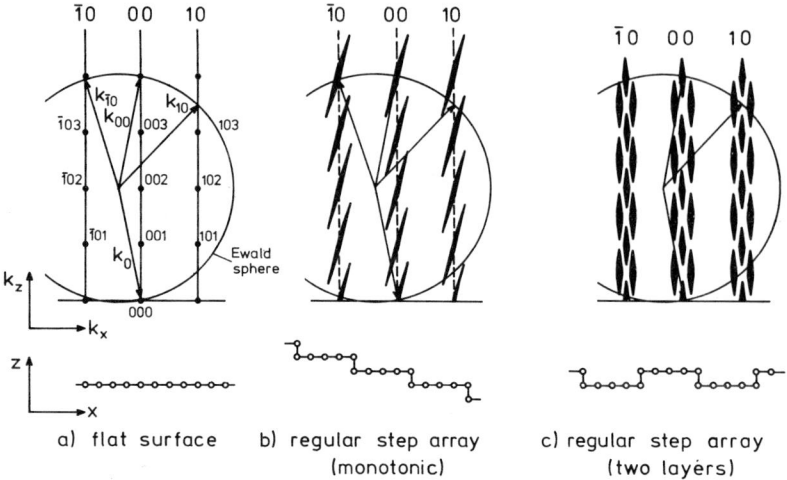

Fig. 4.9a-c Reciprocal space with Ewald sphere for three forms of surfaces. k_o denotes the wave vector of the incident beam, k_{00}, k_{10}, and $k_{\bar{1}0}$ that of the diffracted beams with indices 00, 10, and $\bar{1}0$, respectively. With the shown position of Ewald sphere the 00 beam reflects in-phase condition, the beam out-of-phase condition between adjacent terraces

A more general approach is possible by using the reciprocal lattice and the Ewald construction (Fig. 4.9, [4.16,17]). For a flat low-index plane Fig. 4.9a yields sharp spots for all voltages, since in reciprocal space the rods are sharp in the directions parallel to the surface. If the surface shows a regular array of steps (corresponding to a high-index plane close to a low-index plane) the reciprocal lattice consists of sharp rods normal to that plane. Due to the structure of the flat terraces the rods are important only in the vicinity of the reciprocal rods of the corresponding low-index plane (effect of folding between single terrace and high-index plane). Ewald's construction now yields for each spot a single or double spot depending on the orientation and radius of Ewald's sphere (Fig. 4.9b). In the case of Fig. 4.9b the 00-beam and the 10-beam appear as single and double spot, respectively. If many layers normal to the surface contribute to the scattered intensity (as in X-ray diffraction) the rods in reciprocal space shrink to spots due to the 3rd Laue condition: a three-dimensional reciprocal lattice is formed independent of surface orientation. If the Ewald construction yields a spot of the reciprocal lattice, all atoms, therefore also all surface atoms on all terraces, scatter in phase. In this case the corresponding spot of the LEED pattern is identical for flat or stepped surfaces. The splitting of the spots corresponds to an out-of-phase scattering of subsequent terraces for the exact spot position (for example center of 10 spot in Fig. 4.9b). The single spot corresponds to an in-phase scattering of all terraces.

Therefore the points of three-dimensional reciprocal lattice (as shown by three indices in Fig. 4.9a) are common to the reciprocal lattice of all surfaces no matter which orientation or uniformity, step density or orientation.

The case of a regular array restricted to two atomic layers is treated in Fig. 4.9c. It may be produced by half a monolayer on top of a flat surface forming regular islands. The pattern is identical with that shown in Fig. 4.9b for the exact condition for single and double spots. This is reasonable since the phase shift between the first and second terrace is identical in both cases and that between the first and third layer differs by 2π. For intermediate voltages, however, the two cases are easily distinguished: for the case of Fig. 4.9b two spots of unequal intensity are observed, whereas for the case of Fig. 4.9c three spots with the outer two having equal intensity are expected. If within the LEED beam regular arrays of Fig. 4.9b are found half of them going up, the other half going down over many layers, the diffraction pattern again is identical for the exact in-phase and out-of-phase condition. For the intermediate case now four spots are expected with a characteristic change in intensity and position with electron energy, as easily derived from a superposition of Fig. 4.9b with its mirror image.

The crucial point of the Ewald construction with respect to steps (Fig. 4.9) is the basic property, that the LEED pattern is not affected by steps, if the construction yields a point of the three-dimensional reciprocal lattice. It is, however, affected at maximum, half way between those points. It is therefore important to know in advance those "characteristic" voltages for in-phase or out-of-phase scattering of adjacent terraces. Those voltages depend on the indices of the spot (h,k), on bulk parameters (like step heigth d, lattice vectors $\underline{a},\underline{b}$ and reciprocal lattice vectors $\underline{a}*,\underline{b}*$ -- which are chosen to lie in the plane of a single terrace --, lateral shift of subsequent layers measured as fractions x and y of the lattice vectors $\underline{a}$ and $\underline{b}$).

A general formula for normal incidence (with respect to a single terrace) has been given in [4.16]. For oblique incidence the general formula is [4.17].

$$V_{hk} = \frac{M}{4d^2} \cdot \left[\frac{(S - hx - ky)^2 + (h\underline{a}* + k\underline{b}*)^2 d^2/4\pi^2}{(S-hx-ky)\cos\theta - (h|\underline{a}*|\cos\phi_a + k|\underline{b}*|\cos\phi_b) d \sin\theta/2\pi} \right]^2 \quad (4.7)$$

$$\text{(with } M = \frac{h^2}{2em} = 150 \text{ V\AA}^2 = 1{,}5 \times 10^{-18} \text{ Vm}^2\text{)} \; .$$

This expression includes additionally the angle θ between the direction of the incident electron beam and the normal of a terrace, and the azimuthal angles ϕ_a and ϕ_b between the projection of the incident direction on to a single terrace and the lattice vectors $\underline{a}$ and $\underline{b}$, respectively. If S is an integer, the calculated voltage yields an in-phase condition (that is a sharp and single spot); for half-integer values the out-of-phase condition with split or diffuse spots is obtained. To avoid confusion between transmitted and reflected beams (since with LEED only reflected beams may be observed) S has to be chosen sufficiently large, so that for all S

$$(S-hx-ky)^2 + (S-hx-ky) \times tg\theta(h|\underline{a}^*|\cos\theta_a + k|\underline{b}^*|\cos\phi_b)/\pi \geq (h\underline{a}^* + k\underline{b}^*)^2 d^2/4\pi^2 . \tag{4.8}$$

The angles for the diffracted beams do not have to be specified, since those are automatically determined by the incident direction, the indices h, k of the reflected beam and the electron wavelength via the Ewald construction.

The above derivation allows a simple and accurate evaluation of the LEED pattern.
1) The step height is derived from the sequence of characteristic voltages. Especially for the 00-beam (4.7) yields a simple expression

$$V = \frac{M}{4d^2 \cos^2\theta} \tag{4.9}$$

which allows an easy and accurate determination of the step height (experimental error less than 1% of the step height). 2) The average terrace width is taken from the spot splitting. The ratio of the distance of normal spots to the spot splitting gives directly the terrace width in distances of parallel atomic rows. A more general formula, valid for all spots, however, only for normal incidence, has been given in the literature [4.18]. 3) The orientation of step edges is normal to the spot splitting. 4) The combination of step height and terrace width yields the angle of inclination and the step density. The angle may be also determined directly by observing the point towards which the reflexes move with increasing electron energy, as demonstrated by photographs with varying electron energy [4.13].

The above evaluations are somewhat hindered by the varying intensity of the spots, especially when spots just have zero intensity close to the desired energy. This restriction is serious only for surfaces of high defect density, when spots are visible only for a few voltages. Since many spots with many characteristic voltages exist and the evaluation is equivalent in all cases, any surface which shows a pattern at all may be evaluated in this respect, as demonstrated in many examples.

4.3.2 Irregular Step Arrays

A regular step array is given by three conditions: constant terrace width, step height, and edge orientation. If at least one of the three parameters varies at random (not periodically) an irregular step array is formed. Many irregularities may be described as irregular step arrays. If a surface is not atomically flat over macroscopic distances due to nonuniform cutting or polishing or due to random attack during etching or sputtering, a surface with a certain number of random distributed steps is always formed. Although heat treatment in many cases lowers the number of steps due to surface diffusion, a higher or lower number of steps remains depending on material, orientation and treatment. Only a few exemptions like cleaved mica or cleaved gallium arsenide may have a really negligible number of steps. Most investigated surfaces should show random arrangement of steps. Therefore ways to determine the number and distribution of such step configurations are rather important.

For a quantitative description first a one-dimensional model is discussed. The terraces are described by their width Γ in multiples of the distance of adjacent atomic rows. For a given surface the width may occur with the (normalized) probability $P(\Gamma)$ [4.6]. Also the step height is described by a probability distribution including up and down with equal or not equal probability. Here only monatomic step heights (including up and down) are considered. Whereas $P(\Gamma)$ gives the probability of finding a terrace of width Γ, the function

$$P'(\Gamma) = \Gamma \cdot P(\Gamma) / \sum_{\Gamma} \Gamma \cdot P(\Gamma)$$

gives the probability that an arbitrary surface atom is found on a terrace of width Γ. The function P' rather than P should be used for averaging purposes.

For a rough prediction of the LEED pattern the surface may be approximated by areas with regular step arrays, where the coverage with arrays of terrace width Γ is given by $P'(\Gamma)$. Then the intensity functions of the regular portions within the primary beam are added (incoherently). The single spot is not affected. Its half width is still given by the instrumental function. The double spot, however, reflects all terrace widths by a varying splitting. For a broad mixture of terrace widths with equal probability going up and down therefore a **reciprocal** lattice as in Fig. 4.10a

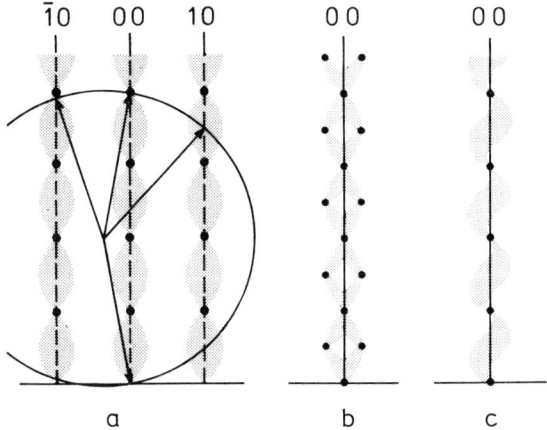

Fig. 4.10a-c Reciprocal space for three arrangements with irregular step arrays: a) both terrace width and direction of step (going up or down) are irregular, b) width regular, step direction (up or down) irregular, c) irregular in width, monotonic step sequence

is expected. With increasing electron energy all spots vary between sharp and diffuse shape periodically. Again the characteristic voltages for on-phase scattering are the same as before. If a constant terrace width is assumed, steps may, however, go up and down with equal probability; a reciprocal lattice as in Fig. 4.10b is expected. If on the other side a broad mixture of terrace widths is admitted with nearly all steps going down, Fig. 4.10c is obtained. By observing the spot shape not just at characteristic voltages, but additionally at voltages in between, a lot of information on sequence and distribution of terraces is obtained. If the spot shape in the two directions is observed, additionally information on symmetry of the step structure is obtained.

Although this model of incoherent superposition of many different areas with regular step arrays describes the LEED pattern qualitatively correctly, it should not be expected that the distribution $P'(\Gamma)$ of terrace widths may be evaluated correctly. The interference of neighboring terraces with different width is neglected. Therefore a random mixture of all possible terrace widths has to be arranged and the corresponding LEED pattern has to be calculated via the average autocorrelation function and Fourier transform as described in Subsection 4.1.2. The first attempt in this respect was made by HOUSTON and PARK [4.6]. They approximated the average autocorrelation function by a folding of the probability distribution of step edges with the autocorrelation of the averaged terrace. Their results show that a fairly broad distribution of terrace widths still results in a splitting of the spots at the appropriate voltages. If just the existence of spot splitting is observed, the step array may be rather irregular. Additionally each of the spots has to show the same sharpness as a single spot so that the existence of a step array regular within the coherence zone may be conclusive.

The above construction of the autocorrelation or pair distribution function of the random stepped surface is only approximately correct. It treats all terraces alike by replacing them with an average. Therefore the pair correlation is not given exactly if a narrow and a wide terrace are situated close to each other. For example the above approximation yields a small nonzero probability that atoms of the second layer are treated as surface atoms which should be avoided. If the terrace widths scatter over a not too wide range, the approximation is very good. Therefore all examples and conclusions given in the paper are accurate.

For a correct pair distribution function the probability that, starting from an arbitrary surface atom, another surface atom is found after going a given number of atomic distances to the side and up or down has to be calculated. A formula similar to that of Houston and Park has not been given so far. It is, however, possible to construct the autocorrelation function by an algorithm, if a terrace width distribution $P(\Gamma)$ or $P'(\Gamma)$ is assumed [4.19]. If the autocorrelation function is calculated as a two-dimensional array (parallel and normal to surface), the instrumental function may be included by multiplication and the spot shape of each spot at any angle of incidence and at any voltage may be calculated by Fourier transform. By comparison with observed spot shapes the terrace width distribution may be evaluated.

If the spot shows a diffuse broad shape without any splitting into two spots (as in Fig. 4.10a), a rather broad mixture of terraces with different widths has to be taken. Additionally the probability $P'(\Gamma)$ has to be chosen appropriately to get a smooth shape without satellite peaks and kinks. Widths from one to more than hundred atomic distances have been included so that some coherence zones show not even a single step. Some recently calculated results are seen in the following figures [4.25]. The calculated spot shape of a diffuse spot (that is for a characteristic voltage according to (4.7) with half integer S) for several terrace width distributions is shown in Fig. 4.11. The abscissa is divided in units of the relative scattering vector: the value one indicates the position of the nearest normal spot. The percentage of step atoms is obtained by the average of the reciprocal terrace width Γ.

$$\langle \frac{1}{\Gamma} \rangle = \sum_\Gamma \frac{1}{\Gamma} P'(\Gamma) \tag{4.10}$$

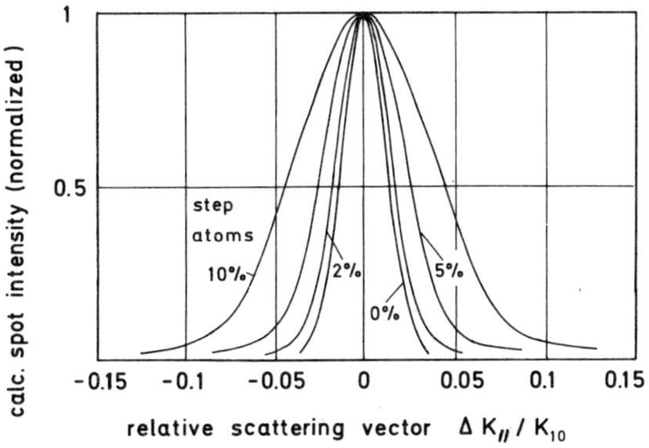

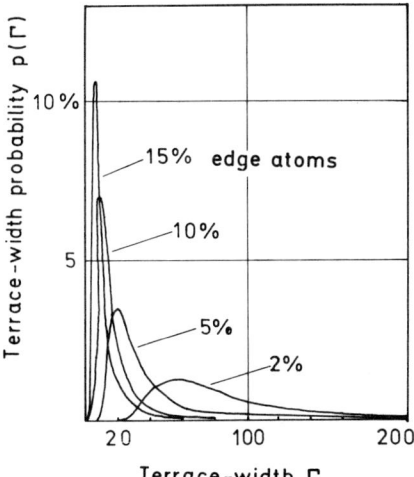

Fig. 4.11 Calculated spot shapes for terrace width distributions chosen to yield approximately a gaussion shape

Fig. 4.12 Terrace width distributions used to calculate the spot shapes of Fig. 4.11

Since the same terrace width distribution is assumed for all directions on the surface, the average of $1/\Gamma$ has been doubled to obtain the fraction of step atoms. Some of the terrace width distributions used are shown in Fig.4.12. From checking some alternate distributions it is assumed that no drastically different distribution would yield a similar spot shape. Including some error bars the given distribution should therefore be unique. From Fig. 4.11 a correlation between step atom density and half width may be derived (Fig.4.13). If the half width is reduced by defolding with the instrumental function (assuming gaussian shape), relative half width and fraction of step atoms are nearly proportional, even nearly identical. In this way step atom densities may be determined quantitatively by spot shape measurements.

If the three methods (incoherent summation of regular arrays, approximate average autocorrelation function, exact average autocorrelation function) are compared with respect to their numerical results they agree very well for constant and nearly constant terrace width. For a mixture of a few close terrace widths the two latter methods yield fairly sharp double spots with a splitting distance corresponding to the average terrace width, whereas broad double spots are obtained with the first method. For diffuse spots (Fig.4.10a) the second method yields a lower intensity of the center of the spot. For very broad distributions the first and third methods agree fairly well. Extensive comparisons, however, have not been made.

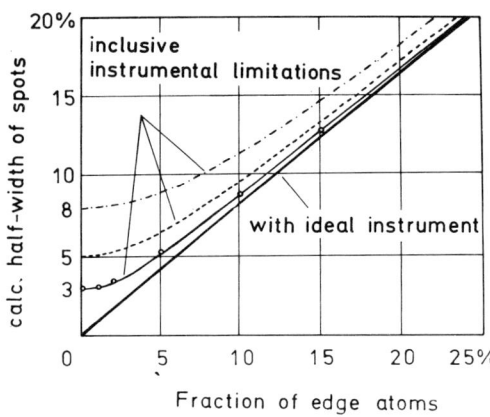

Fig. 4.13 Calculated half width versus edge atom density for several instrumental functions (0, 3, 5, and 8%)

4.3.3 Nonprimitive Lattices

The calculation of the characteristic voltages in Subsection 4.3.1 is based on the assumption that the distance between any two surface atoms is described by linear combination of multiples of three unit vectors. This is obviously true for all primitive lattices with respect to all faces with or without defects. Nonprimitive lattices may be constructed by superposition of several primitive sublattices (e.g., the diamond consists of two identical sublattices displaced in the [111] direction). If a stepped surface consists of atoms of only one primitive sublattice, the description taken from primitive lattices is sufficient. For example close to the (111)-face of silicon and germanium all surface atoms most likely belong to one sublattice, since two subsequent layers of the different sublattice form a close double layer. Even for the polar (0001) faces of ZnO it has been shown by measurement of the step height that all terraces belong to one of the four sublattices. This result is supported by symmetry argument with respect to step edge structure [4.20]. On the other hand in the [100] direction of silicon the two sublattices form a stack of alternate equidistant layers. To predict the LEED pattern, again the Ewald construction of Fig.4.9 is used. Now the three-dimensional reciprocal lattice is modified due to the interference of the two sublattices: some points are missing. For those points the Ewald construction now predicts a diffuse or split spot, whereas for only one sublattice a sharp single spot would be obtained. For other points the sublattices scatter in phase with consequently sharp spots. For a group of spots a complete in-phase or out-of-phase condition is never obtained so that the shape of the spots is not easily predicted, as discussed in detail for the Si (100) face [4.20].

4.3.4 Examples of Stepped Surfaces

Regular step arrays of monatomic step height are produced by cutting single crystals at a definite angle and orientation with respect to a plane with high atom density (that is a low-index plane). At small angles the step density is proportional to the angle of inclination [4.21]. After proper treatment including etching, ion bombardment and annealing and other cleaning procedures for the material under investigation, the LEED pattern of a regular step array has been obtained for metals like Cu [4.15], Pt [4.22], W [4.18], Pd [4.23], and nonmetals like UO_2 [4.13]. This method does not work for silicon and germanium, if clean surfaces are used; annealing makes steps disappear (by forming broad terraces and multiple step heights, not detectable with usual LEED technique).

With cleavage, however, silicon and germanium (111) faces develop portions with regular arrays of monatomic step height [4.14,21]. Cleaved ZnO (0001) shows a step height of two double layers, which has been explained by symmetry arguments [4.21]. For regular step arrays a convenient nomenclature has been given [4.22]; it indicates both width of terraces and orientation and height of steps. For example Pt (S) [6(111) × 1(100)] describes a stepped platinum surface with regular (111) terraces of a width of six atomic distances and step faces with (100) orientation and a width of one atomic distance. Only periodic arrangements are described with this nomenclature. For cleaved silicon and germanium faces and for all other faces, where steps are arranged monotonically over at leat 1 μm, step density and step orientation are easily measured and photographically mapped by a reflection technique [4.21]. Therefore such cleavage planes are especially suited to compare differently stepped portions of the same clean surface with respect to any property (like surface states or adsorption) [4.20]. The case of Fig.4.9c has not been reported so far. The high precision of step height determination has been used to get information on step atom rearrangement [4.20]. Whereas for ZnO both on polar (0001 and 000$\bar{1}$) and nonpolar (10$\bar{1}$0) faces the measured step height agrees precisely (0.8% accuracy) with bulk distances, for Ge (111) and Si (111) a systematic decrease of the step height (about 3%) compared with bulk values has been found. Several rearrangements have been discussed; the most simple model of a uniform depression (or lifting) of all surface atoms (see Fig. 5 of Ref. 4.20) does not work. In this way the step height is still identical with the bulk value. If, however, only the edge atoms are depressed, the terraces are inclined in the average towards the bulk (111) faces. This model has been checked by several experiments and evaluated quantitatively [4.20].

When cleaved silicon surfaces with regular step arrays are heated in ultrahigh vacuum, the step arrays disappear due to surface diffusion by forming fewer steps with multiple step height so that finally the LEED pattern of a flat surface is obtained (that means atomically flat in most coherence zones).

Irregular step arrays were first identified on GaAs (110) faces after ion bombardment and weak heat treatment [4.16].

Immediately after argon ion bombardment germanium, silicon and gallium arsenide show no LEED pattern. Random disorder of the top layers deteriorates the pattern completely. After heat treatment a LEED pattern becomes clearer and clearer with increasing temperature. As soon as a pattern is visible the characteristic change between sharp and diffuse spot shape with increasing electron energy is observed, that is, a random step arrangement is visible. With increasing temperature of heat treatment the diffuse spots get sharper and sharper which corresponds to a decrease of the number of steps (see Fig.4.13)[4.24].

Etched silicon surfaces show the pattern of a stepped surface, if oxide has been removed by fluoric acid and an iodine coverage has been used for transfer to the UHV system. No further treatment within the LEED system is needed. The interface between silicon and thermally grown silicon dioxide has been investigated more carefully. After removing the oxide with fluoric acid and treatment as given above for etched samples the halfwidth of the 00-beam of the LEED pattern was measured as a function of electron energy (Fig.4.14)[4.25]. The oscillating width demonstrates the effect of a random arrangement of steps. The position of maxima and minima corresponds to the predicted positions as seen by indicated S-values (see 4.7). The increasing value from minimum to minimum is due to the increasing instrumental function, as has been checked with flat surfaces. If the effect of the instrumental function is numerically eliminated (by assuming gaussian shape), the maximum values are all approximately the same as predicted by theory. With the results of

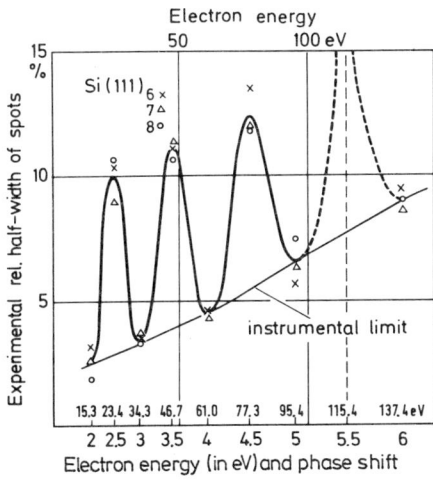

Fig. 4.14 Measured half width of the 00-spot of a LEED pattern of an oxidized silicon (111) face after removal of oxide versus electron energy

Fig. 4.13 it is concluded that the surface has a step edge atom density of about 12%. This density seems to be independent of step density before oxidation. An effect of oxidation parameters so far has not been found. Also a comparison of different faces of silicon is under investigation.

4.4 Domains and Facets

4.4.1 Superstructures and Domains

A frequently observed complication of a LEED pattern is the existence of extra spots, that is spots which are not explained by the substrate atoms in positions as given by bulk structure. Both clean surfaces (e.g., silicon and germanium) and adsorbate structures may show superstructures. In all cases the periodicity of the surface atom arrangement differs from that of the substrate. Its unit vectors $\underline{a}'$, $\underline{b}'$ may be expressed by linear combination of the unit vectors of the substrate $\underline{a}$ and $\underline{b}$. As treated in the literature [4.4] the coefficients may be used to classify superstructures. A superstructure may be due to a regular, however, partial occupation of available surface sites or due to periodic displacements of surface atoms with respect to bulk positions. If the surface is covered with a perfectly periodic superstructure, sharp extra spots are obtained. If on the other hand the superperiodicity prevails only over a limited distance and vanishes or changes after that distance the shape of extra and normal spots is affected by those defects. An area with an arrangement of a superstructure and perfect periodicity within this area is called domain.

Point defects of superstructure may be treated exactly as in Section 4.2. The intensity of the extra spot is reduced and background is formed. Thermal motion of surface atoms different from the substrate may yield a different temperature dependence of extra spot intensities compared to normal spots.

4.4.2 LEED Patterns of Domain Structures

In the following it is assumed that the superperiodicity is maintained within a finite length. If this length is smaller than the coherence length (see Subsec. 4.1.3) and additionally at most one domain in each coherently diffracted area, interference between domains does not occur. Only the diameter of the domain is responsible for the spot shape. The smaller the domain diameter the wider spot diameter as treated in Subsection 4.1.4 and Fig.4.1. The half width of the spot may be evaluated directly to determine the size of the domain. If the diameter of the domain varies with direction (e.g., on a (110) face due to unequal periodicities in the two directions along the surface) an elliptical shape of the spot is obtained.

If the domain diameter exceeds the coherence width considerably, no determination of domain size from spot shape is possible. If several superstructures differ in unit vector orientation or length, different spots are formed and therefore the existence of domains is conclusive. In this case the size of a single domain again from spot shape may not be derived.

A more sophisticated discussion is required if adjacent domains having the same unit vectors are, however, shifted with respect to each other. Since both domains produce the same spot pattern, interference in all spots occurs with all spot shapes possible as discussed in Section 4.1. The phase shift from domain to domain depends

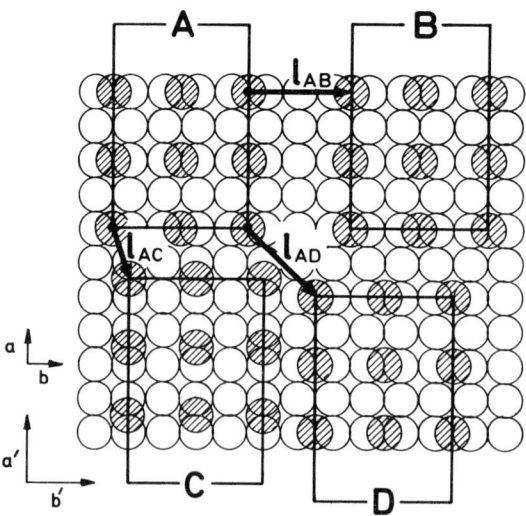

Fig. 4.15 Schematic presentations of four domains A-D of the same superstructure (unit vectors of superstructure are twice the length of substrate vectors) which are shifted against each other by the vectors $\underline{\ell}_{AB}$, $\underline{\ell}_{AC}$, and $\underline{\ell}_{AD}$

both on the connecting translation vector and on spot index. From index dependence possible translation vectors are deduced. As an example in Fig.4.15 four domains A-D of the same superstructure are shown. The substrate square lattice has the unit vectors $\underline{a}$ and $\underline{b}$, the superstructure the unit vectors $\underline{a}'$ and $\underline{b}'$ which are twice the length of the corresponding substrate vectors (in general $\underline{a}'$ and $\underline{b}'$ are given by a linear combination of $\underline{a}$ and $\underline{b}$ with integer coefficients for simple superstructures). The phase shift between the scattered waves of different domains is given by $\Delta\phi = \underline{K} \cdot \underline{\ell}$ with $\underline{K}$ the scattering vector and $\underline{\ell}$ a connecting vector from one domain to another one (see $\underline{\ell}_{AB}, \underline{\ell}_{AC}, \underline{\ell}_{AD}$ in Fig.4.15). The scattering vector $\underline{K}$ may be expressed as usual by $\underline{K} = h\,\underline{a} + k\,\underline{b} + \underline{g}_z$ (with $\underline{g}_z$ an arbitrary vector normal to the surface). The indices h and k have fractional values for the extra spots of the superstructure. The vector $\underline{\ell}$ may be expressed as $\ell_a \cdot \underline{a} + \ell_b \cdot \underline{b}$. The phase shift $\Delta\phi$ is therefore given by

$$\Delta\phi = 2\pi(h \cdot \ell_a + k \cdot \ell_b) \ . \tag{4.11}$$

If the sum in parentheses gives an integer value the two domains scatter in phase; the spot with indices h and k is not degraded in sharpness. If $\underline{\ell}$ is a vector of the substrate (that is ℓ_a and ℓ_b are integers) as in the case of of domain A and B in Fig. 4.15 all normal spots remain completely sharp; some of the extra spots, however, are changed by splitting or broadening depending on domain width and distribution as discussed in detail for step distributions in Section 4.3. Some of the extra spots keep the full sharpness of the instrumental function. If, however the vector $\underline{\ell}$ is not a vector of the substrate (with ℓ_a, ℓ_b nonintegers) as with domain A-C in Fig. 4.15 also some of the normal spots are affected due to noninteger values of the parenthesis (4.11). If finally the vector $\underline{\ell}$ is a vector of the superstructure as with domains A-D in Fig.4.15 the parenthesis (4.11) yields always integer values; no spot is affected by destructive interference. It should be noted that other than integer and half integer values for the parenthesis are possible, resulting in a more complicated spot shape (similar as for stepped surfaces with voltages other than given by (4.7) with integer and half integer values of S). By observing the shape of all normal and extra spots and comparing it with (4.11) the possible vectors connecting different domains may be determined.

4.4.3 Quantitative Description of LEED Patterns

As with doublets due to arrays of terraces, a continuous variation of spot splitting due to domain arrangement has been observed. It was therefore necessary to assume a mixture of domains of different size (in the same way as terraces of different width in Section 4.3). It is not necessary to assume a regular change in domain size to obtain the correct average domain diameter. As shown by HOUSTON and PARK for a one-dimensional arrangement a probability distribution $P(\Gamma)$ may be assumed with Γ the width of a domain in multiples of the superstructure unit vector [4.6]. In nearly exactly the same way as with steps (see Subsection 4.3.2) the broadening and splitting of the spots is calculated for a given distribution P(Γ). Again the result is obtained that the average domain width determines the width of splitting and a rather broad mixture of domains of different width may be tolerated with a still visible splitting of spots. Conversely, the splitting may be used to determine the average domain width, and the shape or broadening of the spot gives information on probability distribution P(Γ).

A quite different, computer-oriented approach has been followed up by ERTL and KÜPPERS [4.26]. A two-dimensional model was used including surface diffusion of an originally randomly distributed adsorbate. With appropriate intermolecular forces a more or less ordered layer is formed after a sufficient number of diffusion steps of each adsorbed atom. The computed LEED pattern between the (0,0) spot and the (0,1) spot is shown in Fig.4.16 (after ERTL and KÜPPERS [4.26]). The intermolecular

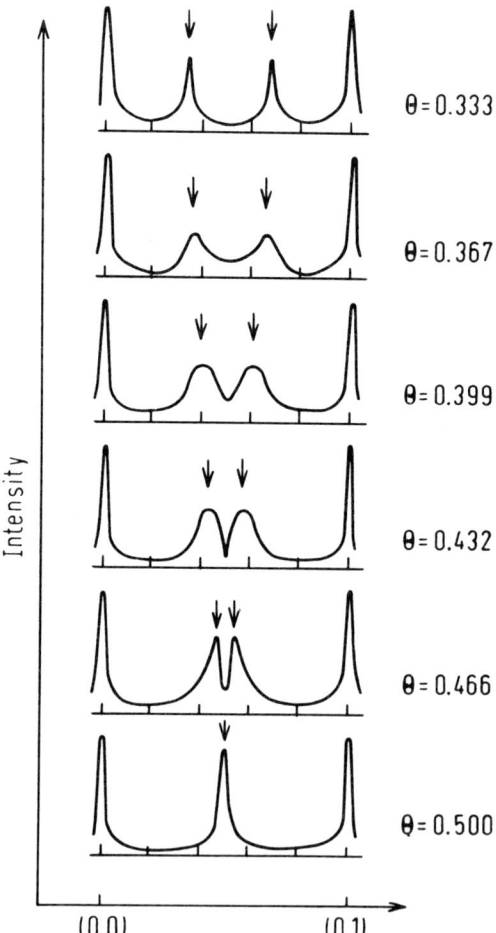

Fig. 4.16 Calculated diffracted intensity profile between (0,0) and (0,1) spot for an adsorbate with indicated coverage, which has been allowed to find equilibrium distribution by surface diffusion. The lowest and highest coverages shown correspond to fairly well ordered structures, the intermediate ones require a minimum number of defects (broadening of extra spots) (taken from [4.26])

energies have been chosen so that a 1 × 2 superstructure yields the lowest energy for half a monolayer. Consequently a nearly completely ordered layer with sharp normal and extra spots is formed. The same sharpness and degree of order is found for one third of a monolayer. For all coverages in between the spot splitting varies continuously between 1/3 and 1/2 positions according to the mixture of arrangements of adatoms with two or three atomic distances. In contrast to the pessimistic interpretation of the authors that this figure demonstrates the insensitivity of LEED against disorder, this figure clearly reflects the disorder in the atom arrangement by the width of the extra spots; all coverages between 1/3 and 1/2 show an increased degree of disorder which may be evaluated quantitatively by analysis of the spot shape.

Both evaluations of LEED patterns show that it is possible to interpret observed patterns by appropriate defect structure in superstructure domain arrangements. The parameters used are different in both models. The concept of probability distribution $P(\Gamma)$ clearly describes the surfaces for coverage close to ordered structures. On the other hand the computer model describes all structures but it does not, however, give an easily available measure of order or disorder. A degree of disorder (as a number of defect sites) has not been given so far.

4.4.4 Facets

Sometimes during processing (like etching or thermal treatment) a different kind of surface defect develops; new planes with mostly low indices, which are formed on part of the surface, are named facets. They have definite, not too small angles against the flat portions. For example etch pits are formed by 2,3 or 4 equivalent faces or facets. If they are fairly flat and in diameter wider than the coherence width, their diffraction pattern is that of an independent inclined face; sharp spots are formed. Their 00-spot identified by a position independent of electron energy may be found at twice the angle of inclination. Inclinations of more than 30° are not seen on the screen of a commercial unit at normal incidence. The motion of all other spots of the facet with increasing electron energy indicates the position of the 00-beam. For an exact determination of the angle of inclination the motion of any spot of the facet pattern close to the 00-beam of flat portions may be observed [4.4,27]. The change of diffracting angle with wave length enables a direct measurement of the angle of inclination. Since frequently several equivalent facets are present simultaneously the reciprocal space is schematically represented as in Fig.4.17c. Each facet forms its own set of rods. These sets, however, are by no means independent. Since all facets are surfaces of the same single crystal, the points of the three-dimensional reciprocal lattice of the substrate are points of constructive interference for all faces.

Therefore each facet has one rod crossing each point of the reciprocal lattice. If the Ewald construction yields such a point, a sharp single spot is found irrespective of the presence of facets (or steps). For other conditions, however, each spot is split into several spots, one for each orientation of a facet. If the voltage is increased starting from a characteristic voltage according to (4.7) with integer S, the originally sharp spot separates into several sharp spots with increasing distance, each spot moving towards the respective 00-beam. If a spot of a facet is followed up from one spot of the flat surface to another one, the orientation of the rod and therefore the angle of inclination of the facet is easily determined from the two voltages of crossing and the positions in the three-dimensional reciprocal lattice. If for example the (20) spot of facet 2 in Fig. 4.17c (its indices have not to be known) is seen at the crossings with the (00) and ($\bar{1}$0) spot of the flat area, from the voltages the crossing points are indexed as (200) and (3$\bar{1}$0), respectively. In this way the orientation of 20-rod is fixed.

A more difficult problem arises when facets are very small, that is comparable with or smaller than coherence width. It is easy to predict that also in this case the characteristic voltages of (4.7) with integer S yield sharp spots. For other voltages the spot is no more split in several sharp spots, several more or less diffuse spots or one broad diffuse spot may arise similar to the cases treated for steps and domains (Sec.4.3 and Subsec.4.4.3). The size and distribution of facets determines the spot shape. A possible maximum intensity at a characteristic voltage should not be used for facet identification, rather the spot shape close to a characteristic voltage yields the information on facets. Finally the same surface may be described by steps or by facets. In some cases one or the other description may be more easily done and is therefore more appropriate. Since there is no clear-cut difference, the LEED pattern especially for very small facets may be also interpreted by a step arrangement. The description by facets should be used for fairly flat nearly perfect portions with low indices, whereas the discription with regular step arrays is appropirate for high index faces. Irregular step arrays finally may be used for description of surfaces which show more roughness on an atomic scale.

4.5 The Interpretation of a LEED Pattern

4.5.1 Parameters to be Observed

Originally LEED patterns were divided into "good" and "bad" ones. Good ones are those with bright sharp spots and low background and have been very much desired as being conclusive for a "good", that is strictly periodic, surface. Bad patterns are those with high background and weak or diffuse spots. They have been disregarded for many years as being an indicator of a disordered and therefore not desirable surface. In any case the existence of normal and extra spots is good for valuable information: existence of long range order. It has been learned, however, that useful information is found in "bad patterns" too and that the length of the ordered range may be less than originally thought. Each spot in the pattern corresponds to a periodicity of the autocorrelation function (see. Sec.4.1) and such a periodicity is maintained even with an appreciable amount of disorder.

It is therefore necessary to observe the other information of the LEED pattern too. As shown in Section 4.2 the background is indicative of point defects. The spot shape (splitting and broadening) gives information on periodic subsections on the surface (like terraces or domains) and their arrangement. The next important parameter is the comparison of spot shapes of all spots of the pattern. If only part of the spots are degraded in sharpness the defect structure still is perfectly periodic with respect to the periodicity of those spots (like normal spots of a superstructure of adatoms in bulk atom positions). Finally the voltage dependence of the spot shape and the background should be observed. Its importance is demonstrated in Fig.4.17. The reciprocal lattice is shown for three different defect structures: domains, steps, and facets. If the Ewald construction yields a spot of the three-dimensional reciprocal lattice (that means a characteristic voltage according to (4.7) with integer S has been chosen) then the normal spots have mostly the sharpness due to instrumental limitations. Another shape may be due to special domain arrangements. If the shape of all spots is not changed with varying voltage, the surface including its defects have to be within an atomic layer. Steps and facets are excluded. If the voltage is changed starting from a characteristic voltage and a splitting in two, three or more spots with increasing distance is observed, the same

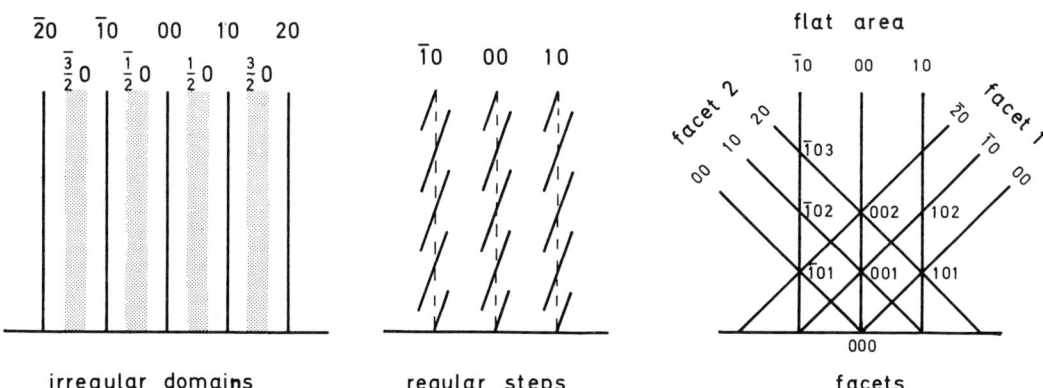

Fig. 4.17 Reciprocal space for three surface structures. Since from voltage dependence of the LEED pattern the reciprocal space may be derived, the LEED patterns enables a distinction between different structures (for details see text)

number of equivalent facets is present. If the special spot shape with two spots of varying intensity or other change in shape is observed as described in Section 4.3 atomic steps are present with the corresponding terrace arrangement. In reality difficulties may occur due to the varying intensity of the spots, especially if the spots are visible only in limited voltage ranges due to high background. If a sufficient number of spots is visible at several different voltages, a distinction of the above cases should be possible in all cases. A single photograph of a LEED pattern, however, is in no case conclusive. The voltage dependence should always be used.

4.5.2 Interrelation of Defects and Effects

The different defects are not always clearly separated; they may be present simultaneously and may therefore make a unique interpretation difficult. On the other hand those interrelations may yield a lot of additional information on surface properties.

Stepped surfaces may show superstructures. An example is the 2×1 superstructure of the cleaved silicon and germanium surface only observed on wide terraces [4.14,28]. Obviously the step edge influences the atom position with respect to superstructure. Adsorption of oxygen on stepped tungsten reveals a correlation of the superstructure with respect to step edge [4.29]; the splitting of the extra spots is also correctly predicted by (4.7). Step edges are not strictly straight, they may show kinks and point defects, which should be visible in the shape of the spots. Models of such arrangements have been studied by optical Fourier transform [4.8]. Point defects may be correlated so that a continuous change to the pattern of a stepped or superstructure surface may occur. A qualitative description with respect to appropriate defect parameters or a quantitative calculation has not been given in general.

If defects of surface structures are known a lot of other properties may be studied as with bulk defects and bulk properties. A few surface defect properties have already been studied. Electronic surface states [4.20], and work function [4.21] have been found to be step dependent; some catalytic and adsorption properties are still controversial (for a recent review see [4.17]).

The frequently published opinion that LEED is very insensitive to surface defects should be revised according to the possibilities and results presented in this chapter. Looking only at the existence of spots of course is not conclusive for a strictly or nearly periodic surface. Knowing the limitations, the evaluation of spot shape and background provides direct and quantitative information on various kinds of surface defects.

The more surface defect structures are observed and measured quantitatively, the more open is the field for the study of further surface properties and for an understanding for the real, in most cases by far not ideal, crystal surfaces.

References

4.1 R. Hosemann, S.N. Bagchi: *Direct Analysis of Diffraction by Matter* (North Holland, Amsterdam 1962)
A. Guinier: *X-Ray Diffraction* (Freeman, San Francisco 1963)

4.2 R.W. James: *The Optical Principles of the Diffraction of X-Rays* (Bell & Sons, London 1962)

4.3 E.G. Bauer: In *Topics in Applied Physics* Vol.4, ed. by R. Gomer (Springer, Berlin, Heidelberg, New York 1975) p. 225

4.4 G. Ertl, J. Küppers: *Low Energy Electrons and Surface Chemistry* (Verlag Chemie, Weinheim 1974)

4.5 R.L. Park: In Proc. 4th Intern. Mat. Symp. *Structure and Chemistry of Solid Surfaces* ed. by G. Somorjai (Wiley, New York 1969)

4.6 J.E. Houston, R.L. Park: Surface Sci. $\underline{21}$, 209 (1970); Surface Sci. $\underline{26}$, 269 (1971)

4.7 R.L. Park, J.E. Houston, D.G. Schreiner: Rev. Sci. Instr. $\underline{42}$, 60 (1971)

4.8 W.P. Ellis: In *Optical Transforms*, ed. by H. Lipson (Academic Press, London 1972)

4.9 C. Kittel: *Quantentheorie des Festkörpers* (Oldenbourg, München 1970)

4.10 R.L. Dennis, M.B. Webb: J. Vac. Sci. Tech. $\underline{10}$, 192 (1973)

4.11 M.B. Webb, M.G. Lagally: Solid State Phys. $\underline{28}$, 301 (1973)

4.12 M.G. Lagally: In *Surface Physics of Materials* ed. by J.M. Blakely (Academic Press, London 1975)

4.13 W.P. Ellis, R.L. Schwoebel: Surface Sci. $\underline{11}$, 82 (1968)

4.14 M. Henzler: Surface Sci. $\underline{19}$, 159 (1970)

4.15 J. Perdereau, G.E. Rhead: Surface Sci. $\underline{24}$, 555 (1970)

4.16 M. Henzler: Surface Sci. $\underline{22}$, 12 (1970)

4.17 M. Henzler: Appl. Phys. 9, 11 (1976)

4.18 K. Besocke, H. Wagner: Surface Sci. $\underline{53}$, 351 (1975); $\underline{52}$, 653 (1975)

4.19 M. Henzler: to be published

4.20 M. Henzler, J. Clabes: Japan. J. Appl. Phys. Suppl. 2 Pt 2, 389 (1974)

4.21 M. Henzler: Surface Sci. $\underline{36}$, 109 (1973)

4.22 B. Lang, R.W. Joyner, G. Somorjai: Surface Sci. $\underline{30}$, 440, 453 (1972)

4.23 H. Conrad, G. Ertl, E.E. Latta: Surface Sci. $\underline{41}$, 435 (1974)

4.24 G. Schulze, M. Henzler: Verh. Dt. Phys. Ges. and to be published

4.25 M. Henzler, F.W. Wulfert: Proc. XIII Intern. Conf. Phys. Semiconductors, Rome 1976

4.26 G. Ertl, J. Küppers: Surface Sci. $\underline{21}$, 61 (1970)

4.27 C.W. Tucker: J. Appl. Phys. $\underline{38}$, 1988 (1967)

4.28 J.E. Rowe, S.B. Christman, H. Ibach: Phys. Rev. Lett. $\underline{34}$, 874 (1975)

4.29 T. Engel, E. Bauer: to be published

5. Photoemission Spectroscopy

B. Feuerbacher and B. Fitton

With 20 Figures

Photoemission plays a special part among electron spectroscopies, in that it combines both optical and electron spectroscopic methods. Stimulation of the sample under observation occurs by absorption of photons, and the electrons emitted as a response are observed. A variety of combinations of variables and observed parameters is offered in such an experiment, leading to a variety of experimental methods sensitive to distinctly different physical properties of the sample. The present chapter aims to give an assessment of such methods in terms of their relative merits for application to the observation of surface related features. The emphasis throughout is on the experimental techniques. Consequently the theoretical aspects are treated somewhat sketchily. An attempt is made to compensate for this shortcoming by providing an extensive bibliography, including a structured list of relevant review articles [5.1,42].

The chapter is laid out as follows. A short overview of the field is presented in the first section. This includes a brief discussion of the basic principles of photoemission, a definition of the various subfields as delineated by characteristic ranges of certain parameters, and a description of the basic processes leading to particular features observable in photoelectron spectra. The following sections are kept fairly autonomous to allow the reader to select individual sections of interest without being bothered by excessive cross references. Section 5.2 deals with instrumentation, discussing in particular the light sources used for photoexcitation. Section 5.3 will present, in a very limited framework, some of the theoretical problems in photoemission. Those will be linked directly to practical aspects such as surface sensitivity, the interpretation of chemical shifts, and adsorbate studies. With this background available, the various methods of photoemission spectroscopy are discussed in Section 5.4. This will cover the classical energy-resolved type of measurement and also the angle-resolved experiments which are recently gaining attention. Yield spectroscopy will be discussed in view of the renewed interest in this technique with the availability of synchrotron radiation as a tunable light source. Finally the use of spin polarization measurements on photoemitted electrons will be briefly discussed under the aspect of applications to surface studies.

5.1 Principles of Photoemission

In photoelectron spectroscopy, photons of well-defined energy are absorbed in a sample by the process of electron excitation. If the photon energy $\hbar\omega$ is sufficiently high, the sample may be excited above the ionization threshold (or the work function in the case of solids), leading to emission of electrons. This process is illustrated, in a simplified single-particle picture, in Fig. 5.1a. The kinetic energy E_{kin} of the emitted electron can be measured, and may be traced back to the initial state energy of the electron before excitation. This step is not always without problems, but we may assume that the observable kinetic energy distribution of photoemitted electrons yields in some way information on the energy states of electrons inside the emitter. At the same time we may observe the emission *direction* of the photoelectrons (Fig.5.1b) in terms of the polar angle ϕ and the azimuth θ. The emission direction will be related to the momenta of exciting photon and excited electron. One could expect that the angular distribution of photoemitted electrons might provide information on the orbital quantum numbers of locally bound electrons or on the wave vector of delocalized electrons in solids. So, in contrast to purely optical experiments that measure only energy differences, photoemission observes the absolute energy of the initial and final electron state and in addition allows selection of particular regions in momentum space.

Surface sensitivity may be achieved in photoemission experiments by various means. The incident light will in general penetrate fairly deep into the solid, in the

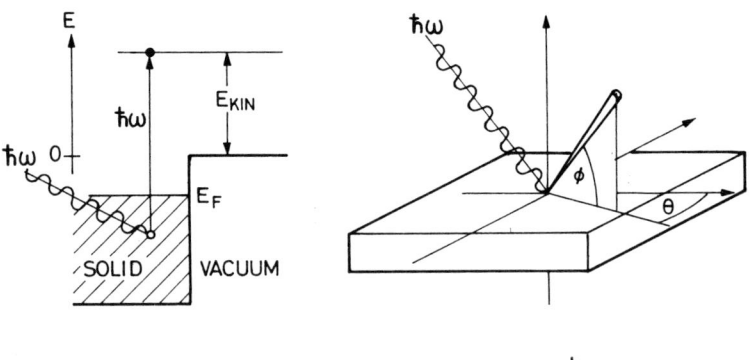

Fig. 5.1 (a) Photoexication of electrons in a crystal by photons of energy $\hbar\omega$ increases their energy above the vacuum level and can lead to emission through the surface. By observing the external kinetic energy E_{kin} it is possible to obtain information on the initial energy states in the solid. (b) Additional momentum information can be obtained from a measurement of polar (ϕ) and azimuthal (θ) emission angles

order of several tens of atomic layers. However, light interacts with the solid via the discontinuity arising from the presence of the surface, exciting optical transitions in the immediate vicinity of the surface by means of a "surface optical effect". On the other hand, the escape depth of excited electrons is usually very short, ranging between 5 and 100 Å as a function of material and electron energy, so the observation of photoemitted electrons intrinsically emphasizes surface contributions. In addition, the surface sensitivity of observed spectra may be enhanced considerably by a proper choice of external parameters such as photon energy, angle of incidence, or emission angle. It is one of the main objectives of the present chapter to provide the experimenter with a guide for the use of these parameters such as to obtain effectively information on the specific surface features in which he may be interested.

Photoemission is a particularly useful tool for the study of adsorbed species on surfaces, since it is much less destructive than electron impact spectroscopies. The cross sections of photodesorption and photodissociation are small compared to the respective cross sections of electron impact, even for high-energy (X-ray) photons. This allows the observation of an undisturbed system, as shown schematically in Fig.5.2. Here the left-hand side shows a clean solid surface and an atom in free space with a filled electronic level of binding energy E_{AT}. After adsorption (right-hand side of Fig.5.2), the atomic energy level might be shifted (or split) and broadened, forming a resonance level E_{ads}. A photoemission experiment may detect such resonance levels as distinct changes in the local density of states near the surface after adsorption. Comparison of the energy levels of the same atom before and after adsorption, that is, comparison of gas-phase photoemission spectra to adsorbate resonance levels, can provide information on the electronic character of the chemisorption bond acting in the adsorbate system. Photoelectrons emitted from adsorbate levels may also be observed as a function of emission angle. The results

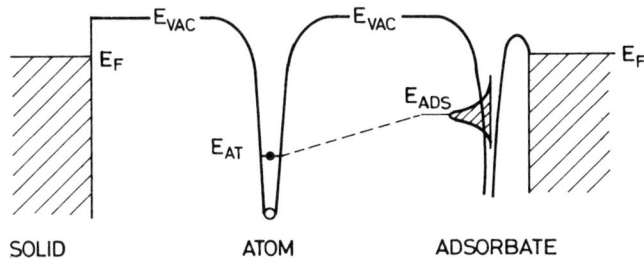

Fig. 5.2 Principle of adsorbate system as seen in a photoemission experiment. For an isolated atom the energy level E_{AT} is sharp. After adsorption it will broaden and may shift and split, forming an adsorbate resonance level E_{ADS} observable as a change in surface local density of states when compared to the clean surface

of such experiments are likely to provide insight into the nature of the
atomic or molecular orbitals involved, or into the geometric arrangement of the
adsorbed atom on a single-crystalline substrate surface.

The theoretical interpretation of photoelectron spectra is somewhat more
involved than that of purely optical spectra, where the dipole approximation of
the interaction of electromagnetic waves with electrons provides a fairly detailed
description of the experiment. In photoemission the important role of the spatial
region close to the emitting surface makes the specific nature of local fields at
the surface an essential element in any theoretical description. The whole problem
becomes even more difficult when reconstructed or adsorbate-covered surfaces are
considered. In additon, the spatial limitation to the surface region accentuates
many-body features in the spectra, leading to troublesome complications especially
for the case of localized adsorbate energy levels. In fact, surface sensitivity
in electron emission spectra is often achieved as a result of these strong many-
body interactions, so that the observer is trading the easy interpretation of
single-particle spectra against increased surface sensitivity.

5.1.1 Parameters and Ranges

Photoemission spectroscopy allows of a considerable number of experimental variables
and observable parameters. As far as the emitted electrons are concerned, one can
measure their number (the photoelectric yield) and their energy, as a function of
two angles, polar and azimuthal, and their spin. The experimenter has at hand the
incident light, the sample under observation, and possibly adsorbates. For the light,
the energy, angle of incidence, and the polarization may be varied. The sample may
be prepared in various thicknesses, single- or polycrystalline, it may expose
various crystal faces, each of which could reconstruct in several different ways.
Finally, various adsorbates may be brought to the surface in known coverages, and
the adsorbates may exhibit distinct phases depending, for example, on temperature.

A simple experimental approach is to avoid energy analysis and measure the *yield*
as a function of photon energy. Such experiments have recently gained interest due
to the availability of tunable high-energy light sources like the synchrotron, but
the identification of surface-derived features is difficult, since no energy analysis is made. Still, this type of spectroscopy has widely been used to verify the
concept of surface photoemission by means of the *vectorial photoeffect*, measuring
the yield as a function of light polarization angle. *Energy analysis* is, however,
the basic measurement in photoelectron spectroscopy. Energy distribution spectra
may be scanned for all emitted electrons or for a restricted sample in an angle-
resolved experiment. While in the past most experiments kept the photon energy constant during a scan, synchrotron sources permit the photon energy to be scanned
while observing through a narrow energy window, which may be either kept fixed
(constant final state spectroscopy) or moved together with the photon energy (constant initial state spectroscopy).

Angle-resolved spectroscopy is a new and promising way of performing photoemission measurements. It does require a high-intensity light source, since the observation is is done on a small fraction of the total number of photoemitted electrons. The measurement may be performed by scanning an energy distribution spectrum for fixed values of the emission angle, or by fixing the energy window at an interesting feature, say an adsorbate level, and varying the polar or azimuthal angle. *Spin polarization* experiments observe the ratio between electrons with spin up and spin down. This ratio, detected in a Mott analyzer, may be measured as a function of magnetization, temperature, or photon energy and polarization.

The *photon energy* is by far the most important parameter in photoemission experiments. Ranging over several decades as shown schematically in Fig.5.3, it encompasses

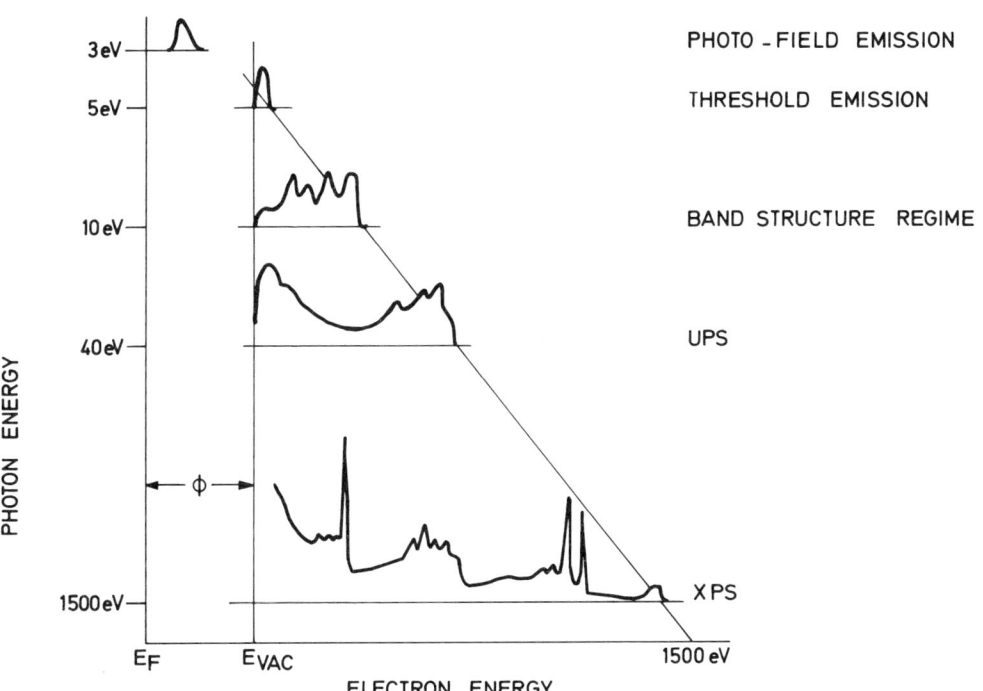

Fig. 5.3 Energy ranges and specialized spectroscopies in photoemission. XPS, excited by soft X-rays, shows spectra of considerable complexity including core level spikes, Auger peaks, valence-band emission and inelastic electrons. UPS has an intrinsically higher resolution and cross section for the valence band. The bandstructure regime, $\hbar\omega \simeq 10$ eV, shows sharp structure arising from bulk selection rules. Threshold emission is generally observed without energy analysis. Subthreshold spectroscopy requires additional means to emit photoexcited electrons over the work function barrier ϕ, such as, e.g., a high electric field

spectroscopies that use different experimental and theoretical techniques and are applied in different fields such as chemistry, surface physics or bulk band structure investigations. At low photon energies, below the emission threshold, additional means are required to emit photoexcited electrons. In *photo-assisted field emission* the surface barrier is made transparent for tunneling by a large electrostatic field. Double excitation is possible by populating normally empty states via thermal excitation, secondary-electron cascades, or by using two-photon effects. Such subthreshold spectroscopies explore the energy region of unoccupied states between the Fermi level and the vacuum level, which are not easily accessible by other means. *Threshold spectroscopy*, using photon energies around 5 eV, is usually performed without energy analysis. It probes the yield in a region where it is dominated by the emission threshold function, and has been used to observe work functions or surface states in semiconductors. At higher photon energies in the 5-20 eV range (*band structure regime*), energy-resolved spectra show rich structure due to k-conservation selection rules or final-state modulation. *Ultraviolet photoemission spectroscopy* (UPS) is commonly understood as excited by resonance light sources in the 16 to 41 eV range. A short electron mean free path in the sample makes those spectra surface sensitive. The energy probing depth is sufficient to cover the valence band of most solids. A large low-energy background of scattered electrons is found in these spectra. Resolution is generally limited by the analyzer, in contrast to X-ray excited spectra (XPS, *X-ray photoelectron spectroscopy*), where the resolution limit is given by the width of the characteristic X-ray line used to photoemit the electrons. Photon energies in the 0.1-5 keV range are sufficient to excite core electrons, giving rise to sharp characteristic lines in the spectra. The accurate observation of core levels makes this type of spectroscopy element specific, providing a powerful tool for chemical analysis. In addition, XPS spectra contain Auger lines and allow the observation of characteristic energy losses.

5.1.2 Basic Processes

To illustrate the basic physical processes giving rise to typical features in a photoemission spectrum we will use the example of an X-ray excited photoemission spectrum, which has a wide energetic range. The upper part of Fig.5.4 shows an XPS spectrum of a metal in the second row of the periodic table. The spectrum has the typical appearance of sloped steps with sharp lines at the high-energy edge of the steps. The total width of the spectrum is $\hbar\omega - \phi$, where $\hbar\omega$ is the exciting photon energy and ϕ the work function of the sample. The lower part of Fig.5.4 shows schematic diagrams of the processes that lead to the characteristic features in the spectra. Diagrams b and d illustrate the primary excitation processes. The photon energy $\hbar\omega$ is absorbed by a core elctron or a valence electron, leading to a photoionization process. The (negative) binding energy of the emitted electron is found

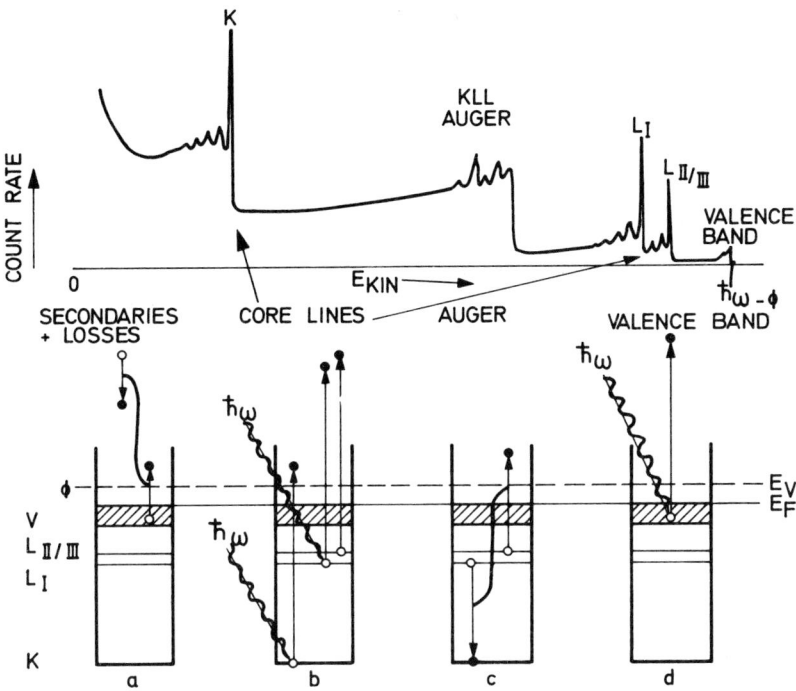

Fig. 5.4a-d The basic processes contributing to an XPS spectrum (top) include direct optical exitation of core electrons (b) or valence electrons (d), hole deexcitation via the Auger process (c), and inelastic losses (a) giving rise to secondary electrons and plasmon satellites

by subtracting the photon energy from the measured kinetic energy of the core line,

$$-E_B^V = E_{Kin} - \hbar\omega \tag{5.1}$$

if energy is referred to the vacuum level. For metals, the binding energy is commonly referred to the Fermi level. In this case

$$-E_B^F = E_{Kin} - \hbar\omega + \phi. \tag{5.2}$$

The binding energy represents, apart from a small correction termed *relaxation shift* (see Subsec.5.3.3), the one-electron energy level of the core shell, so XPS allows a direct observation of core terms. This fact immediately provides *chemical specificity* for the spectra. In a different chemical environment the core levels exhibit a small shift due to the different contributions of the charge density of the valence electrons at the core orbit. This *chemical shift* permits conclusions on the binding of the atom involved, making its observation very attractive in chemistry. The electrons with the highest kinetic energy in the spectra reflect the density of states in the valence band, with possible intensity modulations due to cross-section variations for electrons of different orbital character.

After excitation and emission of a core electron a hole is left in the core shell. This hole can be filled either by radiative deexcitation, giving rise to a characteristic X-ray spectrum, or by an *Auger process*. The electrons emitted by the Auger process, which is the dominant deexcitation mechanism for elements lighter than $Z \approx 35$, are observable in the XPS spectrum. Diagram c in Fig.5.4 schematically shows a KLL Auger process. Here a hole in the K shell is filled by an electron from the L shell. Another L electron carries away the energy arising from this process. Auger lines are independent of the way the core hole has been created, so they are found at the same kinetic energy for photon- and electron-excited spectra. Being element specific, those lines may also be used for chemical identification. Each peak in the spectrum is accompanied, on the low kinetic energy side, by a characteristic loss spectrum. Core electrons may excite collective electron oscillations, plasmons, during their excitation or emission, so bulk and surface plasmon loss lines are found on the low energy side of each sharp structure. This and many other energy loss mechanisms lead to the low-energy tails forming the characteristic overall step shape of the spectra. Figure 5.4a shows the process of secondary electron emission, where a hot electron loses energy by creating an electron-hole pair in the valence band. These secondary electrons give rise to the huge low-energy hump in photoemission spectra.

As shown here, XPS spectra carry a wealth of information. Core lines identify the chemical elements and the bonding states. The state density in the valence band is obtained. Plasmon structure and Auger lines appear in the spectra. One might therefore ask, what use are photoelectron spectra at lower photon energies, when we get all this information from XPS? In fact it turns out that UPS and XPS both have their virtues and disadvantages and we will try to collect some of the most apparent ones in order to see those spectroscopies in perspective, with special emphasis on their use as a tool for surface investigations.

XPS Characteristics

- Large energetic probing depth makes observations of core levels possible. This results in *chemical specificity*.
- XPS is not intrinsically very *surface sensitive*. The spatial information depth is limited by the electron mean free path to some 50 Å. But the combination of chemical specificity and high sensitivity makes it an attractive tool for the study of *adsorbates*.
- *Chemical shifts* on *adsorbates* permit conclusions on chemisorption bonding. However, the interpretation is complicated due to atomic and extra-atomic *relaxation*.

- *Valence states* may be directly observed. Usually the *resolution* is not sufficient to show fine details, unless an X-ray monochromator is used to reduce the line width of the light source.
- *Energy-loss* structure and *Auger lines* are observable in XPS spectra.

UPS Characteristics

- *Intensities* of available light sources are high, permitting relatively easy angle-resolved studies.
- *Resolution* is high and usually limited by analyzer performance.
- *Surface sensitivity* is achieved either by a very short electron mean free path (5-10 Å) for 20-100 eV photon energies or, at lower photon energies, by the surface photoelectric effect.
- *Valence levels*, involved in bonding or not, are directly observed with high resolution.
- *Adsorbate resonances* may be identified.
- Chemical specificity is given in a very limited way by the *"fingerprinting"* technique of observing characteristic features assigned to specific valence orbitals.
- Surface and volume contributions are intermixed and very difficult to distinguish.
- The assignment of adsorbate features by difference spectra is not unique.

5.2 Instrumentation

The hardware required for a photoemission experiment consists of three major parts. A light source produces a monochromatic beam of photons. The sample is kept in a vacuum chamber, which for surface related experiments needs to be of ultrahigh vacuum capability. An electron analyzer, including detection electronics, provides the angular and energetic selection of the emitted electrons. A detailed discussion of the latter part was given in Chapter 2 of this volume. In the present context we will therefore concentrate on the problem of light sources, which are the most important ingredient and from many aspects the limiting factor in a photoemission experiment.

5.2.1 Light Sources

The prime criteria in the selection of light sources for a photoemission experiment are the energy of the photons, their number at the sample position, the energetic band width of the beam, and possibly the compatibility with an ultrahigh vacuum system. If the source offers a continuous spectrum, a monochromator is required in order to provide a beam of defined photon energy.

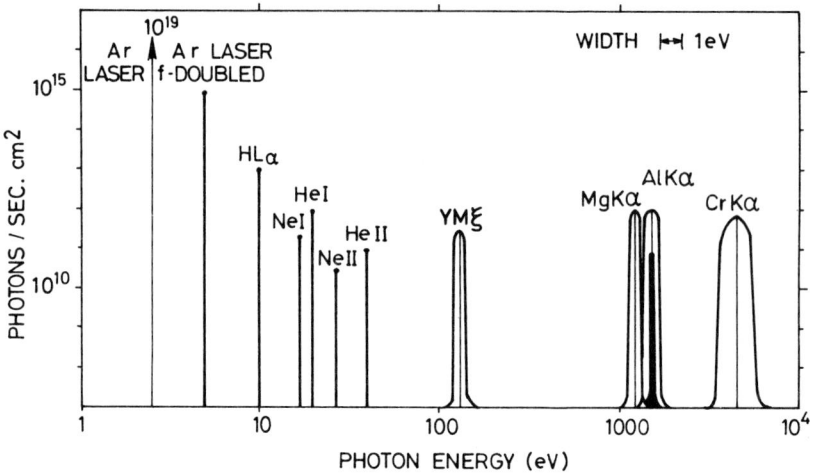

Fig. 5.5 Photon energies and intensities for line sources commonly used in photoemission experiments. Intensities are given for a 10 x 10 mm illuminated area at the sample position. Line widths are indicated on an expanded scale as shown in the top right corner. The AlK$_\alpha$ line is also shown after monochromatization (full peak)

A summary of presently available *line sources* is presented in Fig.5.5. Intensities indicated are typical and expressed in a photon flux "at the sample", i.e., taking into account that the sample usually subtends a small solid angle as seen from the light source. Intense line emission is presented by lasers in the energy range up to about 3.6 eV. Lasers offer a highly collimated, polarized, small band width beam of very high intensity, but the photon energy is small. Even frequency doubling extends energies to no more than about 6 eV. Therefore lasers find their application in photoemission spectroscopies investigating the energy region just above or below threshold. In the 10-40 eV photon energy region gas discharge light sources are most widely used [5.43,44]. The most effective design appears to be a capillary discharge excited by dc or microwave power, using gas pressures in the 1 Torr range. For energies above 11.8 eV window materials are not available (except for thin films), so the capillary light source has to be differentially pumped [5.45-47] to give a reduced gas flow into the vacuum system and a collimated light beam. High intensities can be obtained at 10.2 eV (H Lyman α), 16.8 eV (Ne I), 21.1 eV (He II), and 40.8 eV (He II). The noble gases used in the windowless light sources have very low sticking coefficients on most room temperature surfaces, so pressures in the upper 10^{-9} Torr range may be tolerated even in surface-sensitive experiments. Problems may arise from the beam of neutral gas atoms impinging onto the sample surface together with the photons. Such a bombardment may disturb sensitive adsorbate systems, so it might be necessary to operate light sources with thin-film windows or to find light sources that run at extremely low pressures [5.48]. Gas-discharge light sources emit lines of very small energetic width in the order of 20 meV or less. This is commonly less than the analyzer resolution, which will be in the order of 0.1 eV. In contrast, the natural line width of characteristic X-ray lines is much larger, ranging from 0.5 eV to several eV for heavy elements. The most commonly used X-ray sources [5.20] are the Kα lines of aluminum (1486.6 eV) or magnesium (1253.6 eV) with a line width of 0.85 and 0.7 eV, respectively. The photon energy of these lines is sufficient to allow observation of core levels of most elements, so it is hardly necessary to employ harder radiation with its inherently larger line width. Lower energies are desirable for the study of the energy dependence of photoionization [5.49]. The gap in energies between the gas discharge lines and the Kα X-ray lines is partly bridged by a group of lines between 100 and 200 eV, of which the Be K (108.9 eV), Y Mζ (132.3 eV), Zr Mζ (131.4 eV) and Nb Mζ (171.4 eV) are the most intense ones with typical half widths of 0.5-1.5 eV. A distinct improvement in resolution is obtained by monochromatization of the broad X-ray lines by Bragg reflection in a bent quartz crystal. The inherent loss in intensity reduces the signal-to-noise ratio, but the signal-to-background ratio is usually increased since the background is drastically reduced in a monochromatized beam lacking satellite lines and bremsstrahlung continuum. The intensity loss may be overcome partly by dispersion compensation [5.50], where the dispersion of the monochromator is matched by an electron lens to the dispersion of the analzyer, and a multidetector array increases detection efficiency. In the method of fine focusing [5.51] a very small emitting spot on the anode is focused onto the sample. Line widths of less than 0.2 eV may be achieved in this way, with good intensity if a rotating anode is used.

A comparison of available *continuum light sources* is presented in Fig.5.6. Photon continuum sources always require a monochromator, so a comparison in terms of "photons at the sample" is difficult since the efficiency depends strongly on the monochromator design and wave length range. The data in Fig.5.6 therefore should be regarded as a rough guideline, giving upper limits for optimum monochromator performance at each particular wave length. The most versatile continuous light source is an electron synchrotron or storage ring [5.52-55]. Synchrotron radiation offers a continuum of high intensity ranging from the infrared to hard X-rays. The light is collimated in a plane, so the source is of high brightness, and it operates under high vacuum conditions. The light is polarized, and the time structure permits the application of time-of-flight energy analysis [5.56]. All these properties make the synchrotron a very attractive source, but there are some inherent disadvantages. First of all, it is inconvenient compared to a laboratory source. If the synchrotron is not dedicated as a radiation source, the use will be parasitic with all its problems. Safety requirements may complicate experiments considerably and impose large distances from the source. Still, there is no question that

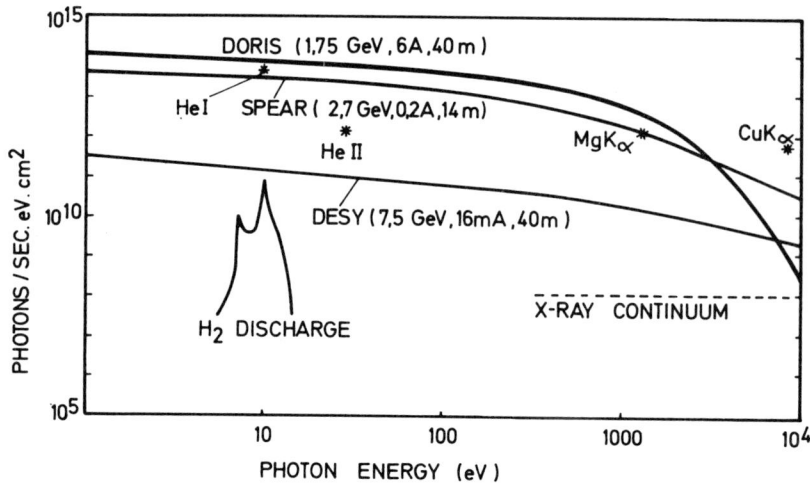

Fig. 5.6 Intensities and photon energies for continuum sources, used with an "optimum" monochromator at each particular wave length, and calculated for a sample size of 10 x 10 mm. Curves are presented for a synchrotron (DESY) and two storage-ring sources (DORIS, SPEAR) at typical operating conditions of electron energy, current, and working distance. For comparison, total intensities of a few line sources are shown by asterisks, taking their intrinsic, unmonochromatized energetic width

synchrotron light is unique and irreplaceable in many experiments. Especially storage rings with their high-current beams and noise-free stable output are very desirable light sources for photoelectric experiments.

A comparison of synchrotron light to conventional light sources, as attempted in Fig.5.6, is difficult due to the different spectral distribution of the intensity. It appears that, whenever a fixed energy can be tolerated, line sources are preferable, since they provide intensities at least comparable to the strongest synchrotron sources. There is, however, always the possibility to sacrifice resolution for intensity in a continuum source. On the other hand, monochromatization of hard radiation from a synchrotron is not an easy problem, and sophisticated monochromator designs have to be used to ensure high efficiency and suppression of higher orders [5.57-59].

5.3 Theoretical and Practical Aspects

No attempt is made here to give a comprehensive review of theoretical work in the field of photoemission. Rather, we will try to describe theoretical aspects in a qualitative way, with ample references to original literature, keeping in mind at each stage the practical applications. Therefore, the electon excitation process will be treated in a conceptually simple and pictorial manner, emphasizing those features that lead to the particular advantages or disadvantages of the various spectral regimes. Reference to the less complex case of gas-phase photoemission will be made, so that the essential problems of solid state photoemission become clearer. On the basis of a simplified understanding of the excitation process, the various ways of achieving surface sensitivity in a photoemission experiment will be discussed. Subsection 5.3.3 is devoted to the discussion of level shifts as found in photoemission, including chemical shifts and relaxation effects. The final part of this section describes the theoretical aspects of adsorbate spectroscopy, including angular photoemission studies on adsorbate systems.

5.3.1 Electron Excitation and Emission

The theory of the interaction of light with electrons in atoms or molecules is well established [5.60], and the same is true for the theory of the optical properties of bulk solids [5.61]. Only recently has the problem of photoelectron emission from solids been tackled in a rigorous way [5.62-68]. The main difficulties in a quantum mechanical description of the photemission process are those of wave matching at the solid-vacuum interface, the inelastic processes leading to a finite electron escape depth, and the correct treatment of the local field effects near the surface [5.69-72]. The present discussion will remain qualitative and the reader is referred to the original papers [5.62-72], or to recent review articles emphasizing theoretical aspects [5.28,32,35,39,40] for a quantitative description of the phenomena.

The quantum mechanical interaction of an electromagnetic field of vector potential $\underline{A}$ with an electron is given by

$$\tau = \frac{1}{2} (\underline{A} \cdot \underline{p} + \underline{p} \cdot \underline{A}) \tag{5.3}$$

where $\underline{p}$ is the momentum operator, and terms of higher order in $\underline{A}$ are neglected (dipole approximation). If this interaction appears as a small perturbation in a quantum mechanical system, the electron transition rate from an initial state $|i\rangle$ to a final state $|f\rangle$ is governed by the matrix element $M_{fi} = \langle f|\tau|i\rangle$. The photoelectron current j, observed through an analyzer set at the energy E with band width dE and accepting a solid angle $d\Omega$, can be described by [5.67].

$$\frac{d^2 j}{dE d\Omega} = 2 \, ev (\frac{m}{2mc})^2 \, (\frac{m}{2\pi \hbar^2})^2 \sum_i |M_{fi}|^2 \, \delta(E - \hbar\omega - E_i) \tag{5.4}$$

where v is the observed electron velocity, and the δ-function ensures energy conservation. The general formula (5.4) for the differential photoemission current is common to all single electron theories. The current is expressed as a sum over all occupied initial states, so an immediate conclusion from this formula is that *the observed photoelectron current is determined essentially by the initial density of states, each state |i> contributing as energy conservation permits, with an intensity given by the square of the matrix element* M_{fi}. In other words, the observed spectrum will reflect the initial state density, provided a sufficient number of final states is available and the matrix element may be regarded constant.

The main difference between gas-phase and solid-state photoemission arises from the form of the wave functions of initial and final states, entering (5.4) via the matrix element. This is illustrated in Fig.5.7, where the left-hand side describes the situation for an isolated atom. The initial state is an atomic orbital, localized at the atom. The final state E_1 lies above the vacuum level E_{vac} in the ionization continuum, and it is a propagating state close to a free-electron wave function. The increased complexity of a photoemission system including a solid is apparent from Fig.5.7b. We consider a semi-infinite emitter, left, and a vacuum half space, to the right. In the vacuum, all states above the vacuum level are allowed free-electron states, while inside the solid there are band gaps, where no allowed electron states exist, and bands, as given by the band structure E-$\underline{k}$ diagram. Wave functions of the total system, vacuum plus solid, are those that fulfill the existence conditions on either side, and are connected by quantum mechanical matching conditions at the interface. The initial state E_0, occupied and therefore below E_F, may be described by a delocalized Bloch state inside the solid, while outside it decays exponentially. Three distinctly different cases may occur for the final states, depending on the selected analyzer bandpass energy. First, consider an energy range where the damping of excited electrons within the solid is small, i.e., the mean free path is large. If the energy is chosen such that it coincides with an allowed band, E_1, the free-electron state in vacuum may be matched to a Bloch state within the solid. The overlap of initial and final states in the matrix element now ranges over large distances in the solid, so the observed electrons carry information on bulk properties, provided the operator τ in the matrix element (5.4) couples efficiently. On the other hand, if the final state energy is chosen to coincide with a band gap, E_2, the wave function will be evanescent in the solid, and the matrix element vanishes everywhere except for a narrow spatial region close to the interface. Photoemission under these conditions, termed *band-gap emission* [5.67], therefore carries information on the immediate vicinity of the interface [5.73-75]. The third case occurs if the electron mean free path is short in the solid. The free electron wave now couples to a damped, decaying wave, E_3, and again the matrix element vanishes everywhere except close to the surface due to limited spatial overlap of initial and final state wave functions. This discussion points to the fact that *solid state photoemission is intrinsically a surface sensitive process*, due to the condition that the initial state wave functions must overlap with final states that connect at the surface to free electron waves in vacuum. The overlap may be spatially limited to the close vicinity of the interface, depending on the surface matching conditions, which are identical to those for LEED wave functions [5.76].

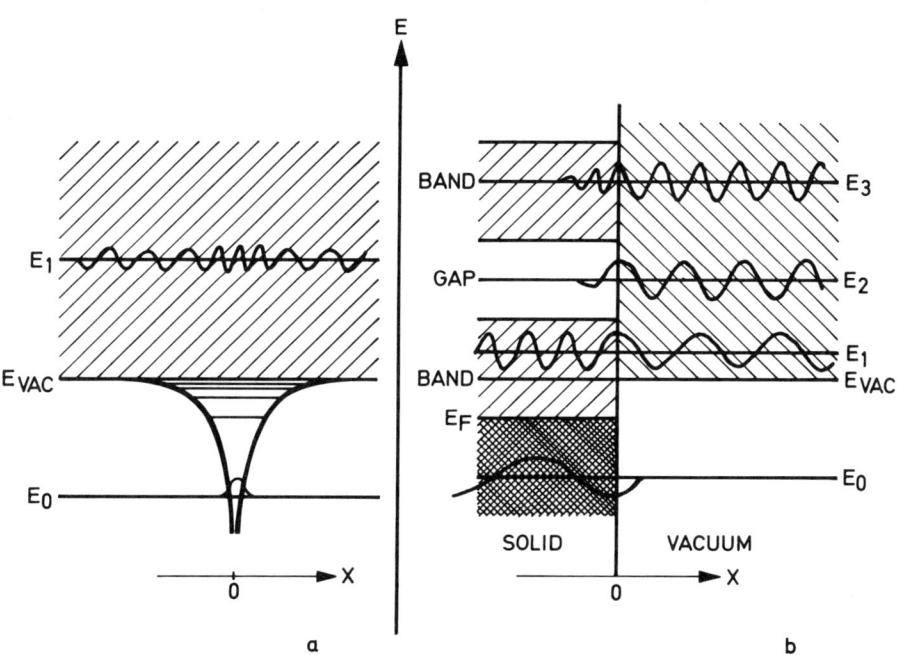

Fig. 5.7 a-b Initial and final state wave functions in photoemission from a free atom (a) or a solid (b). For a gas atom, the final state wave function is in the vacuum continuum and spatially extended. For a solid, extended free-electron wave functions in vacuum may be matched to extended Bloch functions within a band (E_1), to surface localized evanescent wave functions in a band gap (E_2), or to strongly damped wave functions in a band with short electron lifetime (E_3). Depending on photon energy and band structure, photoemission spectra may therefore sample extended regions in the solid or only the immediate vicinity of the surface as determined by the overlap of initial and final state wave functions

The intensity of the emission current, and so the spectral shape of the energy distribution curve, is determined by the coupling strength due to the operator in the matrix element. Possible selection rules that may arise in the matrix element have been discussed in a series of papers by GADZUK [5.40,77], who related the time scales in the photoemission process to the coherence between electron emission from each atom in the ensemble considered. This method allows a unified view of emission phenomena, covering atomic, molecular, solid state, and adsorbate photoemission. In the following we will try to adopt this picture in a simplified way to discuss selection rules and spectral shapes in solid state photoemission.

Consider a system consisting of two atoms. A photoionization process creates a hole at one of these atoms, and this hole may hop from one atom to the other. If the hopping is fast compared to the measurement time, the observer cannot determine where the hole is. Consequently, the two atoms appear as coherent emitters, the *amplitudes* of emission from the two centers add, giving rise to interference effects in the measured photocurrent. On the other hand, if the hole remains localized on one atom, the two atoms appear to emit independently, the *intensities* of the individual currents add, and no interference effects are found in the spectra. This criterion of measurement and hopping time scales may be converted, via the uncertainty principle, into more familiar energy scales. The hopping time t_H is related to the hopping matrix element $V_H = <1|H|2>$ by the relation $t_H = \hbar / V_H$ [5.77]. On the other hand, the observation time is related to the energy resolution ΔE of the experiment by the same relation, so the criterion of rapid hopping becomes

$$t_H = \hbar/V_H \ll h/\Delta E \qquad (5.5)$$

or $\Delta E \ll V_H$, which means that interference effects are observed only if the energy discrimination is sufficient to resolve the level splitting arising from the hopping matrix element.

The principle of photoemission time scales, explained here using the example of a two-atom molecule, may be applied also for solid state photoemission. It has been shown by GADZUK [5.77] that the criterion for incoherent emission from solids is that the initial state band is much narrower than the final state band. In wide valence bands the hole delocalizes rapidly. If observed with good energy resolution, the hole may spread over a large number of atoms during the measurement time. All these atoms will appear to emit coherently, so sharp interference effects are observed, which are identical to the *k-conservation* selection rule familiar from solid state optics. This situation is characteristic for low excitation energies in the *band structure regime*. A photoemission spectrum in this energy range will therefore reflect the energy density of the *joint density of states*, that is, at any final state energy fixed by the analyzer passband the matrix element will couple only to electron states vertically downwards in a band structure diagram. In angle-resolved measurements, the parallel components of the k-vector are fixed so the system is overdetermined. Four selection rules (3 for the k-vector components, one for the energy) act in a system with four parameters (two for the angle of observation, one for the photon energy, and one for the electron energy), so in general no electrons will be observed. During an energy (or angle) scan, a condition with all selection rules fulfilled may occur, so the spectrum shows a narrow δ-peak. Angle-integrated

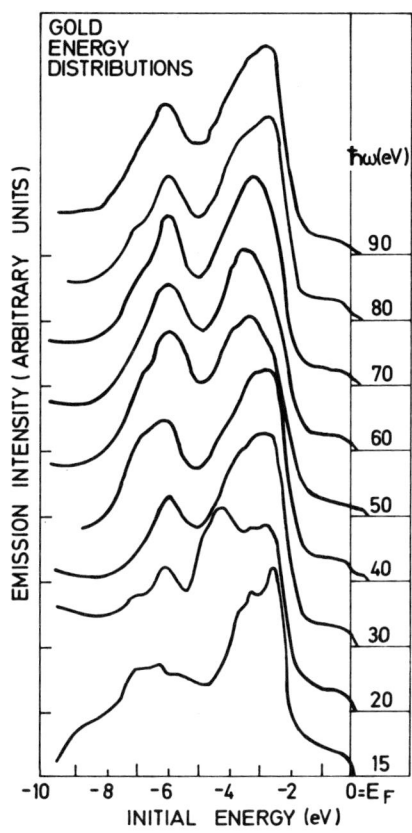

Fig. 5.8 Photoemission spectra gradually change form the rich structured band-structure regime at low excitation energies, where the spectral shape is determined by bulk transition selection rules, to the XPS regime at high energies, where the spectrum reflects the density of electron states in the emitting crystal. The measurements for gold shown here were made using synchroton radiation by FREEOUF et al. [5.80]

spectra, however, will be continuous and show relatively sharp structure, characteristic for the joint density of states of the emitting crystal, and therefore strongly depending on final state energy or exciting photon energy.

As the final state energy increases, many body effects such as electron-electron scattering will decrease the lifetime of the excited electrons, resulting in an effectively decreased energy resolution. Therefore, the number of electrons seen to emit coherently will decrease, and so will the importance of the wave vector selection rule. In the limiting case of vanishing selection rules the matrix element in (5.4) will couple all initial states to all final states with equal probability. Due to the short lifetime structure in the final state bands will not be sharp, so the spectra reflect the *initial density of states* in the high damping regime characteristic for UPS. It should, nevertheless, be kept in mind that the independent particle picture invoked here does not necessarily apply, and the many-body effects required to decrease the final state lifetime may well alter the conclusions derived from single particle theories. At high electron energies, in the *XPS* regime, core levels become observable. *Core levels* are localized so no interference effects are expected and the emission characteristics will be *atom like*. For emission from the valence band the criterion of initial and final state band width may be applied. With essentially free electron-like states at high photoelectron energies one may conclude that interference effects are unimportant, so the *XPS valence emission* will reflect the *density of states* with little or no modulation.

To summarize, we may say that the appearance of photoemission spectra depends a great deal on the photon energy used for excitation. At low photon energies sharp k-vector selection rules lead to well-structured spectra. Angle-resolved measurements are expected to show discontinuous spectra consisting of sharp peaks, and recent experimental results confirm this view [5.78,79]. Wide angle spectra are related to the energy density of the joint density of states, so structure changes rapidly with photon energy. At higher photon energies, the spectra are expected to change gradually into a shape representing the initial density of states in the XPS limit. An experimental example for this statement is presented in Fig.5.8 (from [5.80]).

5.3.2 Surface Sensitivity

Many experimental parameters influence the behavior of spectra with respect to surface contributions. The most generally accepted way to achieve surface sensitivity is by minimizing the escape depth of excited electrons, i.e., the electron mean free path. There are, however, other factors such as the importance of optical surface effects, band gap emission as discussed above, or the variation of photoionization cross sections with photon energy. If the surface can be modified without changing the bulk contribution, the identification of surface emitted electrons in the spectra is facilitated. The difference spectra widely applied in adsorbate spectroscopy take advantage of this fact, but the same methods may be applied to clean surfaces if, for example, several different reconstructed surfaces can be produced. It is obvious that chemical specificity of spectral features allows identification and observation of atoms at the surface once they are different form the bulk atoms. Core-level spectroscopy makes ample use of this possibility.

The Electron Mean Free Path

The mean free path for photons in solids is larger than the scattering length of excited electrons for all energies under consideration here, so the electron mean free path will be the parameter that limits the escape depth of photoemitted electrons. This, however, is not the only effect of a short electron mean free path. Inelastic collisions of hot electrons will result in a loss of energy and a change in momentum, and will tend to invalidate the useful and relatively simple single-particle description of the photoemission process.

A thorough treatment of inelastic effects in photoemission was presented by CAROLI et al. [5.66]. The inelastic processes that define the mean free path of hot electrons are electron-electron, electron-phonon, and electron-impurity scattering.

Electron-electron scattering may lead to pair excitation, a single particle event, or plasmon emission, a collective many-body excitation. Both of these processes are accompanied by a relatively large energy transfer, namely a few eV for pair excitation and $\hbar\omega_p$, the plasmon energy of the order of 10 eV, for plasmon excitation. This implies that it is possible, for parts of the photoelectron spectrum, to experimentally identify unscattered electrons. Those electrons with the highest kinetic energies are likely to have left the solid without electron-electron scattering, while inelastic electrons and secondaries will be removed from the elastic energies and piled up at the low-energy end of the spectrum. The elastic electrons therefore represent a sample that derives from a limited region close to the surface. The "master" curve for the electron mean free path presented in Fig.1.1 shows that photoelectrons in the 20-100 eV range have free path lengths comparable to the lattice spacing, so up to 30% of the total mumber of electrons in the spectra derive from the top layer of atoms. The short means free path leads to a relaxation of sharp k-vector selection rules in the matrix element, as discussed in Subsection 5.3.1, so one expects an energy distribution spectrum closely resembling the initial density of states corresponding to the topmost atomic layers, provided the one-electron picture still holds. For photoelectron kinetic energies above 100 eV the electron mean free path increases, leading to a decreasing surface sensitivity. At energies in the XPS regime the core levels become accessible, and a high sensitivity for adsorbates is achieved due to the chemical specificity of any core-level spectroscopy.

The effect of quasielastic processes such as impurity scattering [5.81] or phonon scattering [5.66] is quite different. The energy transfer in a single phonon scattering event is small, of the order of the Debye frequency (10-100 meV), but the momentum transfer is large, of the order of a reciprocal lattice vector. Electrons inelastically scattered from phonons are therefore much more difficult to identify experimentally. The hot-electron mean free path (Fig.1.1) is limited by quasielastic events for low kinetic energies of emitted electrons. In particular for insulating materials this regime extends from threshold to twice the band gap energy. Here the hot electron does not lose energy within the experimental resolution, but merely changes momentum. The escape depth is large, much larger than the electron-phonon scattering length, and the *energy spectrum* of an angle-integrated measurement will therefore be determined by the *excitation* process, reflecting the joint density of states. The k-space information on the initial state will be lost after phonon scattering, so the *angular distribution* will be determined by the *final state* alone. This has been demonstrated by HIMPSEL and STEINMANN [5.82] in a measurement of the angular distribution of photoelectrons form KCl, where the angular pattern was found to depend only on the final state energy for photoelectrons excited below twice the band gap energy.

Optical Surface Effects

The matrix element governing the photoemission process has been given as

$$M_{fi} = \langle f | \underline{A} \cdot \underline{p} + \underline{p} \cdot \underline{A} | i \rangle \tag{5.6}$$

in Subsection 5.3.1. Surface sensitivity in this matrix element may be achieved in two ways. Either the overlap of initial and final state wave functions $|i\rangle$ and $|f\rangle$ is such that it differs from zero only in the immediate neighborhood of the surface (as in the "band gap case", Subsec. 5.3.1), or the operator $\tau = (\underline{A} \cdot \underline{p} + \underline{p} \cdot \underline{A})$ is such that it couples wave functions of whatever form only in spatial regions close to the surface. The latter process is usually referred to as an optical surface effect. On the other hand, the historic development of photoemission theory has led to the definition of a "classical" *surface photoelectric effect*, derived for free-electron solids [5.83,84], where optical excitation conserving both energy and momentum is possible exclusively at the surface (in a "real" solid G-vectors are supplied by bulk periodicity to permit momentum conservation). Under the assumption of spatial independence, the vector potential may be written in the form $A_0 \underline{e}$, $\underline{e}$ being a unit polarization vector. The second term in (5.6), $\underline{p} \cdot \underline{A}$, then vanishes, and the $\underline{A} \cdot \underline{p}$ term may be transformed into a gradient to the effective potential V_{eff}, using the identity [5.60]

$$\langle f | \underline{e} \cdot \underline{p} | i \rangle = \frac{1}{\hbar \omega} \langle f | \underline{e} \cdot \nabla V_{eff} | i \rangle \quad . \tag{5.7}$$

In a free-electron gas, the only potential discontinuity is provided by the surface, so the matrix element becomes

$$M_s = \frac{A_0}{\hbar \omega} \langle f | \underline{e} \cdot \frac{\delta V_s}{\delta z} | \rangle \quad . \tag{5.8}$$

Since the surface potential gradient is a vector along the surface normal, the surface photoeffect acts only for p-polarized light, where the polarization vector $\underline{e}$ has a nonvanishing component along the surface normal.

It has been realized that the approximations leading to (5.8) are not generally valid. The normal component of the electromagnetic field vector changes rapidly when passing through the surface, so the assumption of a spatially independent vector potential is not justified near the interface. Calculations including the $\underline{p} \cdot \underline{A}$ term have been presented by MAKINSON [5.85], and by ENDRIZ [5.86] with an improved surface potential. Recent work has taken into account the longitudinal nature of the electromagnetic field near the surface (KLIEWER [5.69,70]), and treated field and potential in a self-consistent way (FEIBELMAN [5.71,72]) for "jellium" metal. The results show that the surface photoelectric effect is strongest near

threshold and decreases toward the plasma frequency, where it vanishes. For higher
excitation energies it remains small compared to the threshold values. No calculations are available, to date, for realistic materials including a band structure,
so it is not possible to derive data for the relative values of surface and bulk
effects from first principles. Attempts to do this [5.64] had to rely on phenomeno-
logical photoelectron escape depths. Nor have any theoretical results been published
on optical surface effects at high excitation energies, applicable for XPS ranges.
It is to be expected that those effects will be small, since the optical constants
approach the vacuum values at high energies, so the electromagnetic vector potential
does not change drastically across the surface. On the other hand, the light wave
length becomes comparable to the atomic distances, so local field effects due to
spatial inhomogeneity cannot be neglected.

5.3.3 Relaxation and Chemical Shift

So far we have treated photoelectron emission exclusively in a single electron
picture. The effect of the other electrons in a solid or a molecule was taken into
account only via the one-electron potential, which was a self-consistent potential
formed by the electron under consideration and all other electrons. In addition,
electron-electron inelastic effects have been included phenomenologically by lifetime
and mean-free-path considerations. In such a picture the total energy $\hbar\omega$ of an
absorbed photon is taken up by a single electron, and the energy of this electron
before excitation may be found simply by subtracting the photon energy from the measured elastic photoelectron energy.

A more appropriate point of view is to consider the electron states of an N electron
system and compare them to those of the (N-1) electron system left behind after the
photoemission event. If N is very large, these states will not differ very much,
and we are permitted to use the single particle picture, since apparently the self-
consistent field does not change much as a single electron is removed. This is the
case for solids with wide, overlapping bands. The other extreme is the case of localized energy levels as found in atoms. Here the total number N of electrons is small,
so the self-consistent field is sensitive to the removal of a single electron. The
orbitals of the remaining electrons adjust to the new field, which leads to a relaxation toward lower energies. The *relaxation energy* has to be carried away by the photo-
ejected electron, which appears at a higher kinetic energy. Here is a point that
frequently gives rise to confusion. In solid-state photoemission the terms initial
and final state usually mean *one-electron* states, while in ESCA theory these terms
are generally used to describe the states of the emitting *molecule*. Therefore, if
the final molecular state relaxes to *lower* energy, the emitted photoelectron appears
at a *higher* "final state" energy.

An additional effect occurs with the emitting atom embedded in a polarizable
medium be it a large molecule or a solid. In this case the local hole created by the

photoemission process, acting as a local positive charge, attracts electrons that tend to screen it. The screening results in a lowering of the total energy of the system, and the energy difference again is given to the emitted photoelectron. This latter effect, usually referred to as *extraatomic relaxation* or polarization, will be observed for atoms in large molecules, adsorbed on surfaces or embedded in a solid. It requires a well-localized hole state, so, for an atom in a solid, extra-atomic relaxation might be large for a core hole, but weak for a valence-band hole, while for an adsorbate atom strong relaxation effects may be found also for valence emission, if the binding is such that the hole remains localized, i.e. for narrow adsorbate levels.

A schematic picture of these two types of relaxation effects and their relation to the single particle scheme is shown in Fig.5.9. The first diagram, Fig.5.9a, illustrates the single particle energy diagram with no relaxation present, applicable to UV photoemission from solids. Here the binding energy E_B may be deduced directly from the observed kinetic electron energy E_{Kin}. Figure 5.9b shows a plot of the total energy, not the single particle energy, of a photoemitting atom as a function

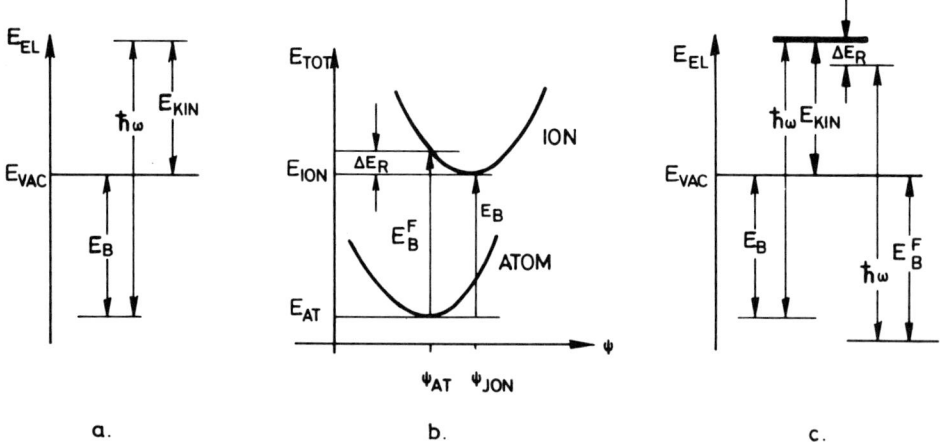

Fig. 5.9 a-c Relaxation effects and the single particle picture. In the single-particle scheme (a) the binding energy E_B is obtained directly by subtracting $\hbar\omega$ from the observed kinetic energy E_{kin}. The total energy of an atom or an ion (b) shows a minimum (ground state) at different values of the parameter ψ describing the total wave function of the system. The assumption of frozen orbitals therefore leads to a binding energy E_B^F that is higher (by the amount of relaxation energy ΔE_R) than the measured binding energy E_B, given by the difference of the ground-state total energies of atom and ion. In a single-particle energy diagram (c) one therefore has to correct the observed kinetic energy by ΔE_R to find the binding energy E_B^F for "frozen" orbitals

of a parameter Ψ describing the total wave function, e.g., by the radii of the individual electron orbitals. The minimum of the Ψ versus E curve represents the energy and wave function values of the ground state. After the ionization process the total energy is higher, and a new total energy curve is found for the ion which is different from that of an atom due to the rearrangement of the orbitals in the new selfconsistent field. The energy E_B^F is the binding energy for the case of "frozen" orbitals, i.e. for constant values of the wave function parameter Ψ. As the orbitals rearrange to the ionic form, the total energy is lowered by the relaxation energy ΔE_R. One will therefore measure $E_B = E_B^F - \Delta E_R$, not equal to the "frozen orbital" energy E_B^F which corresponds to the wave function values of the unperturbed system. The intraatomic relaxation energy ΔE_R cannot directly be observed. If, however, the shift in total energy is due to extrinsic effects, as in the case of extra-atomic relaxation, an energy difference becomes observable by comparing the free-atom binding energy to the adsorbed-atom or embedded-atom binding energy. If the binding energy is calculated directly by subtracting the photon energy from the measured kinetic energy (Fig.5.9c) a too small (absolute) value is obtained. The "true" binding energy E_B^F can only be found if the relaxation correction ΔE_R is known.

The above concept of interpreting X-ray spectra dates back to KOOPMANS [5.87], who noted that the observed ionization energies might be related to the electron orbital energies under the assumption of "frozen" orbitals. Correct binding energies were calculated by BAGUS [5.88] by comparing the energy levels of the N and (N-1) electron system. The problem of extra-atomic [5.89] and intra-atomic relaxation has been described in a number of original [5.90-94] and review articles [5.1,18,19,39, 40,95]. The important role of time scales on relaxation effects in photoionization has been emphasized by GADZUK and ŠUNJIĆ [5.96], following ideas of MELDNER and PEREZ [5.97]. Their consideration of time scales allows a very lucid description of the many-body processes associated with relaxation effects, including the influence of satellite structure.

Once the inherent complications of the *final state* of the photoemitting system, arising from many-body relaxation and polarization effects, are known, the observed kinetic energies of photomitted electrons may be used to deduce information on the initial state of the system and hence of the intrinsic electronic properties of the emitter. One may change the *initial state* by manipulating the average charge distribution, for example by a different chemical environment. This leads to *bonding shifts*, arising from a rearrangement of valence electrons following a chemical reaction, which alters the electrostatic fields in the vicinity of the core and therefore the core orbital energies. These energy shifts are small, in the order of a few percent of the core level energies, but easily observable. But why are changes in the valence structure due to bonding observed in this indirect way, via small effects on the core orbitals instead of by direct spectroscopy on the valence shell? The problem is that the valence shell has a complex structure that rearranges in a complicated way on chemical bonding. XPS shifts, on the other hand, yield a single parameter that may, under favorable circumstances, be directly related to the integrated charge transfer during bonding. This is the basis for the tremendous success of XPS in chemistry, and the original name ESCA for this spectroscopy was coined for that reason. An example of the versatility of chemical shifts is presented in Fig.5.10, which shows the carbon 1s line in the spectrum of ethyl fluoroacetate [5.1]. The carbon atom is bound in this molecule in four different ways, leading to four different states of the carbon valence shell. Indeed the carbon 1s line is found split into four lines of roughly equal intensity. The molecular structure at the top of Fig.5.10 is drawn such that the carbon atoms appear in the same sequence as the corresponding C 1s XPS lines.

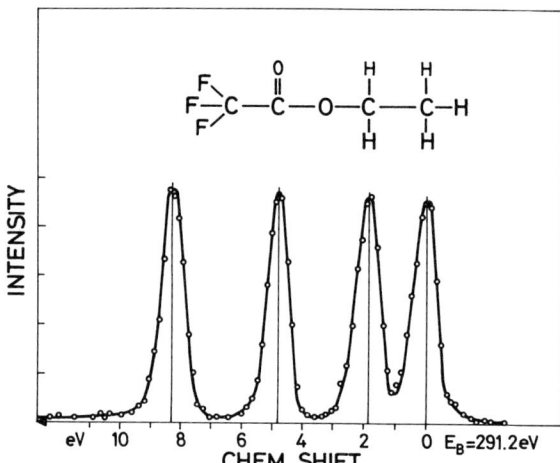

Fig. 5.10 The carbon 1s core line emitted from ethyl fluoroacetate shows four different chemically shifted positions arising from different chemical states of the carbon atom. The molecular structure at the top shows the carbon atoms in the same sequence as the corresponding ESCA lines (after [5.1])

In summmary we may say that small shifts are expected and observed in XPS corelevel line positions as the atomic environment changes. Those *chemical shifts*, of the order of 10 eV, are partly due to initial state effects, arising from bonding changes in the valence-electron charge distribution, and partly due to final state effects arising from a charge redistribution as a reaction to the creation of a local hole. Final state relaxation effects may be split into intra- and extra-atomic contributions, where only the latter are experimentally accessible. The problem of chemical shifts is discussed widely in literature and some comprehensive review articles may serve as a source for further references [5.1,18,19,39,98,99].

5.3.4 Photoemission from Adsorbates

Photoemission spectroscopy is intrinsically surface sensitive and therefore an obvious tool to study adsorption phenomena at surfaces. Two lines of investigation are available for such investigations. Using X-ray excitation one may observe the core levels, which provide chemical identification. The relative intensity of core lines gives a semiquantitative indication of the adsorbate coverage. Bonding information can be deduced from chemical shifts, if relaxation effects are known. On the other hand, UV excited spectra allow direct access to the valence electrons involved in bonding. Much weaker interatomic interaction than that inside a solid is expected in a chemisorption bond, so discrete valence levels may be observable and a shift, splitting, and broadening can be measured.

The interpretation of photoemission spectra of adsorbates in terms of available chemisorption theory is presently restricted, a fact that is merely a reflection of the current state of develepment of photoemission theory rather than of chemisorption theory itself. Calculations are now available that treat the difficult problems of *clean surfaces*, so far limited to idealized surfaces of free-electron solids [5.69, 72]. Very few attempts have made to date to properly assess photoelectric emission from *adsorbate-covered surfaces*. A number of additional difficulties make this problem much harder to solve than emission from a clean surface. First, the local electromagnetic fields at the surface, which are troublesome for clean crystals, can become extremely complex for an adsorbate covered surface. Second, many-body effects and relaxation phenomena are likely to be important for localized adsorbate electrons, so a proper treatment would have to go beyond the single-particle approximation. Third, interference effects of surface excited electrons with bulk excited electrons may occur and their quantitative implications are unknown. Due to those inherent difficulties, the general approach taken to date has been rather pragmatic. It is simply assumed that the contribution to photoelectric spectra from adsorbates reflects the *local density of states* at the surface, or at the adsorbate. With this hypothesis, measured spectra are related directly to calculated adatom level densities derived from chemisorption theory. The justification for such an assumption is obtained from several observations. As shown in Subsection 5.3.1, the single-particle photoelectron current is determined by the initial density of states (5.4), modulated by a matrix element, if the density of final states is sufficiently high to allow transition from all initial states. For adsorbates, we can assume that a reasonable overlap with free-electron final states outside the metal is always present (see, e.g., Fig.5.7, band-gap emission case, E_2). On the other hand, the local field effects near the surface tend to reduce the importance of any sharp selection rules (this is not necessarily true, of course, for selection rules on the parallel k-vector components). Such arguments are intuitively obvious, but it should always be kept in mind that there is no firm theoretical backing for identifying adsorbate-induced photoemission features directly with the local density of states at the adsorbate. There appears to be some experimental justification for this interpretation [5.39,100] derived mainly from field-emission energy-distribution spectroscopy, which is on a much better theoretical footing than photoemission theory [5.100].

The usual method to experimentally distinguish adsorbate-induced structure in photoemission spectra is by means of *difference curves*. Two spectra are scanned, one on the clean surface and one with the adsorbate present, and the difference is taken. Positive contributions are assigned to additional emission from the adsorbate, while negative contributions are ascribed to suppression of emission from the substrate. This method, although common practice, has to be treated with some care. Sometimes a scaling of the clean spectrum appears to be necessary to compensate for an overall reduction of substrate emission after adsorption. No firm guidelines are avaibable at present to remove the arbitrariness from this scaling factor. PENN [5.102] has pointed out that interference effects between substrate and adsorbate emission may lead to peak shifts and even to the formation of an antiresonance (negative contribution). EINSTEIN [5.103] has shown that extended negative contributions to difference spectra are to be expected as a consequence of a sum rule establishing charge neutrality: a large amount of charge density concentrated in a strong adsorbate resonance leads to a depletion in other regions of the spectra. For all these reasons difference spectra are difficult to interpret, and it is only the strong and unambiguous features that should be used for comparison with results of chemisorption theory.

Once structure in difference curves has been identified as being due to an adsorbate resonance level, one would like to compare it to the energy levels of the free atom or molecule of the adsorbate in the gas phase. Here the problem of *referencing* arises [5.104]. As sketched in Fig.5.2, both the atom and the solid are referenced to a "vacuum level", which is set equal to the ionization threshold for the gas and the work function for the solid. The question is, which work function one should take, the value of the clean surface, that of the saturated surface, or the "current" value depending on coverage. In fact, Fig.5.2 is misleading, since a "vacuum level" for a solid is not well defined, in contrast to a free atom, where it is equal to the ionization threshold. It is well known that a crystal exhibits different work-function values on different crystal faces. One can therefore define an ionization threshold, or work function, as the energy necessary to bring an electron from the Fermi level just outside the surface, at rest, in a distance small compared to the crystal face dimensions. The vacuum level, i.e., the energy the electron has far away from the crystal, depends on the geometry, as it is clearly different for an infinite single crystal face and a finite crystal exposing faces with different work function. An additional complication arises with a polycrystalline surface, where the individual crystallites have different work functions, but the observer at a large distance sees an average value. A similar situation is found for the case of a surface partially covered with an adsorbate. Each atom, as it adsorbs, establishes its new, microscopic, ionization level, but at a large distance an average work function is measured, changing with coverage. The correct work function value of the adsorbate system may be observable at full coverage if all atoms are bound in the same way, and consequently the macroscopic work function value is equal to the microscopic one. This is only possible if no coverage-dependent phase changes occur. Experimental observations support this view insofar as clear evidence was found that adsorbate resonance levels do not follow coverage-dependent work function changes. Results of Xe [5.105] or H_2 [5.106,107] on tungsten show that the adsorbate levels appear to be fixed with respect to the Fermi level, while the work function changes by 1.6 eV (for Xe) or 0.9 eV (for H_2) as a function of coverage.

In parallel with the rapid progress of experimental results in photoemission spectroscopy, *chemisorption theory* has developed its own branch dealing with the calculation of adsorbate resonance levels. This field is growing rapidly, and we will give here only an overview on general features relevant in the present context. For detailed information and further references the reader is referred to a number of excellent review articles [5.108-112].

One way to predict energy levels of adsorbed molecules is by means of *cluster calculations* [5.113-115]. Here a large molecule, a cluster, is formed of a representative number of substrate atoms and the adsorbate in an appropriate binding position. The energy levels of this cluster are calculated explicitly, and their density is then directly compared to the observed photoemission spectrum with the adsorbate present. An example of such a calculation is shown in Fig.5.11, which compares the experimental photoemission spectrum of CO adsorbed on nickel [5.116] to the energy levels of a $CO(Ni)_5$ cluster as reported by BATRA and BAGUS [5.117] and the gas-phase energy levels of CO [5.118]. There is considerable difference between the latter two results, and for two of the levels even the energetic sequence is inverted. Apparently, the assignment of observed adsorbate levels to molecular orbitals is a nontrivial task.

A different way of treating the chemisorption problem as a function of the adsorbate-substrate interaction strength is obtained using the self-consistent-field *molecular orbital schemes* based on the early work of GRIMLEY [5.119] and NEWNS [5.120]. Many ideas have recently been injected into this method [5.103,121-128],

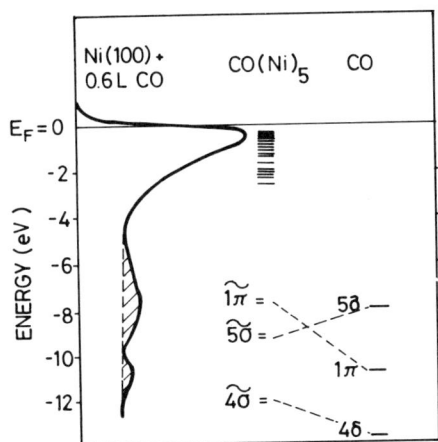

Fig. 5.11 Comparison of the measured energy levels of CO chemisorbed on nickel (100) (cross hatched, from EASTMAN and CASHION [5.116]) with the theoretical levels from a cluster calculation by BATRA and BAGUS, [5.117]. The CO-derived levels in the cluster show a sequence different from that observed in the gas phase [5.118]

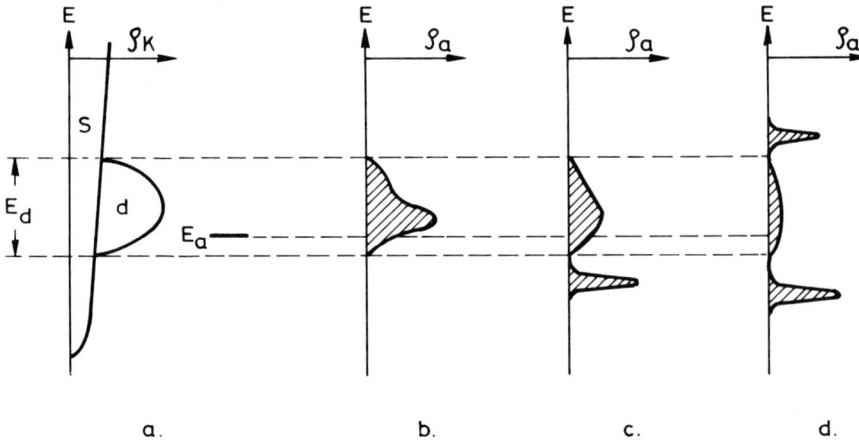

Fig. 5.12 a-d Illustrating possible adsorbate level positions in the Newns-Anderson model of chemisorption. The substrate density of states ρ_k (a) consists of s and d band contributions, the free-atom adsorbate level E_a is assumed to lie within the d band. Three possibilities are found for the adsorbate level density ρ_a. Weak interaction (b) leads to a single, broadened resonance. Strong interaction can give rise to one (c) or two (d) split-off levels, broadened much less by the interaction with the s band (after [5.126])

the basic principle of which is illustrated in Fig.5.12. Consider an adsorbate atom with a single energy level E_a, interacting with a model d-band metal, characterized by a relatively narrow d-band (d in Fig.5.12a) and a wide, low-density s-band. The coupling of the adsorbate state $|a\rangle$ with the substrate states $|k\rangle$ is represented by the "hopping" matrix element $V_{ak} = \langle k|H|a\rangle$ where H is the full Hamiltonian of the combined system. The density of states at the adatom ρ_a may be obtained from the adatom Green's function, and it depends on the values of V_{ak}, the position of the adatom level E_a relative to the d-band, and the d-band density of states. Three "typical" cases are shown schematically in Fig.5.12 [5.126]. If V_{ak}, the hopping interaction, is small compared to the d-band width E_d, i.e., if the adatom is loosely bound, the adsorbate level E_a broadens and shifts, forming a single *resonance level*. This is the situation for a weak physisorption bond. As the interaction strength V_{ak} increases, the width of the resonance level increases, until at some point an adatom level is "split off" the d-band. In the absence of an s-band such a *split-off level* has the shape of a δ-function. However, as shown in Fig.5.12c, the interaction with the s-band will broaden this level, but to a lesser extent due to the low s-band density of states. For values of V_{ak} large compared to E_d one expects two split-off levels, Fig.5.12d, in analogy to the formation of a bonding-antibonding pair of levels due to molecular interaction.

Finally we would like to give a short overview of the effects expected in angle-resolved photoemission studies. Angular structure may arise, as pointed out in Subsection 5.3.1, from the polarization dependence of the operator in the matrix element, from the angular dependence of the wave functions, and from interference effect between adatom emission and substrate emission. Photoemission experiments on gases give a fairly smooth angular distribution, since only the first two effects contribute, and an averaging occurs due to the random orientation of molecules in a gas. However, molecules adsorbed on a surface are usually oriented, so the angular structure of adsorbate emission can be expected to be very pronounced. Theoretical approaches to angular photoemission start from two directions, namely final state effects and initial state effects, as originally introduced by LIEBSCH [5.129] and GADZUK [5.126,130] respectively. The latter has shown that different adsorbate orbitals on the same substrate or on various adsorption sites are expected to give distinctly different angular emission patterns, so the observation of such patterns for different light polarization could provide identification of orbital symmetry and adsorption geometry. Azimuth scans may yield in a simple way the local geometry of the adsorption site, e.g., whether the adatom is in a twofold or fourfold symmetric site. This result is common to the initial state [5.131] and final state approach, while the latter predicts additional features in a final energy scan (in analogy to LEED intensity-voltage curves) or in a polar energy scan, which may allow an experimental determination of the adsorbate-substrate atomic distance.

The above discussion of adsorbate photoemission in terms of initial and final state effects dealt with the case of a single isolated atom on a surface, having well-defined nondispersing resonance levels that can be identified in the spectra. Experimental spectra are usually taken on surfaces with considerable coverages of adsorbates (to improve the signal-to-background ratio), and in most cases a strong chemisorption bond results in island growth or the formation of ordered superstructures due to adsorbate-adsorbate interaction. The identification of resonance levels is not unique, and there is evidence of dispersion with angle (see Fig.5.17). In spite of these complications there is hope that recent theories, treating angular photoemission as a two-dimensional problem in close analogy to LEED theory and permitting both initial and final state effects to be included [5.132,133] may bridge the apparent gap between experimental practice and theoretical approach.

5.4 Measurement Methods

This section is intended to give a general view of the most widely applied methods used in photoemission spectroscopy. Examples will be presented to illustrate some of the topics by typical or recent experimental results. Special emphasis will be placed on surface related effects, where electron states of clean surfaces and adsorbate studies will play an important role. The overall aim of this section is to show how photoemission spectroscopy can be used in an efficient and direct way to provide the surface physicist with the desired information on a specific problem.

5.4.1 Energy-Resolved Spectroscopy

The basic measurement in photoelectron spectroscopy is an energy distribution spectrum of electrons emitted at constant photon energy, integrated over all emission angles. Angular integration tends to average out pronounced final state effects, so the spectra obtained in this way emphasize initial state features. Consequently this is the preferred method in those cases where an overall view of intrinsic bulk or surface properties is desired, for example in density-of-states measurements, or, most important, for the observation of adsorbate resonance levels. As discussed in Subsection 5.3.1, full-angle spectra may contain both bulk and surface information, and under favorable circumstances the bulk part may be identified by its characteristic excitation-energy dependence. Surface contributions to the spectra are expected to be largely independent of excitation energy, except for energy-dependent cross section variations in the transition matrix element. The question of what minimum acceptance angle is required to obtain true angle-integrated spectra cannot be answered uniquely. At low kinetic electron energies even the full $180°$ collection angle may not be representative since the refraction conditions during electron escape [5.32] restrict the escape cone of k-vectors inside the sample. On the other hand, in the X-ray excitation range a few degrees acceptance angle may suffice to observe a representative sample.

Wide-angle photoemission energy-distribution measurements have successfully been used to investigate surface states in the band gaps of semiconductors. Many adsorbate studies have also been performed with this technique, both in the XPS and UPS regime, and experimentalists are now extending to the point of observing chemical reactions on surfaces and exploring the physical basis of the catalytic process. Recent review articles should be consulted for original references. For example, the spectroscopy of clean surfaces has been reviewed by FEUERBACHER and WILLIS [5.32], and results for adsorbate-covered surfaces by PIERCE [5.37], BRUNDLE [5.36,134], PLUMMER [5.39], and MENZEL [5.41]. In the following we will discuss a very limited number of examples that may help to illustrate the basic capabilities of the technique.

Clean Surfaces

The most extensively investigated single-crystal surface is the (111) face of silicon. The existence of localized electron states at this surface, inside the bulk band gap, has been postulated for a long time [5.135] to explain band bending. It was one of the great successes of photoemission spectroscopy in surface physics to directly demonstrate such surface states [5.136,137] on freshly cleaved Si(111) surfaces. The spectra show a hump in the energy-distribution curve, close to the absolute band gap, that remains essentially independent of exciting photon energy, but is very sensitive to small amounts of adsorbed gases. An interesting characteristic of the Si(111) surface is the fact that it has two stable forms of reconstruction. A freshly cleaved surface exhibits a (2 × 1) superstructure, which changes irreversibly into a (7 × 7) structure after annealing to 850°C. One would expect that surface-derived electron states depend on reconstruction. This is indeed so, as shown in Fig.5.13, where the photoemission spectrum observed by ROWE and PHILLIPS [5.138] for the two stable reconstructed Si(111) surfaces is compared to the estimated bulk contribution (shaded). The surface derived features are clearly different, showing one peak for the (2 × 1) and two for the (7 × 7) superstructure. Surface states on silicon are also expected at energies away from the band gap, spreading essentially over the whole occupied valence band. For lower lying states an identification is much more difficult, since strong overlap with bulk states obscures their existence (ROWE and IBACH [5.139]). A short review of measurements on the Si(111) face was presented by ROWE et al. [5.140]; other references on measurements dealing with semiconductor surface states may be found in [5.32].

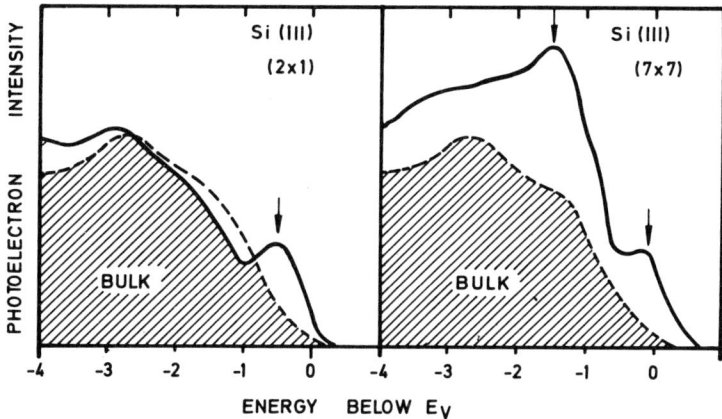

Fig. 5.13 Surface and bulk emission from a silicon (111) surface (ROWE and PHILLIPS [5.138]). The two stable forms of reconstruction have different surface electronic properties, with one surface state on the (2 × 1) surface and two (arrowed) after (7 × 7) reconstruction

Adsorbate Studies

In an XPS experiment it is laboratory routine to check the cleanliness of the sample under observation by a scan over a wide energy range. *Core-level lines* of adsorbates reveal contaminants in the percent coverage range, and identify the chemical nature of the contaminant. Here XPS is used in the same way as Auger spectroscopy. Though the sensitivity is somewhat lower (mainly due to the high current of the Auger exciting electron beam), XPS has the advantage of avoiding beam-induced effects to a large extent. Measurements of the height of an adsorbate core-level peak allow a relative measure of the coverage, since the photoemission cross sections should be independent of coverage. Calibration in absolute terms is possible, e.g., at a known saturation coverage. Therefore sticking coefficients or adsorption and desorption kinetics can be observed by XPS. Furthermore, the energetic position of core lines yields the surface chemical shift by comparison with the gas phase values. As pointed out in Subsection 5.3.3, this surface shift is composed of a relaxation and a bonding contribution, and if one of those can be calculated or is available from other data, the other can be derived. However, a distinction between bonding and relaxation contribution to the surface shift is not necessary to obtain valuable information on the chemisorption process. For example, YATES et al. [5.141,142] have performed extensive X-ray photoemission experiments on gases chemisorbed and physisorbed on tungsten. They found *several* peaks due to a single adsorbate core level, varying strongly in intensity with coverage following heat treatment. These multiple surface shifts could be related to multiple adsorbate phases found in thermal desorption spectra, giving direct evidence that those multiple phases do coexist on the surface and are not produced during the thermal desorption cycle.

X-ray excited photoemission also allows the observation of *valence levels of adsorbates* [5.143]. However, in this realm ultraviolet excited spectra offer distinct advantages of higher intrinsic resolution and increased excitation cross section. Since the first photoemission measurement of adsorbate levels by EASTMAN and CASHION [5.116] a large amount of data has been reported [5.36,37,39,41,134]. Simple adatoms physisorbed or even chemisorbed usually allow the distinction of reasonably well-defined resonance levels. Under favorable circumstances these may be related to energy levels of the free adsorbate molecule (see Fig.5.11). If this is the case, a surface shift may be defined in analogy to core-level shifts. PLUMMER [5.39] has collected data on valence level surface shifts and presented a careful discussion of the phenomenon. Observed shifts have values of 3 eV which should be compared to core-level shifts of typically 5 eV due to the more localized nature of the core levels. DEMUTH and EASTMAN [5.144] have shown that the relaxation and bonding contributions can experimentally be distinguished under minimal assumptions, if a proper system is chosen. These authors observed several valence levels of a hydrocarbon molecule under physisorbed and chemisorbed conditions. In physisorption all levels shift simultaneously by the same amount, which is attributed to extra-atomic relaxation. Chemisorption bonding appears to involve only specific levels, as deduced from their shift being different from that of the other levels. This difference is attributed to a bonding shift and allows the bonding interaction to be estimated. Recently ultraviolet photoemission has been used to directly observe chemical reactions on surfaces [5.144-146]. Ethylene was chemisorbed on nickel, and gentle heating induced dehydrogenation to give acetylene, as evidenced by a drastic change in the photoemission spectrum to a form identical to that obtained by direct adsorption of acetylene. Careful studies allow identification of particular chemical bonds in the spectra [5.146], permitting reactions to be followed through several stages.

As a final example we would like to demonstrate how a close interaction between theory and experiment in photoemission spectroscopy can yield precise and unambiguous information on chemisorption problems. A silicon (111) surface in the stable (7 × 7) form of reconstruction, exposed to atomic hydrogen, exhibits a photoemission spectrum as shown by curve 1 in Fig.5.14 [5.147]. A completely different spectrum

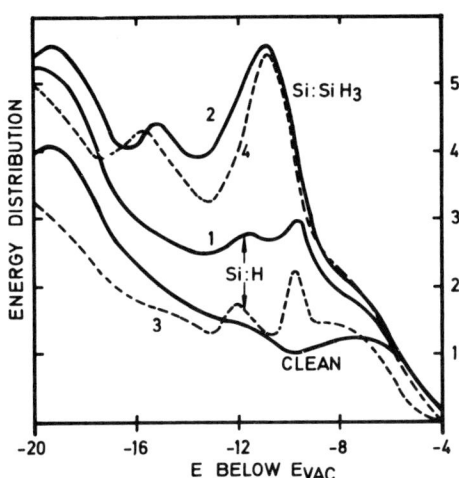

Fig. 5.14 Present chemisorption theory allows identification of adsorbate configurations, as shown for two forms of hydrogen chemisorption on the Si (111) surface. One form is a monohydrate Si : H (Curve 1, measurement, curve 3, theory), the other a trihydrate Si : SiH_3 (curve 2, measurement, curve 4, theory) (after PANDEY et al. [5.147])

is obtained, curve 2, on hydrogenation of a stabilized (1 × 1) surface. The question of the detailed nature of those two different hydrogenated phases of the Si(111) face has been solved by a theoretical calculation of the photoemission spectra. Hydrogen chemisorption on the annealed (7 × 7) surface was shown to give binding of hydrogen atoms to the single dangling bond orbital of each surface Si atom in the form of a monohydride phase Si(111): H. The corresponding calculated spectrum is shown by curve 3 in Fig.5.14. The chemisorption phase formed on hydrogenation of a (1 × 1) surface was identified as being due to SiH_3 radicals bonded to the dangling orbitals of Si(111), a trihydride phase termed Si(111): SiH_3. The electronic energy levels of this phase are strikingly different, and the calculated spectrum (curve 4 in Fig.5.14) shows excellent agreement with the experimental observations.

5.4.2 Angle-Resolved Photoemission

An experiment that measures both the energy and emission angle of a photoelectron simultaneously observes all quantities conserved within the one-electron picture of the photoelectric process: the final state energy $E + \hbar\omega$, and the two k-vector components parallel to the emitter surface. Such a measurement allows a direct comparison with theoretical results, avoiding the tedious integration and interpolation procedures involved in wide angle theoretical predictions. In studies concerned solely with the surface of an emitter the normal k-vector component is either undefined or irrelevant, so an angular emission measurement determines all the important parameters of the electronic system. This is the basis for the predominance of angle-resolved electron emission studies particularly in surface physics [5.129-133]. The spectral shape expected in an angle-resolved energy distribution curve has been discussed in Subsection 5.3.1 and 5.3.4. Bulk emission in the energy range where the electron mean free path is long, i.e., damping is small, leads to a discontinuous spectrum consisting of narrow peaks, with a width inversely proportional to the electron escape depth [5.67]. For surface-related features the situation is different. In regions where there is a continuous electronic state density available, that is within (bulk) bands, surface emission leads to a continuous energy spectrum, reflecting to some degree the local state density near the surface. In contrast, surface states in bulk band gaps have no dispersion normal to the surface, so emission from such states will give rise to sharp features, if damping parallel to the surface is small. Sharp structure arising from either bulk transitions or surface states can be distinguished by measuring spectra at different photon energies, which causes the bulk-derived peaks to move in a characteristic way [5.74,75].

Any interpretation of angle-resolved spectra is complicated by the effects of diffraction and refraction. *Diffraction* originates from interaction of the excited electron with the periodic lattice and gives rise to multiple emission directions termed primary and secondary cones by MAHAN [5.62]. The crystal lattice can provide momentum in discrete amounts of reciprocal lattice vector $\underline{G}$ that may be added or subtracted from the final state k-vector as discussed in more detail in [5.32]. *Refraction* arises from the conservation of energy and the k-vector component parallel to the surface during the electron escape process. The electron observed at the detector can be regarded as free, so the value of the parallel momentum $\underline{k}_\shortparallel$ is obtained from the emission angle θ and the kinetic energy E_{kin} by $\hbar\underline{k}_\shortparallel = (2mE_{kin})^{1/2}\sin\theta$. The normal k-vector component inside the solid is found by cutting the final-state constant energy surface at $\underline{k}_\shortparallel$, a process that may lead to multiple values of $k_\perp$ [5.148]. The difference in $k_\perp$ between the values outside and inside the solid is taken up by the surface, which acts as a continuous source of normal momentum by means of the surface potential discontinuity [5.32]. Refraction is not a serious problem in the investigation of surface properties where only the parallel momentum is of interest. Angular photoemission studies also face a number of technical problems. The most serious one is the light source intensity. The analyzed number of electrons is roughly three orders of magnitude smaller than the total number of emitted electrons. Measurements with reasonable statistics have to be performed within a time that ensures stable conditions at the observed surface,

so the light intensity required is a function of the residual gas pressure at the
sample position. As measurements are performed at low kinetic electron energies, the
reduction of magnetic fields near the sample requires special attention. A field
reduction by passive shielding is in most cases preferable to field compensation.
It is not a good solution to simply increase the kinetic energy by electrostatic
acceleration in a (hemi-)spherical field around the sample, since the dimension of
the flat sample surface is usually not sufficiently small compared to the radius of
the sphere to avoid angular distortion. In no case should one underestimate the
mechanical problems that arise from detector movements and critical alignment
procedures under ultrahigh vacuum conditions.

In principle there are two ways to perform angle-resolved photoemission measurements. A fixed angle may be chosen, and an angle-restricted energy scan is made,
or the analyzer band-pass energy is fixed, and an angular scan is performed along
the polar or azimuthal angle. An energy spectrum appears to be indispensable,
especially if surface contributions to the total spectrum have to be identified. Once
this is done, the analyzer may be locked onto a surface feature. Now an *azimuth* scan
for fixed polar angle will reveal the *symmetry* of the surface feature relative to
the bulk. On the other hand, the experimenter can use known symmetry properties of
the crystal and make *polar* angle scans to investigate the *dispersion* with varing $\underline{k}_{\shortparallel}$.
In the latter case energy distribution curves have to be observed for various points
along the polar arc to relocate the energetic position of the surface features as
they disperse with polar angle.

Clean Surfaces

An excellent example of the power of angle-resolved photoemission measurements in
surface studies has been given by ROWE et al. [5.150]. This experiment also demonstrates the different information content in polar and azimuth angular scans. Energy
spectra of photoelectrons emitted from a Si(111) surface have been observed with
a $4^{\circ} \times 4^{\circ}$ angular resolution for various polar angles along a [112] azimuth. Those
spectra (Fig.5.15a) show a prominent peak just below the Fermi level, which has been
assigned to dangling-bond surface states near the band gap. A comparison to Fig.5.13
demonstrates the improvements in signal-to-background ratio for surface features
achievable by angle-resolved spectroscopy. While in the angle-integrated spectra
(Fig.5.13) the surface state appears as a moderate hump, it is the most prominent
feature in the angle-resolved spectra. Note that the maximum of the surface-state
emission does not occur for the surface normal, as one would expect from the p_z
symmetry of a dangling bond, but rather at an angle of 25° (for the photon energy
of 10.2 eV used in these experiments). Azimuth scans have been performed at the
polar angle of most intense surface state emission, and the results are plotted in
Fig.5.15b. The triangles mark the integrated intensity of the background, which is
found to be nearly isotropic, while the surface state intensity(circles) varies
strongly with azimuth angle, showing three lobes aligned with the [112]

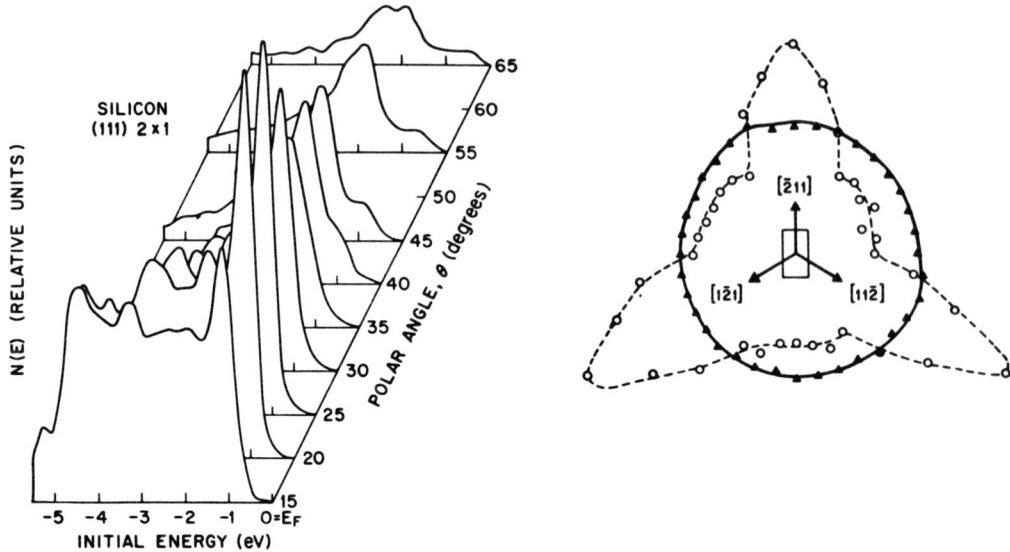

Fig. 5.15 Angle-resolved photoemission experiments applied to the dangling-bond surface state at the Si(111) face. The sharp peak in the energy distribution spectra near -1 eV is due to the surface state and should be compared to the corresponding hump in Fig.5.13 (left side, arrow) for angle-integrated measurements. The azimuth angle distribution of the surface-state emission (open circles) shows a three-lobed structure, while the rest of the spectrum (triangles) is emitted isotropically (after ROWE et al.[5.150])

crystallographic directions. This behavior has been interpreted as an initial state effect [5.150], indicating hybridization of the surface state of cylindrical symmetry with the threefold symmetric back bonds of the top-layer silicon atom. An alternative explanation is also possible in terms of final-state effects, assigning the observed threefold symmetry to diffraction by the crystal lattice.

For metals the absence of absolute band gaps makes the observation of surface state features more difficult than for semiconductors. However metals may exhibit relative band gaps for specific values of k-vector components in the surface. In angle-resolved photoemission experiments both parallel k-vector components are fixed by emission angle and energy, so that these relative band gaps can be observed. An example of this is shown in Fig.5.16 for a measurement on electrons photoemitted *normal* to the (110) face of tungsten [5.73,76]. For emission normal to a low-index crystal face the $k_{\parallel}$ conservation restricts the information to a single symmetry line in the Brillouin zone, which in this particular case exhibits large band gaps in the one-dimensional state density both above and below the Fermi level (first two panels of Fig.5.16). Consequently the technique of band-gap emission may be applied by using appropriate photon energies. Apparently the initial-state band gap around -10 eV is reproduced by the observed spectrum (third panel), but the width of the occupied band is considerable smaller than predicted. A calculated local density

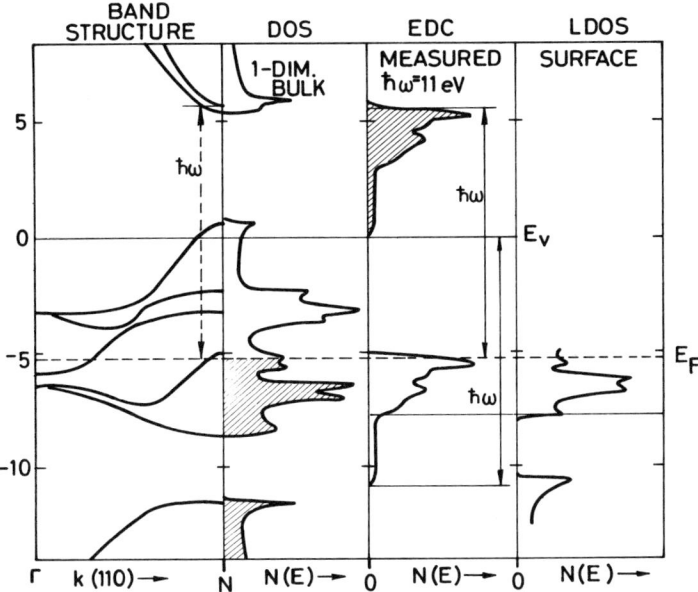

Fig. 5.16 Illustrating the principle of band-gap emission for narrow angle observation normal to a (110) face of tungsten. The photon energy $\hbar\omega$ was chosen such that most of the measured electrons (third panel, shaded [5.75]) have final-state energies in the large one-dimensional gap of the calculated band structure and state density (DOS) [5.74]. The spectrum (third panel, lower curve), reflects the initial-state band gap, and a band having a width close to the calculated surface density of states (LDOS). The peak at the top of the observed spectrum is due to a bulk optical transition (dashed vertical arrow of lenght $\hbar\omega$)

of states at the surface, LDOS, obtained by a rigid compression of the bulk bands taking into account the reduced number of neigbor atoms near the surface [5.74], reproduces the experimental bandwidth fairly well, but the peak near the Fermi level, which is due to a direct, k-conserving transition in the bulk, is of course absent from the LDOS. The results shown in Fig.5.16 could be taken as an indication that the method of band-gap emission may be used to probe the *local density of states* near the surface, which apparently can be different from the bulk values.

Emission from localized states on metal surfaces was observed for the tungsten (100) face [5.151,152], where a prominent structure is seen at energies of about 0.4 eV below the Fermi level, that has been assigned to a surface state in the bulk spin-orbit band gap [5.153]. Experimentally the assignment is based on the observation that the structure is fairly independent of photon energy, even right through final state band gaps [5.74,75] and also on the sensitivity to gas adsorption [5.151]. Similar features have recently been observed for platinum [5.154] and copper single crystal surfaces [5.155]. It appears likely that surface states are in fact a fairly common feature in metals but their observation requires either angle restricted experiments or else the possibility of varying the surface symmetry.

Adsorbates

Few experimental results on the angular distribution of photoemission from adsorbates are available to date. This is in spite of the interesting and new information one can expect to obtain from such experiments. Technical difficulties which are even more severe than in measurements on clean surfaces are the reason for the slow experimental progress in this area. Adsorbate studies require two measurements in order to allow identification of those features due to the adsorbate. Severe signal-to-background ratio problems are faced in addition to the signal-to-noise ratio difficulties present in any angle-resolved photoemission experiment. Still, it is expected that this kind of spectroscopy will gain increased attention in the near future.

A two-dimensional display system to observe the angular features of adsorbate-covered surfaces has been proposed by PLUMMER and WACLAWSKI [5.156], and GADZUK [5.126,130] has shown that such a system allows direct identification of the chemisorption bond geometry. At present, however, only preliminary data are available [5.157]. Polar angle measurements have been performed by EGELHOFF and PERRY [5.158] on the chemisorption system of hydrogen on (100) tungsten. After identification of adsorbate-induced features by difference curves, these authors succeeded in observing the dispersion of adsorbate levels along a symmetry direction, allowing them to plot a one-dimensional absorbate-level band structure diagram. Measurements at different photon energies indicated that refraction effects can be properly taken

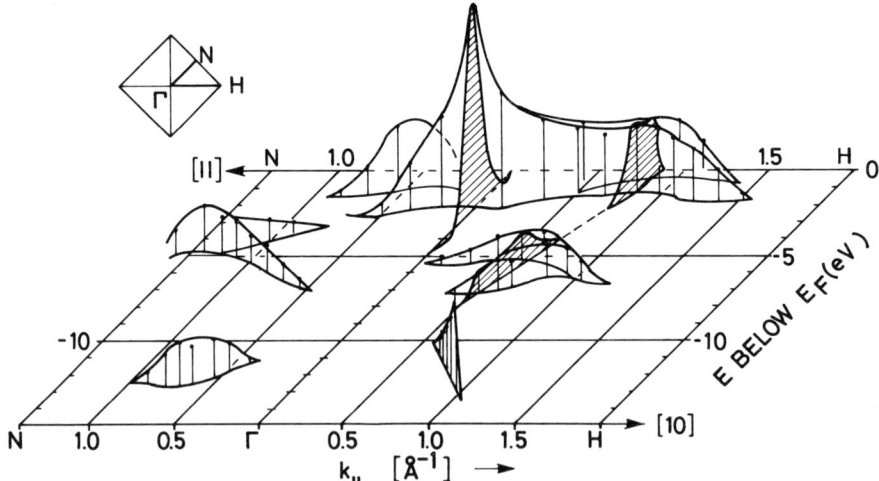

Fig. 5.17 Angle-resolved photoemission spectroscopy on the adsorbate system H on W(100). The position of adsorbate resonance peaks is plotted as a function of the parallell k-vector along two symmetry lines of the two-dimensional Brillouin zone (heavy lines in the inset), forming a two-dimensional band structure in the horizontal plane. The intensity of adsorbate structure is then plotted along the vertical axis. Three adsorbate-induced bands are seen at about -2 eV, -5 eV, and -12 eV, that disperse and split with observation angle

into account, and that the surface features are insensitive to photon energy, as expected. If the excitation can be made independent of observation angle, and if general excitation conditions are chosen (nonnormal incidence, unpolarized light), the intensity of an adsorbate peak may be introduced as an additional parameter. This leads to a pseudo-three-dimensional plot as shown in Fig.5.17 [5.159] where the results of a measurement for hydrogen chemisorbed on a tungsten (100) face at saturation coverage are shown. The horizontal plane in this diagram represents a two-dimensional band structure plot along the ΓH and ΓN symmetry lines of a two-dimensional Brillouin zone (shown in the inset) while the intensity of the adsorbate-induced peaks is plotted along the vertical axis. The narrow peak along the Γ axis indicates strong, narrow hydrogen emission in the surface-normal direction at 2.3 eV below the Fermi level. This peak falls off sharply for observation away from the normal direction, but is present for relatively large angles along the [11] azimuth at energies closer to the Fermi level, where it broadens into a double structure as shown by the hatched area for $25°$ observation angle. Two other bands of adsorbate-induced features are found at lower initial state energies, splitting and dispersing in a way that is qualitatively similar but quantitatively different for the two symmetry directions. The figure clearly demonstrates that, in a restricted-angle measurement of adsorbate features, what one sees depends on how one looks. At normal observation, looking along the $\Gamma(k_\| = 0)$ line, only one peak at -2.3 eV is found [5.107]. In a spectrum taken off normal, say at $45°$, one will see three peaks or two peaks, depending on whether the azimuth is along [10] or [11]. So, if an overall picture of adsorbate induced structure is required, angle-integrated measurements are a necessity. Angle-resolved spectra reveal much more detail, but complete coverage of symmetry directions is required for any interpretation.

5.4.3 Yield Spectroscopies

Measurement of the total flux of photoelectrons emitted from a sample as a function of the incident photon energy is experimentally the simplest form of photoelectron spectroscopy. The sample is held as the photocathode with an applied field such that the emitted photoelectrons are collected at an anode under saturation conditions as $\hbar\omega$ is varied. The spectral dependence of the yield per incident photon is given approximately by $Y_{inc} = A(E) \cdot (1-R) \cdot P(E)$, where $A(E)$ is the escape function describing the probability of electron emission through the surface, R is the optical reflectivity, and $P(E) = \alpha(\omega)\ell(E) [1 + \alpha(\omega)\ell(E)]^{-1}$ is the probability of an electron of energy E and mean free path ℓ reaching the surface when the optical adsorption coefficient is α [5.32,160].

The intrinsic threshold for photoemission is equal to the work function $E_{vac} - E_F$ in the case of metals, and to the energy difference between the top of the valence band and the vacuum level for the case of semiconductors with flat bands to the surface.

Close to the threshold the mean free path is large, comparable to the light penetration depth $1/\alpha$. Therefore, since $P(E) \simeq 1$ and $A(E)$ is smoothly increasing with E, structure in the *threshold yield spectrum* $Y_{inc}(\omega)$ will primarily be due to changes in reflectivity through the (1-R) term. The absorption coefficient enters only insofar as changes in α lead to significant changes in the ratio of the population of those final states lying above the vacuum level to those lying below. Changes in yield due to this latter effect can be used to obtain information on states lying between the Fermi and the vacuum level [5.160]. With a large electron mean free path, the major contribution to the total yield spectrum will be made by electrons originating in the bulk. However, surface contributions to the spectrum due to direct excitation out of filled intrinsic and extrinisc surface states can be important. In addition, the influence of the surface on the yield at threshold enters directly through the escape probability of the electrons at the surface barrier. The barrier is a consequence of the decay of the electron wave functions into the vacuum extending beyond the positive ion cores. Therefore the surface barrier height, and hence the threshold, differs for different crystal planes of the same material and is a sensitive measure of the changes in surface electron distribution resulting from structural changes, adsorbed atoms and external fields at the surface [5.161].

In metals, the photoelectric threshold ϕ or work function is usually determined using the method developed from FOWLER'S theory [5.162] of the yield spectrum near threshold for a free-electron-like solid. Despite the simple model, Fowler's prediction that the yield follows a square law near threshold is generally obeyed over a wave length range sufficient to allow absolute values of ϕ to be determined [5.163]. Changes in work function due to the adsorption of atoms on metal surfaces have been extensively studied [5.164], although the photoelectric method has been of limited application for this purpose since relative measurements are often sufficient, and faster methods are available to measure work function differences. The photoelectric threshold yield data for semiconductors are generally expressed as power laws in photon energy, $Y(\omega) \propto (\hbar\omega - \hbar\omega_0)^n$, where $\hbar\omega_0$ is the threshold energy. Calculations by KANE [5.165] one the basis of energy band expansions about the threshold point showed that the exponent n may have values ranging from 1 to 5/2 depending on the electron excitation and scattering mechanisms operating either in the bulk or at the surface of the semiconductor. BALLANTINE [5.166] has extended these calculations in order to include the effects of phonon scattering.

An example of the use of *threshold yield measurements* to study surface effects is the work of GOBELI and ALLEN [5.167] on cleaved silicon (111). They found an indirect photoelectric threshold of 5.15 eV but their work function measurements, which determine $E_{vac} - E_F$, gave values of ϕ which varied only between 4.75 and 4.82 eV despite the fact that a range of doping levels was used which moved the Fermi level across the the whole $\sim$ 1.1 eV band gap in the bulk. This apparent "pinning" of the Fermi level at the surface and the consequent bending of the bands was interpreted as the first experimental evidence for surface states in semiconductor band gaps. These authors [5.168] were also able to calculate the surface state density profile knowing the degree of the band bending as a function of impurity doping. Yet despite these early successes with this type of measurement and the subsequent refinements [5.169,170], accurate energy distribution measurements have tended to become the prime source of information on surface features derived from photoemission, since energy-resolved spectra allow an easier identification of bulk and surface contributions.

At final state energies above about 10 eV, the electron mean free path ℓ is generally less than 10 Å in both metals and insulators. This is considerably less than the attentuation length of the light $1/\alpha$, so that the yield is essentially determined by the number of photons absorbed within a distance ℓ from the surface. Therefore, in contrast to the situation at threshold, we can expect that structure in the higher energy *total yield spectra*, measured per *absorbed* photon, will mirror directly the absorption spectrum, although exceptions to this situation can occur [5.171]. The short mean free path may also enhance the surface sensitivity of the yield measurement and has been used to reduce uncertainties in energy level determinations in semiconductors where there is appreciable band bending at the surface [5.170,172]. The early studies in this regime were made using a conventional light source and consequently were severely restricted in maximum energy range. The present use of synchrotron radiation in total yield spectroscopy has removed this restriction, allowing measurements from threshold to the X-ray region and leading to renewed interest in this type of spectroscopy. An example of a total yield spectrum obtained with synchrotron light is shown in Fig.5.18, for the case of a polycrystalline

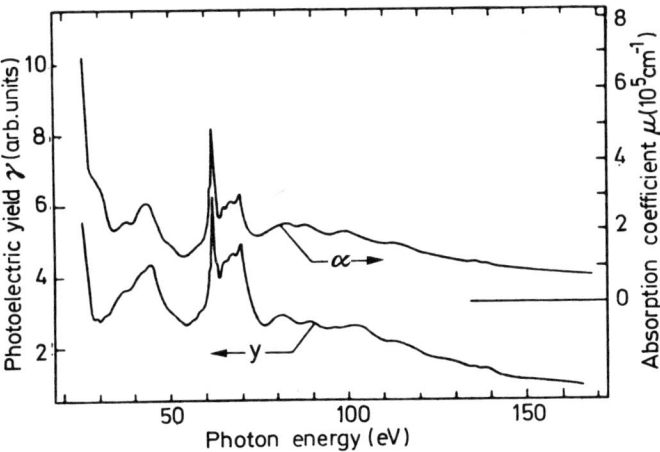

Fig. 5.18 At high photon energies the spectra of total photoelectric yield (Y) and absorption coefficient (α) look very similar, as shown from a measurement on LiF by GUDAT and KUNZ [5.196]

LiF layer, measured by GUDAT and KUNZ [5.173]. The agreement between the absorption spectrum and the total yield spectrum is striking. The yield spectra generally have a better resolution than the absorption spectra, due to the higher signal-to-noise ratio, since a higher absorption leads to a higher electron yield but a lower reflected or transmitted light intensity. Excitation out of a well-defined core state locates the initial state energy and removes to some extent the problem of lack of a reference energy point which is a fundamental limitation in total yield spectroscopy at lower energies. There are, however, a number of important differences between yield spectra obtained in the high-energy regime and the earlier studies at lower energies. Above about 30 eV the optical spectrum is no longer strongly modulated by matrix element effects, final states become available for exciting electrons from all initial states and they are significantly lifetime broadened (see, for example, Fig.5.8). Relaxation shifts, as discussed in Subsection 5.3.3, can lead to an apparent change in the observed core-level energy. Indeed, the correlation between the observed yield spectrum and the absorption spectrum of Fig.5.18 does not arise in quite the same manner as that discussed earlier for lower excitation energies. At these energies, the primary excited electron is strongly scattered, and the total yield spectrum is dominated by the contribution from the secondary electrons. This introduces a serious drawback in the use of yield spectra at these high energies, as far as surface studies are concerned. The mean free path of the low-energy secondary electrons which form the major part of the observed yield is typically in the range 30 to 100 Å for metals and insulators, respectively. Although this is small compared to the optical attenuation length, the emitted current is primarily of bulk origin. Therefore, despite the fact that the *primary* electron mean free path is of the order of 5 to 10 Å, most of the measured secondaries originate well away from the surface and carry relatively little information on surface properties.

A way out of this problem can be to select a higher final state energy, for which the secondary electron mean free path is small, and to measure the current in that particular energy window as $\hbar\omega$ is varied, taking care in the choice of energy ranges to avoid measurement of primary electrons directly excited from a core state or from the valence band. This leads to *partial yield spectroscopy*, as first introduced by EASTMAN and FREEOUF [5.174,175] in their study of the empty surface state bands of a number of III-V semiconductor surfaces. The principle is illustrated in Fig.5.19a. A core vacancy is created by photoexcitation to a final state energy just above the Fermi level. The vacancy may be deexcited by an Auger process involving the valence band, creating a hot electron which is either emitted directly or gives rise to a cascade distribution of secondaries by inelastic processes. Observation occurs in a selected energy band at external energies E^*, so that a particular sample of the *secondary electrons* created by these processes is detected. The secondaries are a direct consequence of the core excitation, and the intermediate steps are normally sufficiently complicated that matrix element effects tend to be smoothed out. Consequently

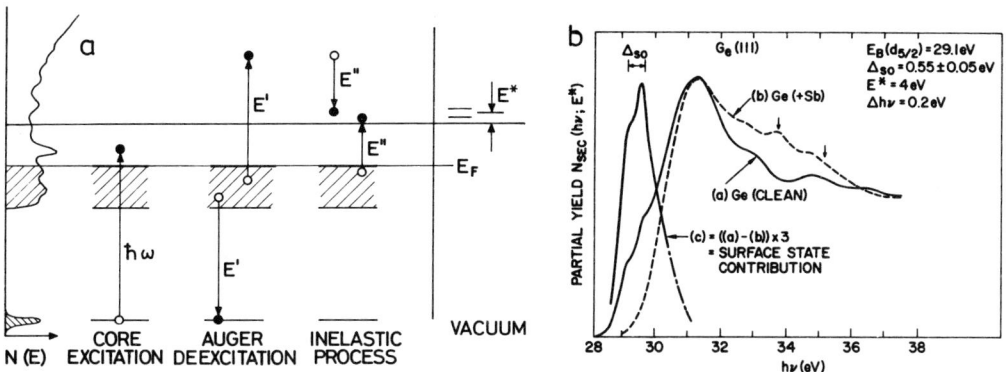

Fig. 5.19 a,b Illustrating the principle of partial yield spectroscopy. (a) A core hole is created by excitation to empty states just above the Fermi level. Deexcitation and inelastic processes lead to low-energy secondaries, which are observed in the energy window E^*. As the photon energy $\hbar\omega$ is varied, the emitted intensity at E^* reflects the number of core holes created and therefore the density of unoccupied states just above the Fermi level. A typical example of the observation of empty surface states on a Ge(111) surface is shown in (b) (from [5.31])

the number of electrons observed in the energy window E^* is proportional to the number of core vacancies created within a secondary electron escape length from the surface. This number of core vacancies is in turn proportional to the number of final states available. As the excitation energy is varied, the spectrum observed therefore reflects the unoccupied density of states just above the Fermi level [5.31]. Typical experimental results for Ge(111) are shown in Fig.5.19b for excitation from the 3d core levels [5.174]. In this figure curve (a) shows a surface state feature at about 29 eV (the doublet structure results from the 0.55 eV spin-orbit splitting of the core levels). Curve (b) is the result after removal of most of these intrinsic surface states by a monolayer of Sb. The difference between curves (a) and (b) leads to the spectral density of surface states (curve (c) enlarged), containing the doublet feature still due to primary excitations from d 5/2 and 3/2 states.

LAPEYRE and ANDERSON [5.176] have introduced an alternative type of partial yield spectroscopy in their study of GaAs (110). In this *constant initial state spectroscopy*, the "initial" state is kept fixed by varying the photon energy and detector band pass E^* synchronously such that, for primary optical excitation, a fixed initial state is observed. Their GaAs spectra exhibit two sharp peaks with the typical separation of the Ga 3d-core levels, occuring at photon energies corresponding to an optical transition from the core levels to an unoccupied surface state as seen by EASTMAN and FREEOUF [5.174]. These features are fairly independent of the constant initial state energy chosen, so that it appears that the energy of the core level-surface state transition is simply transferred to valence electrons. Consequently Lapeyre and Anderson proposed a direct-recombination process, in which the core hole and the electron recombine and transfer their total energy to valence electrons.

5.4.4 Spin-Polarized Photoemission

One observable parameter in a photoemission experiment has not been considered at all in the previous sections: the spin of the emitted electrons. It is not expected that the spin direction is severely affected by the photoejection process, since the time scales involved in the emission process are small compared to spin relaxation times. Consequently a preferential spin orientation, that is a spin polarization, should occur either in photoemission from magnetic materials, with a magnetic axis imposed by an external magnetic field, or in photoemission by circularly polarized light from materials with strong spin-orbit coupling, where optical selection rules allow excitation selective to the spin parameter. Most experimental and theoretical work has focused on bulk effects in spin-polarized photoemission, and reviews on this topic are available [5.42,117]. Recently the applicability of such methods has also been extended to surface investigations.

Spin-polarized photoemission experiments on magnetic solids can be expected to provide information on the magnetization of the initial states from which the electrons originate. The electron spin polarization P may be defined as the expectation value of the Pauli spin operator σ which, for say the z direction, is $\langle\sigma_z\rangle = p = (n_\uparrow - n_\downarrow)/(n_\uparrow + n_\downarrow)$, where $n_\uparrow$ and $n_\downarrow$ are the number of electrons with magnetic moment aligned, respectively, parallel and antiparallel (spin up, spin down) to the magnetic field applied along the z direction. By measuring both P and the total photoyield Y as a function of the optical excitation energy with constant magnetic field H, it is possible to correlate, for example, changes in the yield curve with the onset of transitions from particular bands, provided that they are well separated in energy. Associated changes in P can then be used to identify q in a rather simple way, the magnetic and nonmagnetic electron states [5.178]. It would be preferable, of course, to measure directly the energy distribution of the photoelectrons together with their spin polarization, since the spin of the initial state could then be assigned in a detailed way. So far experimental difficulties have prevented such a development. In these experiments on magnetic materials [5.42], the sample is held well below the Curie temperature and is magnetized perpendicular to the illuminated face so that the resulting photoemitted electrons are longitudinally spin polarized. A subsequent 90° electrostatic deflection of the electron beam then translates this into a transverse spin polarization so that, after acceleration to ~ 100 keV, the degree of polarization can be determined by Mott scattering from a gold foil.

Measurements of the spin polarization P as a function of the external magnetic field H determine, in the absence of spin-flip scattering, the relative magnetization of the sample for a region within the photoelectron escape depth from the surface. These photoelectric magnetization curves can therefore be used to determine departures from bulk magnetization behavior in the surface region. Further, if the polarization is measured as a function of temperature, with the photon energy and the magnetic

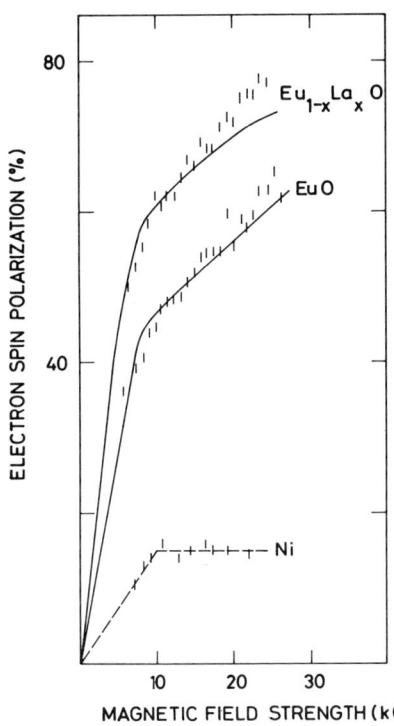

Fig. 5.20 Example of spin-polarized photoemission in surface spectroscopy. The degree of spin polarization of photoemitted electrons from EuO and Ni is measured as a function of the external magnetic field. Whereas for Ni the polarization saturates at the bulk saturation field, this is not the case for EuO, where spin-exchange scattering takes place in a paramagnetic surface layer (from [5.42])

field fixed, it is possible to observe the effect of thermal disorder on the magnetization. The application of these various types of measurement can be illustrated using the example of EuO. Fig.5.20 shows the photoelectron magnetization curve for EuO and, for comparison, that of Ni [5.179]. The polycrystalline Ni was magnetized perpendicular to the plane of the thin film. The photoelectron spin polarization is found to increase with the magnetic field, saturating at about 10 kG as anticipated on the basis of the known bulk magnetization and the demagnetization factor for the geometry of the foil sample. In contrast, the spin polarization of EuO does not saturate, although the discontinuity around 10 kG suggests that the bulk EuO has saturated. The field dependence of P resembles the magnetization curve of a paramagnetic surface layer on a ferromagnetic substrate. However, for the excitation energies used ($\hbar\omega < 6$ eV), the photoelectrons have a mean free path of the order of 50 Å in EuO. In this case therefore, the photoelectric magnetization curves should reflect primarily the bulk ferromagnetic properties rather than those of the one or two paramagnetic surface layers.

What appears to be happening in the saturation region, at these photon energies, is that fully polarized electrons are being photoexcited into the bulk conduction band out of the half filled 4f band of Eu^{2+}, located some 2 eV below the Fermi level [5.180, 181]. They have a large mean free path since they have insufficient energy to cause pair production by inelastic scattering with the oxygen 2p valence electrons located in a 3 eV wide band starting about 4 eV below the Fermi level. These spin-polarized photolectrons from the 4f band do, however, suffer strong spin-exchange scattering with the nonsaturated 4f spins of the paramagnetic surface layer [5.182]. They are then emitted together with those photoelectrons excited directly in the surface layer, the ratio of the bulk emitted electrons to those from the surface being simply the ratio of the thickness of the unsaturated surface layer to the electron mean free path. Assuming that the unsaturated surface layer is one monolayer thick, only about 10% of the photocurrent will be due to emission of partially polarized electrons from this layer. The major contribution to the depolarization is therefore the spin-flip scattering of the bulk electrons at the surface. The spin-flip rate decreases with increasing magnetization of the surface layer, as indicated by the continuing rise in P with H of Fig.5.20. It can also be decreased by increasing the surface magnetization which may be achieved by doping with lanthanum to give an increase in the magnetic coupling of the localized Eu^{2+} 7/2 spins.

Measurements of the spin polarization, photoelectron magnetization and yield can be extremely useful in determining the spin assignment of electron states in magnetic materials [5.178], in determining the effect of disorder [5.183] and the behavior of the bulk and surface magnetization when the photoelectron mean free path is used as a parameter [5.184]. Some care is however necessary in the interpretation of spin photoemission data of magnetic materials. Discrepancies between theory and experiment on both the degree and sign of the polarization are noticeable in the case of the 3d transition metal ferromagnets [5.185,186]. Although electron-magnon scattering cannot be ruled out as a source of some of these discrepancies, the modification of the bulk band structure due to the surface and the possible presence of intrinsic surface states are factors which cannot be ignored [5.187]. Finally, the magnitude and shape of the spin-polarized yield curves may be distorted by the contributions to the total yield from the intrinsic surface photoelectric effect discussed in Subsection 5.3.1. This may be particularly important in the threshold region, which is that region where, for measurements on Fe, Co, and Ni, the major discrepancies between theory and experiment appear to lie.

The possibility of observing spin-polarized photoemission from nonmagnetic solids was investigated after the prediction by FANO [5.188] of a spin polarization (> 85%) in the photoionization of alkali atoms by circularly polarized light and its experimental confirmation [5.189]. In this case the polarization is due to the dependence of the optical absorption on the photon angular momentum direction and the spin-orbit coupling in the p continuum final states. In solid alkali films the polarization was found to be less than 5% when the sample was illuminated by circularly polarized light [5.190,191]. KOYAMA and MERZ [5.192] used a group theoretical approach to determine the selection rules for the spin states of photoelectrons emitted from solid cesium. They concluded that a high degree of polarization should be obtained for particular transitions. The experimental observation of low values of polarization was probably due to the inability to measure specific final state energies and specific $\underline{k}$ values. The same type of measurement, using circularly polarized light and spin-orbit split bands, has been performed on negative electron affinity GaAs, in which spin polarization of up to 45% was obtained at photoyields of a few percent. Maximum polarization was obtained at threshold, where the negative electron affinity allows photoemission directly from the conduction band minimum at the Brillouin zone centre. In this particular case, both the kinetic energy and the region of k-space that is sampled are restricted. A high degree of polarization was maintained for sample temperatures up to 77 K, indicating that GaAs is likely to find application as a spin-polarized electron source, particularly since the polarization direction can be changed simply by changing the optical polarization. Athough the main use of high intensity spin-polarized electron sources has so far been in high energy accelerators, applications such as in polarized electron-magnon scattering experiments [5.194] and spin-polarized LEED studies of surfaces via the spin-orbit coupling or by exchange interactions [5.195] are possible. Furthermore, the study of changes in the spin-orbit splittings due to the surface by means of spin-polarized photoemission measurements of the type described above should be possible provided that energy selection of the final state electrons is performed.

References

Edited Volumes and Monographs

5.1   K. Siegbahn, C. Nordling, A. Fahlman, R. Nordberg, K. Hamrin, J. Hedman, G. Johansson, T. Bergmark, S. Karlsson, I. Lindgren, B. Lindberg:*"ESCA Atomic, Molecular and Solid State Structure Studied by Means of Electron Spectroscopy"*, Nova Acta Reg. Soc. Sci. Upsaliensis IV. $\underline{20}$, 1967.

5.2   K. Siegbahn, C. Nordling, G. Johansson, J. Hedman, P.F. Hedén, K. Hamrin, U. Gelius, T. Bergmark, L.O. Werme, R. Manne, Y. Baer: *ESCA Applied to Free Molecules* (North-Holland Publ. Co., Amsterdam 1969)

5.3   W. Dekeyser, L. Fiermans, G. Vanderkelen, J. Vennik (Eds.): *Electron Emission Spectroscopy* (Reidel Publ. Co., Dordrecht 1973)

5.4   J.T. Devreese, A.B. Kunz, T.C. Collins (Eds.): *Elementary Excitations in Solids, Molecules, and Atoms* (Plenum Press, London 1974)

5.5   R. Gomer (Ed.): *Interactions on Metal Surfaces* (Springer, Berlin, Heidelberg, New York 1975)

5.6   E.G. Derouane, A.A. Lucas (Eds.): *Electronic Structure and Reactivity of Metal Surfaces* (Plenum Press, London 1976)

5.7   G. Ertl, J. Küppers: *Low Energy Electrons and Surface Chemistry* (Verlag Chemie, Weinheim 1974)

5.8   T.A. Carlson: *Photoelectron and Auger Spectroscopy* (Plenum, New York 1975)

Conference Proceedings and Special Journal Issues

5.9 D.A. Shirley (Ed.): *Electron Spectroscopy* (North-Holland, Amsterdam 1972)
5.10 J. Electr. Spectr. 5, 1-1136 (1974)
5.11 "Proceedings of the Second International Conference on Solid Surfaces, Kyoto 1974", Japan. J. Appl. Phys., Suppl. 2, Pt. 2 (1974)
5.12 *Photo-Effects in Adsorbed Species*, Faraday Discussions of the Chemical Society 58 (1974)
5.13 E.E. Koch, R. Haensel, C. Kunz (Eds.): *Vacuum Ultraviolet Radiation Physics* (Pergamon Vieweg 1974)
5.14 Physics Today 28, Nr. 4 (1975)

Review Papers

Mainly XPS

5.15 K. Siegbahn, in [Ref. 5.9, p. 15]
5.16 D.W. Langer: Festkörperprobleme XIII, 193 (1973)
5.17 D.T. Clark, in [Ref. 5.3, p. 373]
5.18 C.S. Fadley, in [Ref. 5.3, p. 151]
5.19 K. Siegbahn: J. Electr. Spectr. 5, 3 (1974)
5.20 H. Fellner-Feldegg, U. Gelius, B. Wannberg, A.G. Nilsson, E. Basilier, K. Siegbahn: J. Electr. Spectr. 5, 643 (1974)
5.21 C.S. Fadley, R.J. Baird, W. Siekhaus, T. Novakov, S.A.L. Bergström: J. Electr. Spectr. 4, 93 (1974)
5.22 C.S. Fadley: J. Electr. Spectr. 5, 725 (1974)
5.23 S.B.M. Hagström, C.S. Fadley: In *X-Ray Spectroscopy*, ed. by L.V. Azaroff (McGraw-Hill, New York 1974) p. 379

Mainly UPS

5.24 D.E. Eastman, in [Ref. 5.9, p.487]
5.25 N.V. Smith: CRC Crit. Rev. Solid State Sci. 2, 45 (1971)
5.26 D.E. Eastman: In *Techniques in Metal Research VI*, ed. by E. Passaglia (Interscience, New York 1972) p. 413
5.27 W.E. Spicer: Comments Solid State Phys. 5, 105 (1973)
5.28 N.W. Ashcroft, in [Ref. 5.13, p.533]
5.29 D.E. Eastman, J.E. Demuth: Japan. J. Appl. Phys. Suppl. 2, Pt. 2, p. 827 (1974)
5.30 D.E. Eastman, M.I. Nathan: Phys. Today 28, 44 (1975)
5.31 J.L. Freeouf, D.E. Eastman: CRC Crit. Rev. Solid State Sci. 5, 245 (1975)
5.32 B. Feuerbacher, R.F. Willis: J. Phys. C: Solid State Phys. 9, 169 (1976)
5.33 W.E. Spicer, K.Y. Yu, I. Lindau, P. Pianetta, D.M. Collins: In *Surface and Defect Properties of Solids*, Vol. 5, ed. by J.M. Thomas M.W. Roberts, in preparation

XPS and UPS

5.34 C.R. Brundle: J. Vac. Sci. Technol. <u>11</u>, 212 (1974)
5.35 J.W. Gadzuk: Japan. J. Appl. Phys. Suppl. <u>2</u>, Pt. 2, 851 (1974)
5.36 C.R. Brundle: Surface Sci. <u>48</u>, 99 (1975)
5.37 D.T. Pierce: Acta Electronica <u>18</u>, 1 (1975)
5.38 D.E. Eastman, in [Ref. 5.13, p. 417]
5.39 E.W. Plummer: In *Interactions on Metal Surfaces*, ed. by R. Gomer (Springer, Berlin, Heidelberg, New York 1975) p. 143
5.40 J.W. Gadzuk, in [Ref. 5.6]
5.41 D. Menzel: J. Vac. Sci. Technol. <u>12</u>, 313 (1975)

Spin-Polarized Photoemission

5.42 M. Campagna, D.T. Pierce, K. Sattler, H.C. Siegmann: J. de Physique <u>34</u>, C6-87 (1973)

Original References

5.43 J.A.R. Samson: *Techniques of Vacuum Ultraviolet Spectroscopy* (Wiley, New York 1967)
5.44 J.K. Cashion, J.L. Mees, D.E. Eastman, J.A. Simpson, C.E. Kuyatt: Rev. Sci. Instr. 42, 1670 (1971)
5.45 C.R. Brundle, M.W. Roberts, D. Latham, K. Yates: J. Electr. Spectr. <u>3</u>, 241 (1974)
5.46 J.E. Rowe, S.B. Christman, E.E. Chaban: Rev. Sci. Instr. <u>44</u>, 1675 (1973)
5.47 R.T. Poole, J. Liesegang, R.C.G. Leckey, J.G. Jenkin: J. Electr. Spectr. <u>5</u>, 773 (1974)
5.48 F. Burger, J.P. Maier: J. Electr. Spectr. <u>5</u>, 783 (1974)
5.49 F. Wuilleumier, M.O. Krause: J. Electr. Spectr. <u>5</u>, 921 (1974)
5.50 K. Siegbahn, D. Hammond, H. Fellner-Feldegg, E.F. Barnett: Science <u>176</u>, 245 (1972)
5.51 U. Gelius, E. Basilier, S. Svensson, T. Bergmark, K. Siegbahn: J. Electr. Spectr. <u>2</u>, 405 (1972)
5.52 R. Haensel, C. Kunz: Z. Angew. Phys. <u>23</u>, 276 (1967)
5.53 R.P. Godwin: Springer Tracts Mod. Phys. <u>51</u>, 1 (1969)
5.54 F.C. Brown: Solid State Phys. <u>29</u>, 1 (1974)
5.55 C. Kunz, in [Ref. 5.13, p. 753 and other papers in chapter 10 of Ref. 5.13]
5.56 R.Z. Bachrach, S.B.M. Hagström, F.C. Brown, in [Ref. 5.13, p. 795]
5.57 H. Dietrich, C. Kunz: Rev. Sci. Instr. <u>43</u>, 434 (1972)
5.58 F.C. Brown, R.Z. Bachrach, S.B.M. Hagström, N. Lien, C.H. Pruett, in [Ref. 5.13, p. 785]
5.59 K. Thimm: J. Electr. Spectr. <u>5</u>, 755 (1974)

5.60 H.A. Bethe, E.E. Salpeter: In *Handbuch der Physik* Vol. 35 (Springer, Berlin, Heidelberg, New York 1957) p. 88
5.61 G.F. Bassani: In *The Optical Properties of Solids*, ed. by J. Tauc (Academic Press, New York 1966) p. 33
5.62 G.D. Mahan: Phys. Rev. B 2, 4334 (1970)
5.63 W.L. Schaich, N.W. Ashcroft: Solid State Commun. 8, 1959 (1970)
5.64 W.L. Schaich, N.W. Ashcroft: Phys. Rev. B 3, 2452 (1971)
5.65 H. Hermeking: J. Phys. C 6, 2989 (1973)
5.66 C. Caroli, D. Lederer-Rozenblatt, B. Roulet, D. Saint-James: Phys. Rev. B 8, 4552 (1973)
5.67 P.J. Feibelman, D.E. Eastman: Phys. Rev. B 10, 4932 (1974)
5.68 H. Hermeking, R.P. Wehrum: J. Phys. C: Solid State Phys. 8, 3468 (1975)
5.69 K.L. Kliewer: Phys. Rev. Lett. 33, 900 (1974)
5.70 K.L. Kliewer, Phys. Rev. B 14, 1412 (1976)
5.71 P.J. Feibelman: Phys. Rev. Letters 34, 1092 (1975)
5.72 P.J. Feibelman: Phys. Rev. B 12, 1319 (1975)
5.73 B. Feuerbacher, B. Fitton: Phys. Rev. Lett. 30, 923 (1973)
5.74 N.E. Christensen, B. Feuerbacher: Phys. Rev. B 10, 2349 (1974)
5.75 B. Feuerbacher, N.E. Christensen: Phys. Rev. B 10, 2373 (1974)
5.76 J.B. Pendry: *Low Energy Electron Diffraction* (Academic Press, London 1974)
5.77 J.W. Gadzuk: *Time Scales and Electron Relaxation Processes*, preprint
5.78 P.O. Nilsson, L. Ilver: Solid State Commun. 17, 667 (1975)
5.79 D.R. Lloyd, C.M. Quinn, N.V. Richardson: J. Phys. C: Solid State Phys. 8, L 371 (1975)
5.80 J. Freeouf, M. Erbudak, D.E. Eastman: Solid State Commun. 13, 771 (1973)
5.81 D.C. Langreth: Phys. Rev. B 3, 3120 (1971)
5.82 F.J. Himpsel, W. Steinmann: Phys. Rev. Lett. 35, 1025 (1975)
5.83 K. Mitchell: Proc. Roy. Soc. A 146, 442 (1934)
5.84 I. Adawi: Phys. Rev. 134, A 788 (1964)
5.85 R.E.B. Makinson: Proc. Roy. Soc. A 162, 367 (1937)
5.86 J.G. Endriz: Phys. Rev. B 7, 3464 (1973)
5.87 T. Koopmans: Physica 1, 104 (1933)
5.88 P.S. Bagus: Phys. Rev. 139, A 619 (1965)
5.89 D.A. Shirley: Chem. Phys. Lett. 16, 220 (1972)
5.90 P.H. Citrin, D.R. Hamann: Chem. Phys. Letters 22, 301 (1973)
5.91 L. Ley, S.P. Kowalczyk, F.R. McFeely, R.A. Pollak, D.A. Shirley: Phys. Rev. B 8, 2392 (1973)
5.92 P.H. Citrin, D.R. Hamann: Phys. Rev. B 10, 4948 (1974)
5.93 J.W. Gadzuk: J. Vac. Sci. Technol. 12, 289 (1975)
5.94 L. Ley, F.R. McFeely, S.P. Kowalczyk, J.G. Jenkin, D.A. Shirley: Phys. Rev. B 11, 600 (1975)

5.95   L. Hedin, S. Lundqvist: Solid State Phys. 23, 1 (1969)
5.96   J.W. Gadzuk, M. Šunjić: Phys. Rev. B 12, 524 (1975)
5.97   H.W. Meldner, J.D. Perez: Phys. Rev. A 4, 1388 (1971)
5.98   U. Gelius: Physica Scripta 9, 133 (1974)
5.99   D.A. Shirley: J. Electr. Spectr. 5, 135 (1974)
5.100 E.W. Plummer, A.E. Bell: J. Vac. Sci. Technol. 9, 583 (1972)
5.101 D.R. Penn, E.W. Plummer: Phys. Rev. B 9, 1216 (1974)
5.102 D.R. Penn: Phys. Rev. Lett. 28, 1041 (1972)
5.103 T.E. Einstein: Phys. Rev. B 12, 1262 (1975)
5.104 H.D. Hagstrum: Surface Sci. 54, 197 (1976)
5.105 B.J. Waclawski, J.F. Herbst: Phys. Rev. Lett. 35, 1594 (1975)
5.106 B. Feuerbacher, B. Fitton: Phys. Rev. B 8, 4890 (1973)
5.107 B. Feuerbacher, M.R. Adriaens: Surface Sci. 45, 553 (1974)
5.108 T.B. Grimley: J. Vac. Sci. Technol. 8, 31 (1971)
5.109 J.R. Schrieffer: J. Vac. Sci. Technol. 9, 561 (1972)
5.110 S.K. Lyo, R. Gomer, in [Ref. 5.5, p. 40]
5.111 J.R. Schrieffer, P. Soven: Physics Today 28, 24 (1975)
5.112 J.W. Gadzuk: In *Surface Physics of Materials*, Vol.II ed. by J.M. Blakely (Academic Press, New York 1975)
5.113 R.P. Messmer, C.W. Tucker, K.H. Johnson: Surface Sci. 42, 341 (1974)
5.114 I.P. Batra, D. Robaux: Surface Sci. 49, 653 (1975)
5.115 J. Harris, G.S. Painter: Phys. Rev. Lett. 36, 151 (1976)
5.116 D.E. Eastman, J.K. Cashion: Phys. Rev. Lett. 27, 1520 (1971)
5.117 I.P. Batra, P.S. Bagus: Solid State Commun. 16, 1097 (1975)
5.118 D.W. Turner, A.D. Baker, C. Baker, C.R. Brundle: *Molecular Photoelectron Spectroscopy* (Wiley, London 1970)
5.119 T.B. Grimley: Proc. Phys. Soc. 92, 776 (1967)
5.120 D.M. Newns: Phys. Rev. 178, 1123 (1969)
5.121 W. Brenig, K. Schönhammer: Z. Physik 267, 201 (1974)
5.122 J.W. Gadzuk: Surface Sci. 43, 44 (1974)
5.123 S.K. Lyo, R. Gomer: Phys. Rev. B 10, 4161 (1974)
5.124 E.A. Hyman: Phys. Rev. B 11, 3739 (1975)
5.125 A. Madhukar, B. Bell: Phys. Rev. Lett. 34, 1631 (1975)
5.126 J.W. Gadzuk: Phys. Rev. B 10, 5030 (1974)
5.127 T.L. Einstein, J.R. Schrieffer: Phys. Rev. B 7, 3629 (1973)
5.128 F. Cyrot-Lackmann, M.C. Desjonquères, J.P. Gaspard: J. Phys. C: Solid State Phys. 7, 925 (1974)
5.129 A. Liebsch: Phys. Rev. Lett. 32, 1203 (1974)
5.130 J.W. Gadzuk: Solid State Commun. 15, 1011 (1974)
5.131 T.B. Grimley: J. Phys. C (in press)
5.132 J.B. Pendry: J. Phys. C: Solid State Phys. 8, 2413 (1975)

5.133 B.W. Holland: J. Phys. C: Solid State Phys. 8, 2679 (1975)
5.134 C.R. Brundle, in [Ref. 5.6]
5.135 J. Bardeen: Phys. Rev. 71, 717 (1947)
5.136 D.E. Eastman, W.D. Grobman: Phys. Rev. Lett. 28, 1378 (1972)
5.137 L.F. Wagner, W.E. Spicer: Phys. Rev. Lett. 28, 1381 (1972)
5.138 J.E. Rowe, J.C. Phillips: Phys. Rev. Lett. 32, 1315 (1974)
5.139 J.E. Rowe, H. Ibach: Phys. Rev. Lett. 32, 421 (1974)
5.140 J.E. Rowe, H. Ibach, H. Froitzheim: Surface Sci. 48, 44 (1975)
5.141 J.T. Yates, T.E. Madey, N.E. Erickson: Surface Sci. 43, 257 (1974)
5.142 J.T. Yates, N.E. Erickson: Surface Sci. 44, 489 (1974)
5.143 J.C. Fuggle, T.E. Madey, M. Steinkilberg, D. Menzel: Phys. Lett. 51A, 163 (1975)
5.144 J.E. Demuth, D.E. Eastman: Phys. Rev. Lett. 32, 1123 (1974)
5.145 D.E. Eastman, J.E. Demuth: Japan. J. Appl. Phys. Supp. 2. p.2, 827 (1974)
5.146 E.W. Plummer, B.J. Waclawski, T.V. Vorburger: Chem. Phys. Lett. 28, 510 (1974)
5.147 K.C. Pandey, T. Sakurai, H.D. Hagstrum: Phys. Rev. Lett. 35, 1728 (1975)
5.148 R.M. More: Phys. Rev. B 9, 392 (1974)
5.149 R.F. Willis, B. Feuerbacher: Surface Sci. 53, 144 (1975)
5.150 J.E. Rowe, M.M. Traum, N.V. Smith: Phys. Rev. Lett. 33, 1333 (1974)
5.151 B.J. Waclawski, E.W. Plummer: Phys. Rev. Lett. 29, 783 (1972)
5.152 B. Feuerbacher, B. Fitton: Phys. Rev. Lett. 29, 786 (1972)
5.153 R. Feder, K. Sturm: Phys. Rev. B 12, 537 (1975)
5.154 H.P. Bonzel, C.R. Helms, S. Kelemen: Phys. Rev. Lett. 35, 1237 (1975)
5.155 P.O. Gartland, B.J. Slagsvold: Phys. Rev. B 12, 4047 (1975)
5.156 E.W. Plummer, B.J. Waclawski: private communication
5.157 B.J. Waclawski, T.V. Vorburger, R.J. Stein: J. Vac. Sci. Technol. 12 301 (1975)
5.158 W.F. Egelhoff, D.L. Perry: Phys. Rev. Lett. 34, 93 (1975)
5.159 B. Feuerbacher, R.F. Willis: Phys. Rev. Lett. 36, 1336 (1976)
5.160 C.N. Berglund, W.E. Spicer: Phys. Rev. 136, A 1030; A 1044 (1964)
5.161 J.R. Smith, in [Ref. 5.5, p. 1]
5.162 R.H. Fowler: Phys. Rev. 38, 45 (1931)
5.163 D.E. Eastman: Phys. Rev. B 2, 1 (1970)
5.164 L.D. Schmidt, in [Ref. 5.5, p. 63]
5.165 E.O. Kane: Phys. Rev. 127, 131 (1962)
5.166 J.M. Ballantine: Phys. Rev. B 6, 1436 (1972)
5.167 G.W. Gobeli, F.G. Allen: Phys. Rev. 127, 141 (1962)
5.168 F.G. Allen, G.W. Gobeli: Phys. Rev. 127, 150 (1962)
5.169 See for example, T.E. Fischer: Surface Sci. 10, 399 (1968); 13, 30 (1969); R.M. Broudy: Phys. Rev. B 1, 3430 (1970)
5.170 C. Sebenne, D. Bolmont, G. Guichar, M. Balkanski: Phys. Rev. B 12, 3280 (1975)

5.171 N. Schwentner, M. Skibowski, W. Steinmann: Phys. Rev. B $\underline{8}$, 2965 (1973)
5.172 F.G. Allen, G.W. Gobeli: J. Appl. Phys. $\underline{35}$, 597 (1964)
5.173 W. Gudat, C. Kunz: Phys. Rev. Lett. $\underline{29}$, 169 (1972)
5.174 D.E. Eastman, J.L. Freeouf: Phys. Rev. Lett. $\underline{33}$, 1601 (1974)
5.175 D.E. Eastman, J.L. Freeouf: Phys. Rev. Lett. $\underline{34}$, 1624 (1975)
5.176 G.J. Lapeyre, J. Anderson: Phys. Rev. Lett. $\underline{35}$, 117 (1975)
5.177 G. Busch, M. Campagna, C.H. Siegmann: J. Appl. Phys. $\underline{41}$, 1044 (1970)
5.178 F. Meier, W. Eib, D.T. Pierce: Solid State Commun. $\underline{16}$, 1089 (1975)
5.179 K. Sattler, C.H. Siegmann: Phys. Rev. Lett. $\underline{29}$, 1565 (1972)
5.180 D.E. Eastman, F. Holtzberg, S. Methfessel: Phys. Rev. Lett. $\underline{23}$, 226 (1970)
5.181 P. Cotti, P. Munz: Phys. Cond. Matter $\underline{17}$, 307 (1974)
5.182 J.S. Helman, H.C. Siegmann: Solid State Comm. $\underline{13}$, 891 (1973)
5.183 U. Bänninger, G. Busch, M. Campagna, H.C. Siegmann: Phys. Rev. Lett. $\underline{25}$, 585 (1970)
5.184 D.T. Pierce, H.C. Siegmann: Phys. Rev. B $\underline{9}$, 4035 (1974)
5.185 H. Alder, M. Campagna, H.C. Siegmann: Phys. Rev. B $\underline{8}$, 2075 (1973)
5.186 N.V. Smith, M.M. Traum: Phys. Rev. Lett. $\underline{27}$, 1388 (1971)
5.187 D.G. Dempsey, L. Kleinman, E. Caruthers: Phys. Rev. B $\underline{13}$, 1489 (1976)
5.188 U. Fano: Phys. Rev. $\underline{178}$, 131 (1969)
5.189 U. Heinzmann, J. Kessler, J. Lorenz: Phys. Rev. Lett. $\underline{25}$, 1325 (1970)
5.190 U. Heinzmann, J. Kessler, B. Ohnemus: Phys. Rev. Lett. $\underline{27}$, 1696 (1971)
5.191 U. Heinzmann, K. Jost, J. Kessler, B. Ohnemus: Z. Physik $\underline{251}$, 354 (1972)
5.192 K. Koyama, H. Merz: Z. Physik B $\underline{20}$, 131 (1975)
5.193 D.T. Pierce, F. Meier, P. Zürcher: Phys. Lett. $\underline{51A}$, 465 (1975)
5.194 R.E.De Wames, L.A. Vredevoe: Phys. Rev. Lett. $\underline{23}$, 123 (1969)
5.195 R. Feder: Surface Sci. $\underline{51}$, 297 (1975)
5.196 W. Gudat, C. Kunz, in [Ref. 5.13, p. 392]

6. Electron Energy Loss Spectroscopy

H. Froitzheim

With 39 Figures

Electron energy loss spectroscopy of solids (ELS) has been developed mainly by the experimental work of RAETHER et al. [6.1], BOERSCH [6.2] and GEIGER [6.3] and theoretically among others by the work of PINES [6.4], BOHM [6.5], FERREL [6.6], WATANABE [6.7] and RITCHIE [6.8]. These papers were mainly concerned with bulk properties. So the experiments were carried out by transmitting electrons of about 50 keV primary energy through thin layers. At this time the experiments were primarily understood in terms of the classical "Dielectric Theory" which is under certain conditions still valid today. Besides some excitations in the bulk several surface effects could be observed in these experiments, but the interest in these effects did not arise before the interest in surface physics was developed. Consequently the development of ELS of solid state surfaces was strongly connected with the reproducible preparation of well-defined solid surfaces which is possible with today's UHV techniques. Because of the scattering mechanism, and not because of the penetration depth of electrons as is often believed, primary energies of less than 1 keV are particularly advantageous when studying surface effects. This means that one has to work in the reflection mode. As the topic of this chapter is ELS of surfaces, only reflection kinematics will be discussed here.

This chapter is organized as follows. In Section 6.1 we first discuss the general model considerations of the "Dielectric Theory", the best available description of ELS up to now. Here we discuss the cross section for each type of excitation of interest in ELS. We finish this section with a discussion about quantitative data reduction. As Section 6.1 contains only little information about experimental data it is of interest only for a better theoretical understanding of the presented experiments. Section 6.2 deals with selected experiments of ELS of collective vibrational states of clean semiconductor surfaces. In Section 6.3 ELS experiments of vibrational states of gas-covered surfaces are presented. Sections 6.4 and 6.5 describe experiments of ELS concerning the excitation of the electronic structure of clean and gas-covered semiconductor surfaces and metal surfaces, respectively. The experiments were selected with regard to the fundamental differences between several experimental techniques in use today and with emphasis on the difference between the materials and methods. The discussions of the experiments are based on the theoretical background given in Section 6.1 but they may be understood also without knowledge of this section.

6.1 Definition of ELS

For the scattering of electrons from a solid state surface with a two-dimensional periodic lattice one has to consider the energy and momentum conservation which are for the case of reflection of the form

$$E^S(\underline{K}^S) = E_0(\underline{K}) - \hbar\omega \quad \text{(energy conservation)} \tag{6.1}$$

$$K^S_{\shortparallel} = K_{\shortparallel} - q_{\shortparallel} \pm G_{\shortparallel} \quad \text{(wave number conservation)} \tag{6.2}$$

$E_0(\underline{K})$ is the energy of the primary electron with momentum $\hbar\underline{K}$, $E^S(\underline{K}^S)$ the energy of the reflected electron with momentum $\hbar k^S$. $\hbar\omega$ means the energy loss of the electron or the energy transfer to the solid. For the reflection from a surface the momentum or wave number conservation is valid only in two dimensions, e.g., parallel to the surface because of the cutoff of periodicity normal to the surface. This is indicated by "parallel" indices. Therefore $q_{\shortparallel}$ measures the momentum transfer along the surface and $G_{\shortparallel}$ means any reciprocal lattice vector determined by a surface lattice unit mesh. Because of (6.1) and (6.2) one can define three independent measurements.

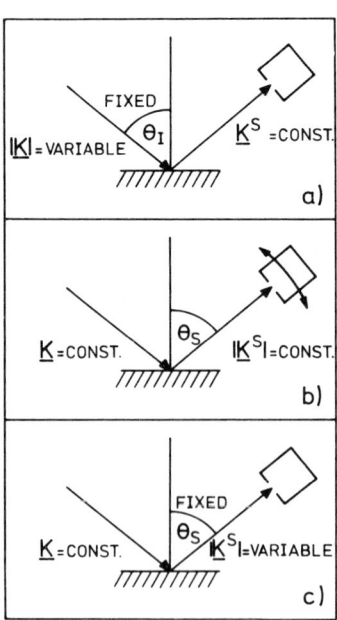

Fig. 6.1a-c Scheme of the three independent measurements which as the whole define ILEED

a) Measuring a spectrum of backscattered electrons with fixed angle of incidence θ_I, angle of reflection θ_s and $\hbar\omega$ while the primary energy E_0 (see Fig.6.1a) is changed.

b) Measuring a spectrum with fixed $\underline{K}, E_0$ and $\hbar\omega$ but variable $\underline{K}^s$. This can be done by moving the analyzer whereas all the other parameters are held fixed (see Fig.6.1b).

c) Measuring a spectrum with fixed $\underline{K}$, E_0 and θ_s but with variable $\hbar\omega$ (Fig.6.1c). The whole set of these three measurements is known under the term ILEED (see Chap.3 of this volume). The present chapter deals only with the measurement defined by c). Although the angular resolution is mentioned above, it is of interest only under certain circumstances, e.g., if one is interested in a quantitative analysis of the loss spectra.

6.2 Theory of Inelastic Scattering

A theory of electron scattering should provide a complete description of all effects which may occur in the set of measurements defined in Section 6.1. The development of such a first principle theory has not been completed up to now. However, numerous papers dealing with the dynamic LEED theory have influenced the theoretical description of ELS. The most important works of the former are those of DAVISSON and GERMER [6.9], DUKE et al. [6.10] and WEBB et al. [6.11]. In these publications the two step model has been developed involving either an energy loss followed by subsequent diffraction or v. vs. This plays an essential role in the quantum mechanical treatments of the ELS problem. In the following section the concept of the classical dielectric theory is presented and afterwards a recent quantum mechanical description is discussed in more detail. The latter is the only theory which permits a fair description of experimental results. Under certain conditions even a quantitative analysis of the measured data is possible with this theory (compare Subsecs. 6.3.2 and 6.5.3).

6.2.1 The Classical Theory (Concept of the "Dielectric Theory")

Despite great progress in the development of a dynamical theory of the scattering of electrons from solid surfaces, it seems that the dielectric theory is still the best available description of ELS experiments. This theory was developed by FERMI [6.12], HUBBARD [6.13], FRÖHLICH [6.14] and PELZER [6.15] and reviewed by RAETHER [6.1] and GEIGER [6.3]. In this theory the scattering cross section of electrons is related to the dielectric response of the system. The dielectric constant $\varepsilon(q,\omega) = \varepsilon_1(q,\omega) + i\varepsilon_2(q,\omega)$ is a function of q and ω and characteristic for each solid. In the case that the "transverse" dielectric constant is identical to the "longitudinal" dielectric constant which comes into play when electric waves of longitudinal character act on the system of atoms, it describes also the response to a perturbation by a (nonrelativistic) moving electron. The interaction can be described by the following model: The Coulomb field accompanying the moving (scattering) electron during its approach to the solid interacts with the electron gas of the solid via long-range dipole fields. This interaction creates a space- and time-dependent polarization field which in turn damps the motion of the electron. The polarization field, on the other hand, can be decomposed by a Fourier expansion into plane waves which are damped inside a dielectric medium proportional to ε_2. Because of this process the energy gained from the relaxation of the moving electron will be transfered to the medium. In this model it is easy to derive an expression for the energy loss of a scattered electron which of course does not consider the kinematics of the scattering. Within the bulk of a dielectric medium the amplitude of the field of an electron is screened by a factor of $1/\varepsilon$, the intensity by a factor $1/|\varepsilon|^2$. If the field moves through a medium then its damping will be proportional to ε_2. Thus, for the energy loss one obtains

$$W_b(q,\omega) \propto \frac{\varepsilon_2}{|\varepsilon|^2} = -\mathrm{Im}\left\{\frac{1}{\varepsilon}\right\} = \frac{\varepsilon_2}{\varepsilon_1^2 + \varepsilon_2^2} \quad . \tag{6.3}$$

In the case of reflection from a dielectric half space the field inside the medium which is set up by a moving electron outside is screened by a factor of $1/\varepsilon+1$ by polarization. It then follows that the energy loss of the reflected electron is proportional to

$$W_s(q,\omega) \propto \frac{\varepsilon_2}{|\varepsilon+1|^2} = -\mathrm{Im}\left\{\frac{1}{\varepsilon+1}\right\} = \frac{\varepsilon_2}{(\varepsilon_1+1)^2 + \varepsilon_2^2} \quad . \tag{6.4}$$

The two functions $-\mathrm{Im}\{1/\varepsilon\}$ and $-\mathrm{Im}\{1/\varepsilon+1\}$ are called the volume loss function and the surface loss function, respectively. They entirely describe the fundamental structure of the loss spectra, because they contain the whole information of the dielectric behavior of the medium. As we shall see later the q-dependence of $\varepsilon(q,\omega)$ is of no importance as long as $E_o \gg \hbar\omega$. As can be seen from (6.3) and (6.4) the loss functions may exhibit maxima which derive either from ε_2 or from the denominator ($\varepsilon_1 = 0$ and $\varepsilon_1 = -1$, respectively). This behavior can be interpreted by the fact that a longitudinal plasma wave traversing a crystal produces density fluctuations (see Fig.6.2) of its charges (in the direction of propagation) thus bringing into play the long-range Coulomb forces between positive and negative charges (collective effect) accompanied by the excitation of collective oscillations like phonons or plasmons in the case of a free electron gas. If there are single excitations possible like intra- and/or interband transitions the plasma frequency is changed by the eigenfrequencies of the crystal electrons. The electron loss spectrum thus contains excitations of collective oscillations as well as additional single electron excitations. Their intensity is given by the maxima of ε_2 reduced by the factor $1/|\varepsilon|^2$ and their position in the loss

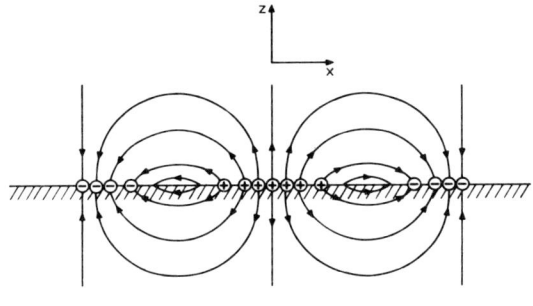

$E_x = +E_0 \sin(qx - \omega t) e^{-q|z|}$

$E_z = \pm E_0 \cos(qx - \omega t) e^{-q|z|}$

Fig. 6.2 Electric field produced by a long wave length surface charge fluctuation wave. Inside the material the arrows indicate the direction of polarization. The shape of the field strongly depends on the wave vector. The above shape corresponds to $q_{\shortparallel} \simeq 0$. For larger $q_{\shortparallel}$ the field becomes unbalanced [6.89-90]. Surface waves are transverse in the sense that $\nabla \times E = 0$; however, they interact with charged particles because of the finite size of the probes and because of retardation effects [6.31]. They may also be excited via bulk waves with higher q-vector [6.90]

function curve is therefore slightly displaced compared to the ε_2 curve (i.e., to the optical adsorption). The fact that loss functions calculated from optical data agree well with those observed in electron scattering experiments at a momentum transfer of $q \simeq 0$ demonstrates the identity of the "transverse" and the "longitudinal" dielectric constant when the approximation $\varepsilon(q,\omega) \simeq \varepsilon(0,\omega)$ is valid (see Subsecs. 6.5.3-5).

In the dielectric theory it is shown that the eigensolutions of longitudinal polarization waves are given by the condition $\varepsilon(q,\omega) = 0$ in the bulk and $\varepsilon(q,\omega) = -1$ for waves localized in the surface region of semifinite solids. These waves contain all possible excitations of the solid as a response to an external electromagnetic perturbation in the form of characteristic dispersion relations. According to this model the scattering electron cannot "distinguish" between the special excitation which is demonstrated by the fact that the dependence of the cross section on the kinematical parameters is exactly the same for collective and single excitations (compare with Subsecs. 6.2.3 and 6.2.5). It should also be mentioned that the loss functions exhibit poles exactly where the above conditions for longitudinal excitations are fulfilled, that means at frequencies where $\varepsilon(\omega) = 0$ and $\varepsilon(\omega) = -1$ for bulk and surface waves, respectively. It is clear that this classical description in the framework of dielectric theory in which the moving electron is treated as a point charge reflected from the surface is only valid in the limit K >> q as in transmission experiments. In fact it nearly describes these experiments quantitatively [6.16,17].

The first semiclassical description of reflection was presented by LUCAS et al. [6.18]. In his treatment the electron is dealt with also as a particle and the crystal surface as an infinitely repulsive barrier with a reflection coefficient of unity. Thus the influence of the elastically scattered electrons on the scattering efficiency is neglected. The wave vector conservation with its resulting kinematical features has to be introduced artificially into this treatment because it is contained naturally only in a full quantum mechanical treatment as a consequence of the lattice periodicity. In addition this theory describes only the excitation of a single collective plasmon or phonon mode by a fast moving electron. The expression "fast electron" is a relative term which shall be discussed later on. However, it should be mentioned that the theory of LUCAS et al. [6.18] describes IBACH'S [6.19] measurements (excitations of phonons by 7 eV electrons) quantitatively.

6.2.2 Quantum Mechanical Description of the Dielectric Theory

A full quantum mechanical treatment of the dielectric theory has been presented by EVANS et al. [6.20,21] and MILLS [6.22,23] in a series of papers. These authors started with a very general Schrödinger equation, and there are no restrictions to the nature of the elementary excitations. The only requirement is that excitations have to set up long-range field fluctuations outside the crystal according to the concept of the dielectric theory. With $\hbar = 1$ the equation is of the form

$$\left[-\frac{\nabla^2}{2m} + V_0(\underline{x}) - e\varphi(\underline{x},t) \right] \Psi(\underline{x},t) = i \frac{\partial \Psi}{\partial t}(\underline{x},t) \quad . \tag{6.5}$$

$V_0(\underline{x})$ is a complex optical potential experienced by the electron in the crystal. Its imaginary part describes the absorption of the electrons inside the solid and is therefore related to the mean free path and the dielectric bulk response function (see Chap.1). It is assumed that $V_0(\underline{x})$ depends only on Z, in the spirit of the jellium model. It has a value different from zero only inside the crystal. With $\varphi(\underline{x},t) = 0$ one would have an incident and specular beam only where the intensity of the latter would be decreased according to the value of the imaginary part of $V_0(\underline{x})$. When a two-dimensional lattice periodicity would be included into $V_0(\underline{x})$ the above equation would then describe the LEED problem. Because of the long-range character of the excitations a microscopic picture of $V_0(\underline{x})$ is of little interest. $\varphi(\underline{x},t)$ is a time-dependent electrostatic potential which is experienced by the incoming electron outside the crystal. It may be related to the charge density fluctuations $\rho(\underline{x},t)$ in the crystal. After a Fourier expansion of the Schrödinger equation and conversion to an integral equation which may be solved in an iterative fashion by introducing a Green's function that satisfies the appropriate boundary conditions, the authors represent in the first Born approximation an expression for the scattering cross section

$$\frac{d^2S}{d\omega d\Omega} = \frac{m^2 e^2}{2\pi \cos\theta_I} \left(\frac{K^S}{K}\right) \frac{1}{q_\parallel^2} P(q_\parallel, \omega)$$

$$\times \left[\frac{1}{q_\parallel + iK_z^S + iK_z} + \frac{R_S}{q_\parallel + iK_z - iK_z^S} + \frac{R_I}{q_\parallel + iK_z^S - iK_z} + \frac{R_I R_S}{q_\parallel - iK_z^S - iK_z} \right] \tag{6.6}$$

The integration yielding the above equation was carried out only in the vacuum region. Thus, the scattering of electrons penetrating into the solid is completely neglected. In the above equation θ_I is the angle of incidence, K_z and K_z^S the perpendicular component of the wave vector of the incoming and the scattered electrons, respectively, and R_I and R_S are the complex reflectivity coefficients of the electrons before and after suffering an energy loss. Therefore R_I describes the diffraction-loss and R_S the loss diffraction process. The function $P(q_{\shortparallel},\omega)$ is the surface analog of the structure factor $S(q,\omega)$ appearing in the theory of neutron scattering [6.24]. It describes the effect outside the crystal of the charge density fluctuation inside (compare Fig.6.2). The meaning of the four contributions on the right-hand side of (6.6) is demonstrated in Fig.6.3. According to Fig.6.3a and b each of two terms proportional to R_I and R_S represents a two-step scattering process, namely the loss diffraction and the diffraction-loss process, respectively. The nature of the process illustrated in Fig.6.3c is a one-step process where the electron is reflected directly by a wide angle scattering process, which is independent to the reflection coefficient. Finally the term proportional to $R_I R_S$, which is shown in Fig.6.3d, represents a three-step process, namely a diffraction-loss-diffraction process. The relative magnitude of these four terms can easily be estimated in the case where de Broglie wave length of the electrons is small compared to that of the excited field fluctuations, which is one of the underlying presumptions of the "dielectric theory". In this case one has $\hbar\omega/E_0 \simeq \hbar\omega/E_S \ll 1$ and therefore $K_z^S \simeq K_z \gg q_{\shortparallel}$. It is clear then that only the two-step processes give large

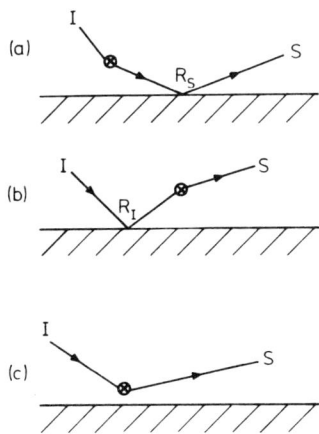

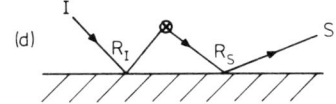

Fig. 6.3a-d Schematic representation of (a) the second term, (b) the third term, (c) the first term and (d) the fourth term of the sum of (6.6)

contributions to the cross section which peak sharply about the specular direction or any other Bragg reflection. The third and fourth term give rise to a broad background, which varies smoothly with scattering angle. The broad background is weak compared to the intensity due to small angle scattering. Thus retaining only the two-step terms and noting that for scattering near the specular reflection, one has

$$K_z^S - K_z = \frac{1}{v_\perp}(v_\| q_\| - \omega) \tag{6.7}$$

and (6.6) may be written

$$\frac{d^2 S}{d\omega d\Omega} = \frac{m e^2 v_\perp}{2\pi \cos\theta_I} \frac{K^S}{K} \frac{|v_\perp q_\|(R_S + R_I) + i(R_I - R_S)(\omega - v_\| q_\|)|^2}{q_\|^2 [v_\perp^2 q_\|^2 + (\omega - v_\| q_\|)^2]^2} \times P(\omega, q_\|) \quad . \tag{6.8}$$

$v_\|$ and $v_\perp$ are the parallel and normal components of the electron velocity with respect to the surface. The properties of the material under study influence the cross section by function $P(q_\|, \omega)$ and by the two reflectivity coefficients. The latter are a characteristic of the quantum mechanical treatment of the problem. The appearance of these two amplitudes means a severe complication with the use of (6.8) as they are not directly measurable. What can be measured directly is only the intensity. The physical origin of their appearance as amplitudes and not as intensities is that the "loss-diffraction" and the "diffraction-loss" processes interfere coherently.

According to (6.8) the cross section decreases rapidly with increasing $q_\|$. The reason for this behavior is that the extending length of the field related to the particular excitation is proportional to $(q_\|)^{-1}$ (see Fig.6.2). This value is therefore a measure of the scattering strength collected in a slab of the thickness proportional to $(q_\|)^{-1}$.

To study the angular distribution of the scattered electrons about the specular direction it is useful to rewrite the expressions $v_\|, v_\perp$ and $q_\|$ in terms of the directly observable parameters E_0, $\hbar\omega$, Ψ and θ_I whose meaning is illustrated in Fig.6.4. Considering wave vector and energy conservations yields

$$q_\| x = (K - K^S) \sin\theta_I + K^S \left[\sin\theta_I (\frac{\vartheta}{2} + \frac{\psi}{2}) - \vartheta \cos\theta_I \right]$$

$$q_\| y = -K^S(\Psi \sin\theta_I + \vartheta\Psi \sin\theta_I) \tag{6.9}$$

$$v_\perp = v_0 \cos\theta_I, \quad v_\| = v_0 \sin\theta_I \quad \text{and} \quad \frac{v_\|}{v_\perp} = \text{tg}\theta_I \quad .$$

From this one has

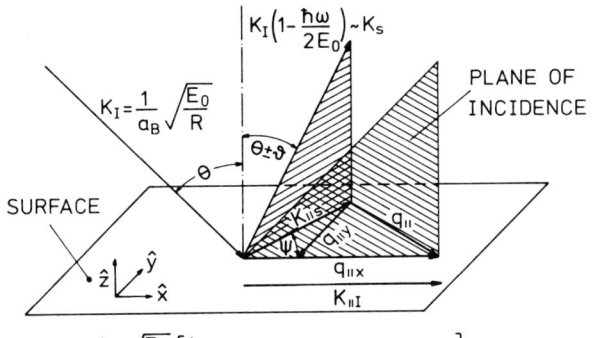

$$q_{\shortparallel x} \sim \frac{1}{a_B}\sqrt{\frac{E_0}{R}}\left[\frac{\hbar\omega}{2E_0}(\sin\theta + \vartheta\cos\theta) - \vartheta\cos\theta\right]$$

$$q_{\shortparallel y} \sim \frac{1}{a_B}\sqrt{\frac{E_0}{R}}\left(1 - \frac{\hbar\omega}{2E_0}\right)\psi\sin\theta$$

Fig. 6.4 Schematic representation of the parameters defining the kinematics of the scattering

$$q_{\shortparallel}^2 + \frac{1}{v_\perp^2}(\omega_s - v_{\shortparallel}q_{\shortparallel})^2 - K^2\left[\vartheta^2 + \left(\frac{\hbar\omega}{2E_0}\right)^2\right] \quad . \tag{6.10}$$

This is the expected result which clearly shows that the scattered electrons are sharply peaked around the specular reflex or any Bragg reflex because there would be no change in the kinematics if in $V_0(Z)$ a lattice periodicity had been introduced. The angular half width is given by $\vartheta_E = \hbar\omega/2E_0$. This is a characteristic angle for the scattering from long-range field fluctuations. Its physical origin is the strong interaction between the electron and the fluctuation wave when the phase velocity of the latter is of the same order of the parallel component of the electron velocity which is equivalent to $\omega/q_{\shortparallel} = v_0$. To what extent (6.8) is valid depends mainly on the value of ϑ_E and has to be determined and controlled in every particular experiment. This question shall be discussed in more detail in Section 6.5.

6.2.3 The Excitation of Optical Surface Phonons in Infrared-Active Material

(A review of the literature about phonons is given in [6.25,35]). Until now the special form of the function $P(q_{\shortparallel},\omega)$ has not been considered. For an isotropic semi-infinite half space $P(q_{\shortparallel},\omega)$ is of the form [6.22]

$$P(q_{\shortparallel},\omega) = \frac{2q_{\shortparallel}}{\pi}\,\text{Im}\left\{\frac{-1}{\varepsilon+1}\right\} \tag{6.11}$$

where $\varepsilon(\omega)$ in this case has the form $\varepsilon(\omega) = \varepsilon_0 + \omega_p^2/(\omega_{TO}^2 - \omega^2 - i\omega\gamma)$ [6.23] with $\omega_p^2 = \omega_L^2 - \omega_{TO}^2$ being the ion plasmon frequency; ω_{TO} an ω_L are the transverse optical phonon and the longitudinal phonon frequency of the bulk, respectively. Substituting this dielectric constant into (6.11) and (6.8) yields after some algebra and integration over the solid angle $d\Omega$ and the energy the total cross section

$$\frac{S}{|R|^2} = \frac{2\omega_p^2 e^2}{\pi\omega_s(1+\varepsilon_0)^2} \int \frac{d^2 q_{\shortparallel} v_{\perp}^2 q_{\shortparallel}^2}{[v_{\perp}^2 q_{\shortparallel}^2 + (\omega_s + v_{\shortparallel} q_{\shortparallel})^2]^2} = \frac{\pi \omega p^2}{\omega_s^2 (1+\varepsilon_0)^2 v_{\perp}} \quad . \tag{6.12}$$

In the above equation the approximation $|R_I - R_S|/|R_I + R_S| \ll 1$ and thus $R_I \simeq R_S = R$ has been introduced. In the case of multiphonon excitation EVANS et al. [6.20] derived an expression for the probability of exciting n optical surface phonons in one scattering process which has in the limit $kT \ll \hbar\omega_s$ the form

$$P_n = |R|^2 \frac{S^n}{n!} \exp(-S) \quad . \tag{6.13}$$

In this limit the probability to excite n surface phonons follows the well-known Poisson distribution and the n surface phonon emission process may be viewed as a series of n random uncorrelated events [6.36]. As will be shown later on, this limit is already fulfilled in ZnO at room temperature. Finally it should be mentioned that the coefficients $\exp(-S)s^n/n!$ obey the sum rule

$$\sum_{n=-\infty}^{\infty} \frac{S^n}{n!} \exp(-S) = 1 \quad . \tag{6.14}$$

Thus for strong coupling between the electrons and the surface phonons (as is in ZnO) the multiphonon emission reduces the strength of the specular beam remarkably.

6.2.4 Excitation of Optical Surface Phonons on Noninfrared Active Substrates

In this section the function $P(q_{\shortparallel},\omega)$ shall be discussed for a material for which the dipole moment vanishes because of reasons of symmetry. Although the effective charge vanishes inside the volume of the material a dipole active surface layer may exist induced by the lower symmetry near the surface. The physical reason may be a layer of adsorbed atoms or molecules or a reconstruction of the surface of the substrate. This case can be treated in a collective model by a modified dielectric constant regarding the different electrical behavior in the uppermost layer as has been done in the case of surface states (see Subsec.6.2.6). But in order to gain the information which is of interest especially in the case of adsorption studies a microscopic model is more advantageous.

EVANS et al. [6.21] found that in this case the potential extending into the vacuum may be described by the expression

$$\varphi(\underline{x},t) = e^{i(q_{\shortparallel}\underline{x} - \omega_s t)} e^{-q_{\shortparallel} z} \frac{4\pi n_0 \varepsilon}{1+\varepsilon} (P_{\perp} - \frac{i}{\varepsilon} q_{\shortparallel} P_{\shortparallel}) \quad . \tag{6.15}$$

The derivation of (6.15) was based upon the assumption that when the optical phonon is excited there will be a two-dimensional wave-like dipole layer induced in the surface region. The thickness of this layer is microscopic, of the order of a few lattice constants at most, and independent of the wave length of the surface phonon, in contrast to the situation in ionic crystals. At first the authors considered the potential in the vacuum generated by a single dipole with a dipole moment $\underline{P} = \underline{P}_\perp + \underline{P}_{\shortparallel}$ placed on the surface of a dielectric substrate with dielectric constant ε. Summation of all dipole moments of each lattice cell results in the above expression. n_0 is the density of lattice cells. The term containing $\underline{P}_{\shortparallel}$ is lowered by a factor of $|\varepsilon|^{-1}$ compared with $\underline{P}_\perp$. This effect can be explained easily by means of the image potential known from simple electrostatics. A dipole placed perpendicular to the surface has an image dipole within the material which is directed parallel to the virtual dipole but its potential is screened by a factor $\varepsilon-1/\varepsilon+1$. Summation of the potentials of both dipoles yields

$$\varphi(P_\perp) = P_\perp(1 + \frac{\varepsilon-1}{\varepsilon+1}) = P_\perp \frac{2\varepsilon}{\varepsilon+1} \quad . \tag{6.16}$$

In contrast, the image dipole of a real dipole which is oriented parallel to the surface is antiparallel oriented. Thus the sum of both potentials in the vavuum is given by

$$\varphi(P_{\shortparallel}) = P_{\shortparallel}(1 - \frac{\varepsilon-1}{\varepsilon+1}) = P_{\shortparallel} \frac{2}{\varepsilon+1} \quad . \tag{6.17}$$

This means that vibrations which induce dipole moments parallel to the surface scatter electrons less by a factor $|\varepsilon|^{-2}$ than those with perpendicular dipole moments. Thus if studying materials with large dielectric constants the term containing $\underline{P}_{\shortparallel}$ can be neglected completely. This provides an important selection rule for dipole scattering of electrons from a layer of adsorbed molecules. In this case EVANS et al. [6.21] find for the cross section with $q_{\shortparallel} \approx 0$ the expression

$$\frac{dS}{d\Omega} = \frac{8mn_0 e^2 q_\perp^2 \varepsilon^2 |R|^2}{M\omega_s E_0 \cos\theta_I (1+\varepsilon)^2} \frac{\psi(\cos\varphi \cos\theta_I - \psi_E \sin\theta_I)^2 + \psi^2 \sin^2\varphi}{(\psi^2 + \psi_E^2)^2} \quad . \tag{6.18}$$

with $q_\perp$ the effective dipole charge of an optical surface phonon with wave vector $q_{\shortparallel} = 0$, $\Psi_E = \hbar\omega/2E_0$ and M the atomic mass. n_0 is the number of dipole active surface atoms. According to the original paper the direction of the scattered electrons is described by the three angles denoted θ_I, ψ, and φ. The x and y axis are oriented such that the specular trajectory lies in the xz plane. The angle between $\underline{K}$ and $\underline{K}^S$ is ψ. The angle between the plane which contains $\underline{K}$ and $\underline{K}^S$ and the xz plane is denoted by φ.

The distinct character of the surface waves of infrared active and noninfrared active materials is given by the different $q_{\shortparallel}$-dependence of the decay length. In the first case it is proportional to $(q_{\shortparallel})^{-1}$ and in the latter it is independent of the wave length. This behavior is reflected in an different angular distribution about the specular reflection and in a different dependence of the cross section on the primary energy. For $\psi > \psi_E$ the loss intensity decreases proportional to ψ^{-2} in the case of infrared active material, in case of noninfrared active material proportional to ψ^{-1} [6.21]. A comparison of the energy dependences is shown in Fig. 6.5. Thus an investigation of the loss spectra will give information about the character of the excited waves.

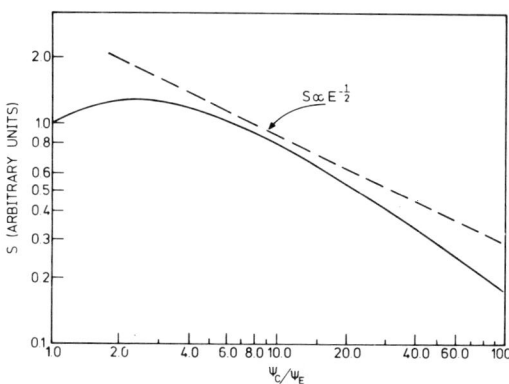

Fig. 6.5 Energy dependence of the one-phonon cross section. S is plotted vs ψ_c/ψ_E with ψ_c the aperture angle of the spectrometer. The values are calculated for electrons incident on a (111) surface of silicon at an angle θ_I of 45° (full line). The dashed line shows the result of the $E_o^{-1/2}$ characteristic appropriate for instance to ZnO (after MILLS [6.21])

6.2.5 Excitation of Plasma Waves

As already mentioned the excitation of plasmons and interband transitions as well as the excitation of long wave optical phonons is of the same nature. Therefore the functions $P(q_\parallel,\omega)$ for the two cases are identical and after substitution of (6.11) into (6.8) one finds with $\hbar^2 K^2 = 2mE_o$;

$$\hbar^2 K_S^2 = 2m(E_o-\hbar\omega); \quad |R_S-R_I|/|R_S+R_I|\ll 1; \quad a_B=\hbar/me^2 \text{ and } \vartheta_E = \hbar\omega/2E_o$$

$$\frac{d^2S}{|R|^2 d(\hbar\omega)d\vartheta d\psi} = \frac{2(1-\vartheta_E)}{\pi^2 a_B K E_o \cos\theta_I} \frac{N^{1/2}}{(N+M)^2} \operatorname{Im}\left\{\frac{1}{\varepsilon+1}\right\} \tag{6.19}$$

where $N = (\vartheta_E \sin\theta_I + \vartheta \cos\theta_I)^2 + (1-\vartheta_E)^2 \psi^2 \sin\theta_I$ and $M = (\vartheta_E \cos\theta_I - \vartheta \sin\theta_I)$; θ_I, ϑ and ψ are defined in Fig.6.4. Eq.(6.19) describes only scattering in the vacuum region (see Subsec. 6.2.2), that means penetration of the electrons is not considered. To what extent this and the approximation $|R_I-R_S|/|R_I+R_S| \ll 1$ are valid shall be discussed by means of experimental data later on.

6.2.6 Electronic Surface Transitions

There is a certain analogy between the excitation of phonons on noninfrared active material and the excitation of electronic surface transitions. A high density of surface states may cause the dielectric constant for the surface to be different from that of the bulk. This situation can be described by a spatially dependent surface dielectric constant $\varepsilon^s(\omega,z)$ with z measuring the distance from the surface. This interesting case has also been considered in the theory of MILLS [6.37]. Similar to models known form ellipsometry [6.38,39] the function $\varepsilon(\omega,z)$ is replaced by a step function. Thus, in this approximation one considers a dielectric half space with $\varepsilon(\omega) = \varepsilon^b(\omega)$ which is covered by a thin layer of thickness d with $\varepsilon'(\omega) = \varepsilon^s(\omega)$. According to Mills one gets the scattering cross section for such a surface active layer if one replaces $\varepsilon(\omega)$ by

$$\varepsilon(\omega,q_{\shortparallel}) = \varepsilon^s(\omega) \frac{1-\Delta \exp(-2q_{\shortparallel}d)}{1+\Delta \exp(-2q_{\shortparallel}d)} \tag{6.20}$$

in (6.19) with $\Delta = [\varepsilon^s(\omega)-\varepsilon^b(\omega)]/[\varepsilon^s(\omega)+\varepsilon^b(\omega)]$.

6.2.7 Data Reduction

In order to perform a quantitative comparison of theory and experiment it is necessary to integrate (6.19) over the solid angle up to a finite value which is given by the aperture of the spectrometer used. This integration yields

$$\frac{1}{|R|^2} \frac{dS}{d(\hbar\omega)} = \frac{2}{\pi a_B K \cos\theta_I} \frac{1}{\hbar\omega} F(\theta_I, \alpha = 2\frac{\vartheta_E}{\Delta\vartheta_C}, \gamma = 2\frac{\vartheta_E}{\Delta\psi_C}) \, \mathrm{Im}\left\{\frac{-1}{\varepsilon+1}\right\} \tag{6.21}$$

with $F = \frac{\vartheta_E}{\pi}(1-\vartheta_E) \displaystyle\iint_{\Delta\vartheta_C \Delta\psi_C} \frac{N^{1/2}}{(M+N)^2} d\vartheta d\psi$

and $\Delta\vartheta_C, \Delta\psi_C$ being the aperture angles of the spectrometer. In Table 6.1 some values of F are listed for several parameters $\alpha = \gamma$ and several angles of incidence. In the case of phonon excitations (2 $E_0 \approx 10$ eV, $\hbar\omega \approx 50$ meV and $\Delta\vartheta_C = \Delta\psi_C \approx 1°$) F approaches $\pi/2$ nearly independent of θ_I and becomes identical to the function $F(\alpha,\gamma)$ defined in the paper of LUCAS [6.18].

In the case of surface transitions it may be possible under certain conditions concerning the material under study to factorize (6.20) which yields an explicit $q_{\shortparallel}$-dependence of $\varepsilon(\omega,q_{\shortparallel})$. Thus also in this case an integration becomes possible [6.40,41]. The resulting function F_s scales similarly to F.

θ \ α	0.1	0.3	1.0	3.0	10
50	$\approx \pi/2$	1.406	0.553	0.101	0.0099
60	$\approx \pi/2$	1.318	0.584	0.113	0.0111
70	1.533	1.267	0.604	0.122	0.0120
80	1.475	1.253	0.614	0.127	0.0126

<u>Table 6.1</u> Typical values for the finite-aperture-size correction $F(\theta,\alpha=\gamma)$. As can be seen there is only a small θ dependence of F

It is worthwhile to mention how the cross section defined in (6.21) is related to the observed intensities. The recorded intensity of the elastically scattered beam is a convolution of the energetic width of the primary beam with the analyzer window:

$$I_{sel}d(\hbar\omega) \simeq I_{sel}\Delta E_{1/2sel} \propto |R|^2 \Delta E_{p1/2} \Delta E_{a1/2} \quad .$$

The signal of the inelastically scattered intensity about $\hbar\omega$ is formed by the same processes and is thus proportional to

$$\frac{dS(\hbar\omega)}{d(\hbar\omega)} \Delta E_{p1/2} \Delta E_{a1/2} \quad .$$

Comparison of the two expressions yields

$$\frac{1}{|R|^2} \frac{dS(\hbar\omega)}{d(\hbar\omega)} = \frac{I_{in}(\hbar\omega)}{I_{sel}d(\hbar\omega)} \frac{I_{in}(\hbar\omega)}{I_{sel}\Delta E_{1/2\ sel}} \quad . \tag{6.22}$$

6.2.8 Anisotropic Effects of ELS

So far we have always presumed isotropic $\varepsilon(\omega)$. In fact there exists no specified description of the scattering from surfaces of anisotropic materials. On the other hand, effects due to anisotropy have already been demonstrated experimentally by FROITZHEIM et al. [6.42] in a paper describing ELS experiments on ZnO. Fig.6.6 shows the loss spectra of ZnO at different orientations of the crystal surface (c-axis) with respect to the plane of incidence. In the case c normal to the plane of incidence (Fig.6.6a) the loss spectrum exhibits only one peak at $\hbar\omega = 3.9$ eV, in the case c parallel to the plane of incidence (Fig.6.6b) three losses can be observed, namely at $\hbar\omega = 3.9$, 5.5 and 7.75 eV. It has also been noticed [6.42] that the relative intensities of the maxima strongly depend on the angle of incidence. The reason is that the function F is anisotropic in the sense that the number of the measured electrons which have excited plasma waves with a wave vector parallel to the plane of incidence is larger compared to those that have excited waves with wave vector perpendicular to this plane. See [6.42] for further details.

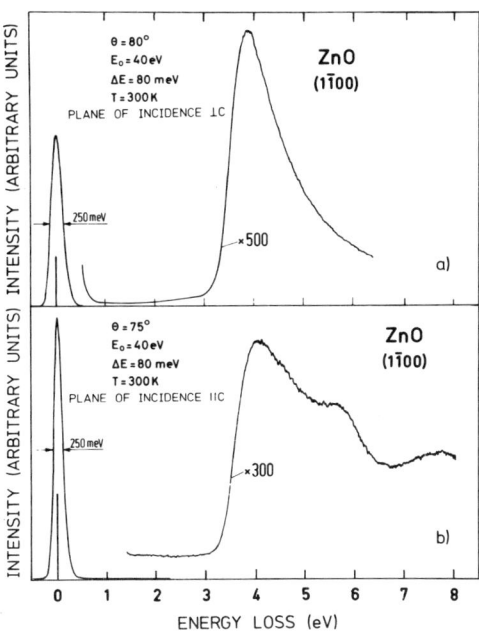

Fig. 6.6a,b Energy-loss spectrum of 40 eV electrons specularly reflected from a (1̄100)-ZnO-surface. The plane of incidence was oriented normal a) and parallel b) to the c-axis

6.3 Experimental Studies of Surface Vibrations (Clean Surfaces)

6.3.1 The Apparatus

In order to perform experiments of scattering of low-energy electrons from surface optical phonons the use of spectrometers with an overall resolution higher than 30 meV is required according to the order of magnitude of quantum energies of the phonons to be excited. The angular resolution plays an important role only when dispersion effects become of interest. In this respect it should only be mentioned that according to (6.9) a 10 eV primary electron will be deflected by an angle $\vartheta \simeq 10°$ from the specular direction after exciting a phonon $\hbar\omega = 50$ meV and wave vector $q_{\shortparallel} \simeq 0.2$ Å^{-1} (about one tenth of a Brillouin zone). With regard to special details concerning spectrometer construction the reader is referred to Chapter 2 of this volume. As an example Fig.6.7 represents schematically a spectrometer which has been used in the most phonon excitation experiments and some experiments of electronic excitations which shall be discussed later. The main parts of this spectrometer are the two cylindrical deflectors with one of them following the electron gun. This in connection with an accelerating electrostatic lens system forms the source of highly monochromized primary electrons. The second deflector works as an analyzing element. The energetic overall resolution can be varied between 7 and 150 meV independent of the primary energy which can be varied between 1 and 70 eV. The angular resolution is about 1.5° in the horizontal and vertical directions with respect to the plane of incidence. The angle of incidence can be varied between 90° and 45°.

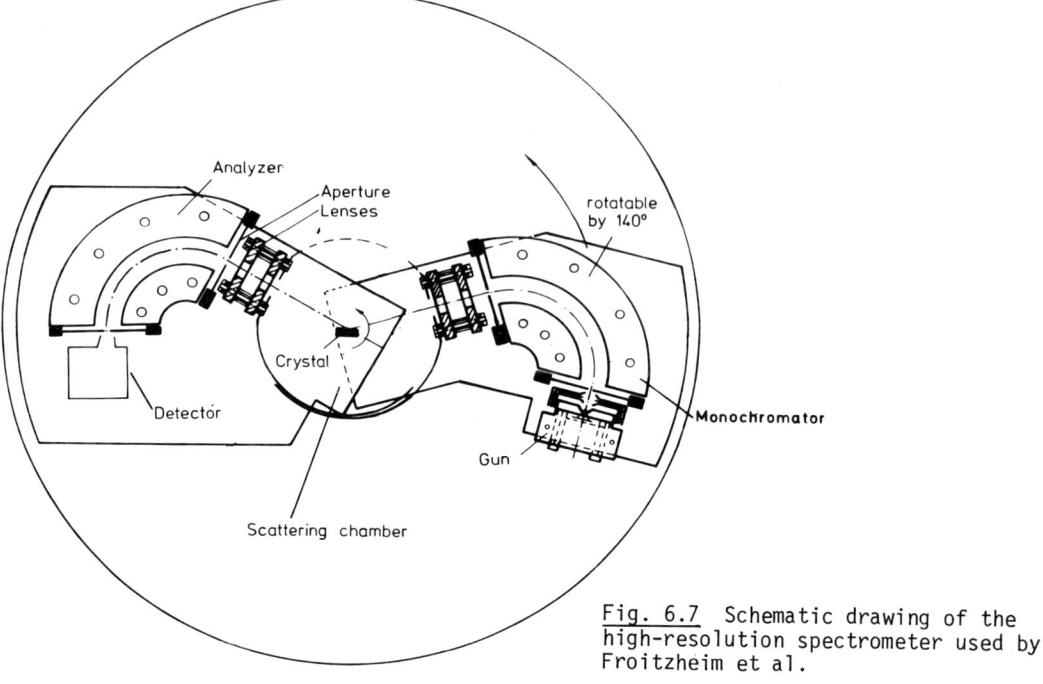

Fig. 6.7 Schematic drawing of the high-resolution spectrometer used by Froitzheim et al.

6.3.2 Infrared Active Material

The first experimental investigations on the inelastic scattering of low-energy electrons from cleaved ZnO (1$\bar{1}$00) surfaces were carried out by IBACH [6.43-44]. A typical loss spectrum is represented in Fig.6.8. The loss peaks are energetically separated by 68.8 ± 0.5 meV. The calculated value yields 69 meV. Contrary to the results of several transmission experiments the excitation of longitudinal optical

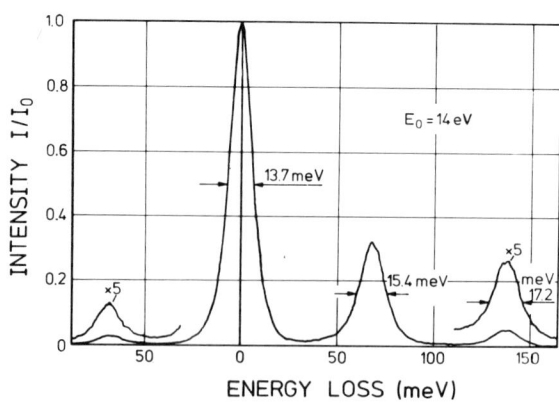

Fig. 6.8 Energy loss spectrum of 14 eV electrons specularly reflected from a (1$\bar{1}$00) surface of ZnO, $\theta_I = 45°$

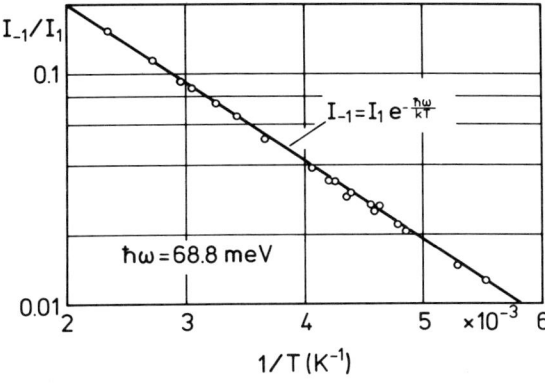

Fig. 6.9 Measured intensity ratio of the one-phonon energy gain I_{-1} to the one-phonon loss I_{+1} vs. temperature

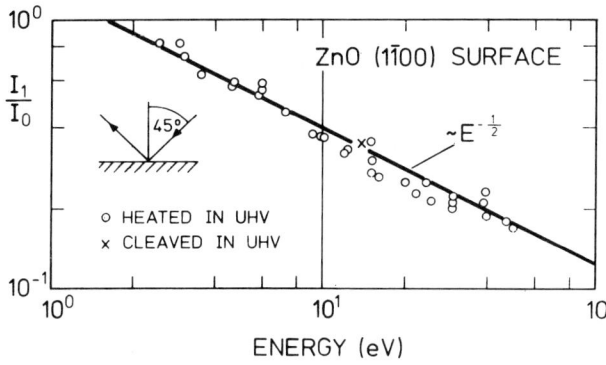

Fig. 6.10 Relative intensity of the one-phonon loss vs. impact energy

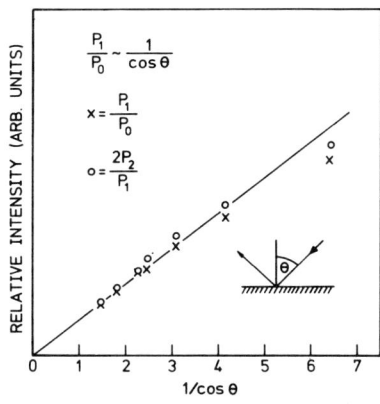

Fig. 6.11 Relative loss intensities of the first and second phonon loss vs. angle of incidence ($\theta_I = \theta_S$)

bulk phonons could not be observed ($\hbar\omega$ = 73.1 meV) (see remarks in Subsec.6.5.2). Fig.6.9 shows the experimental temperature dependence of the one-phonon absorption and excitation probability P_{-1}/P_{+1}. As can be seen it agrees quantitatively with theory. The same agreement exists with the energy and angle dependence (Fig.6.10 and Fig.6.11 [6.45]).

6.3.3 Noninfrared Active Material

A short time after finishing the work on ZnO IBACH presented a paper [6.46] on the scattering of low-energy electrons ($\simeq$ 5 eV) from Si(111) cleaved surfaces. Fig.6.12 shows a loss spectrum of 7 eV [6.47] electrons. The spectrum exhibits an energy gain peak at 56 meV and a loss peak at 56 meV. The ratio of the intensity of the energy gain and the energy loss is related to the Boltzmann factor for 300 K. There are no exact data given about the energy dependence of the relative loss intensity in Ibach's paper, but there are indications that it agrees at least qualitatively with theoretical predictions (see Fig.6.5). In any case the angular distribution about the specular direction supports the assumption of EVANS et al. [6.21] that

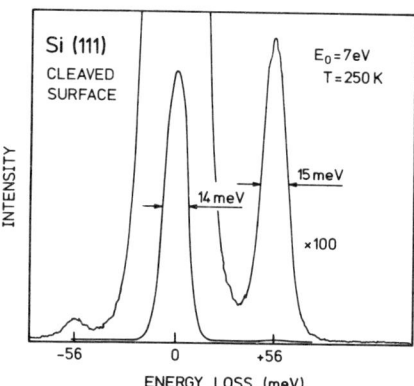

Fig. 6.12 Energy loss spectra of a clean cleaved Si(111) surface

also in this case Coulomb scattering is dominant because the experimentally observed angular half width corresponds well to the theoretical value with respect to the angular resolution of the spectrometer ($\vartheta_E \simeq 0.23°$). A similar agreement of theory with experiment is found for the angle dependence. Fig.6.13 represents the result [6.48] of angle dependence measurement. The solid line indicates the theoretical values. The different scale of the theoretical and the measured curve may be explained by the absence of an angle independent contribution in the theoretical curve arising from surface roughness or volume scattering which is neglected in the theory (see Subsec.6.2.2 and 6.5.2). Using IBACH'S [6.46] data EVANS et al. [6.21] determined a value of the effective dipole charge of $q_\perp = 0.35\ e_0$.

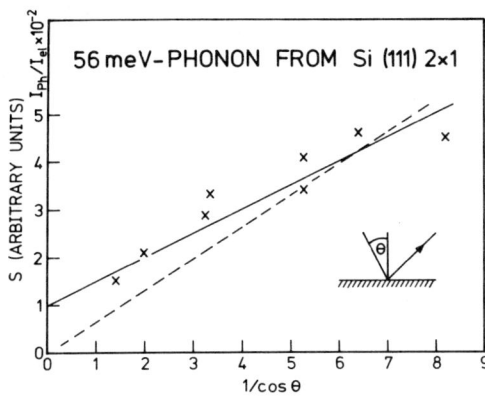

Fig. 6.13 Relative intensity of the surface phonon loss from Si(111) 2 × 1 surface vs. $(\cos\theta)^{-1}$. The dashed curve indicates the exact $(\cos\theta)^{-1}$ characteristic. The experimental deviation from this behavior may be due to volume losses and/or surface roughness $(\theta_I = \theta_S)$

In this connection it should be mentioned that the existence of optical surface phonons on Si(111) is strongly coupled with the 2 × 1 reconstruction of the surface. On the Si(111) 7 × 7 surface no surface phonons could be found. Therefore one may consider the surface dynamic effective charge as a measure of the surface stability and the surface becomes dynamically unstable when the surface effective charges assume values the order of 0.5 e_0 as calculations of TRULLINGER et al. [6.49] show.

6.4 Vibrational Modes on Gas-Covered Surfaces

6.4.1 Apparatus

With regard to the apparative equipment for investigations of vibrational states of adsorbed atoms (localized states), the same conditions have to be fulfilled as in the case when studying clean surfaces. Experience shows, however, that the combination and correlation of results obtained with other experimental methods is often necessary in order to interpret ELS results. The most important methods are LEED, Auger, flash filament and mass spectroscopy. It is therefore advantageous to have all these methods combined with ELS in one apparatus. Simultaneous *in situ* measurements would give an instantaneous picture of the adsorption state from several points of view. If one is interested in measurements under equilibrium conditions of a catalytic system this can be accomplished by differential pumping in the scattering chamber or by cooling the crystal. Fig.6.14 represents an apparatus containing the above mentioned methods which was constructed by the author recently.

6.4.2 Information

Vibrations of adsorbed atoms, e.g., gases on solid surfaces, are well known from infrared spectroscopy. As has already been discussed in Subsection 6.2.4, the incident electrons interact mainly with the long-range fields set up in the vacuum by the oscillating dipoles. We showed in Subsection 6.2.4 that in vacuum a dipole

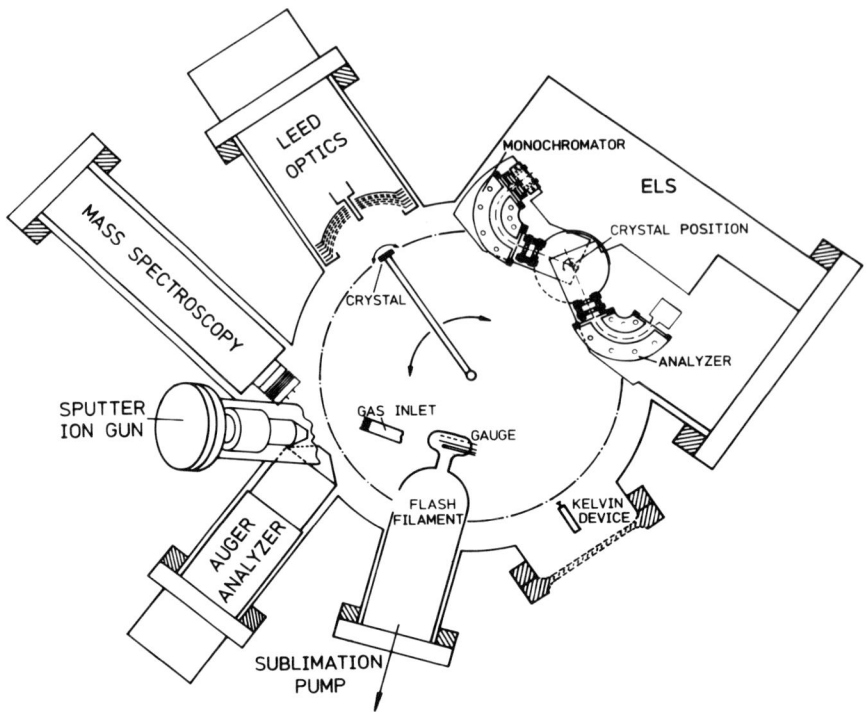

Fig. 6.14 Apparatus design containing all common methods for studying adsorption systems. The sample can be moved on the dashed line and is rotatable by itself. In addition it can be moved in the usual x-y-direction by about 30 mm and in the Z-direction by about 80 mm. The ion gauge, the Auger spectrometer and the mass spectrometer are arranged so that SIMS measurements are possible, too

field which is created by a normal vibration of an adsorbate is screened by a factor of $|\varepsilon|^{-1}$ if the dipole is oriented parallel to the surface of a dielectric medium, whereas this is not the case if the dipole is oriented perpendicular to the surface. This leads to a condition which states that only those vibrations can be excited which create dipoles with a component perpendicular to the surface. Thus in ELS exists the same kind of selection rule as is known for infrared spectroscopy [6.50]. This rule yields additional information and is generally an important help in the interpretation of the loss spectra.

In order to illustrate the selection rules, Fig.6.15 shows several fundamental types of adsorption with their corresponding loss spectra. For simplicity we assume the mass of the adsorbate atoms to be much lighter than the mass of the substrate atoms. Example a) represents a chemisorbed molecule in a bridge position. In this

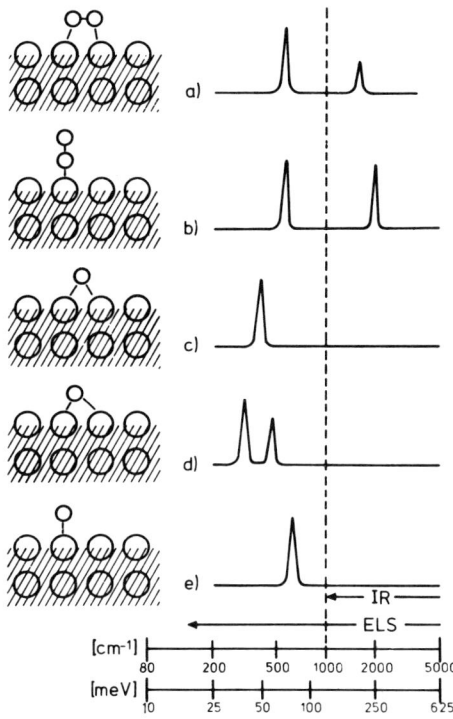

Fig. 6.15 Schematic representation of some fundamental types of adsorption states with their possibly corresponding vibrational spectrum. The vertical dashed line indicates the lowest limit of the spectral range of infrared spectroscopy today

case the corresponding loss spectrum exhibits two loss peaks. The low-frequency mode derives from the vibration of the isolated atoms perpendicular to the surface, the high-frequency mode from the stretching vibration of the quasimolecule. In the case of different masses, e.g., CO, the frequencies are split proportional to the square-root of the mass ratio. The magnitude of the perpendicular component of the dipole moment created by the stretching vibration is a function of the force constant between the two adatoms and the angle γ [6.51,52] as is demonstrated in Fig.6.16.

Example b) shows an adsorbed molecule in a stretched position with two loss peaks in the loss spectrum. The low-frequency mode derives from the vibration of the whole molecule against the substrate. The frequency is low because of the large total mass moving in this mode and the general weaker coupling to the surface. The high-frequency loss corresponds to the stretching vibration of the molecule.

Example c) shows a doubly bonded atom in a symmetric bridge position. This complex has only one normal vibration with a perpendicular component.

Example d) shows that this is not the case for an adsorption site of lower symmetry. In this case a second mode with a perpendicular component occurs.

Example e) shows an atom adsorbed in an on-top position. In this case one has only one loss peak in the loss spectrum. The vertical dashed line crossing the loss spectra indicates about the lower limit of the spectral range currently obtainable in infrared spectroscopy.

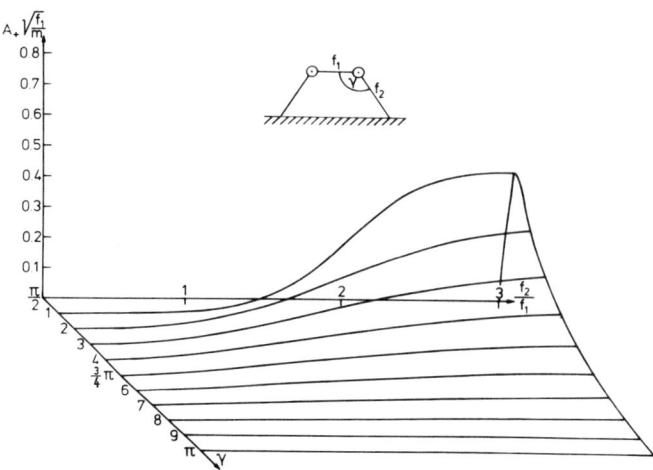

Fig. 6.16 Relation for a bridge position between force constants, the loss intensity, and the angle γ as calculated from [6.52]. A_+ indicates the perpendicular component of the amplitude of the stretching vibation. m is the atomic mass

A conclusion about the information given from ELS studies of adsorption systems can be made as follows: ELS yields information on the microscopic structure of the adsorbate (information about the long-range order may be given from LEED). It also inherently contains information on the type of adsorption, whether dissociative or molecular. One further gains knowledge of the binding potential and the affect of coadsorption. A study of the kinetics of adsorption is possible from intensity versus coverage measurements. Finally measurements of the contact potential can be carried out by the retarding potential method to a high accuracy because of the required high resolution of the spectrometer. This manifold information together with the fact that ELS does not destroy the system under study (current density $< 10^{-12}$ Acm^{-2}, $E_0 \simeq 1$-10 eV) make this method favorable for fundamental studies on catalysis.

6.4.3 Oxygen Adsorption on Si(111) 2 × 1

The vibration spectrum of adsorbed oxygen on Si(111) cleaved surfaces was studied by IBACH [6.47]. Fig.6.17 shows loss spectra with several oxygen exposures. For coverages lower than $\theta \simeq 0.2$ the structure due the surface phonen is unchanged, however, two additional losses at 90 and 125 meV occur, which have to be attributed to localized modes of the adsorbed oxygen. At $\theta \simeq 0.6$ the surface phonen nearly vanished and the two localized modes shifted to 94 and 130 meV. An additional shoulder at 175 meV is also observed. Because of the relatively large dielectric constant (ε_0=12) of silicon, only excitations of vibrations with perpendicular dipole moments can be accepted. Thus one has to look for a configuration which possesses at least three normal vibrations with dipole moments perpendicular to the surface. Contrary to Fig.6.15 the assumption of the substrate being a rigid wall is not valid in the case of the present system. Therefore one has to add to the dynamical system the atoms of the substrate bound to the oxygen.

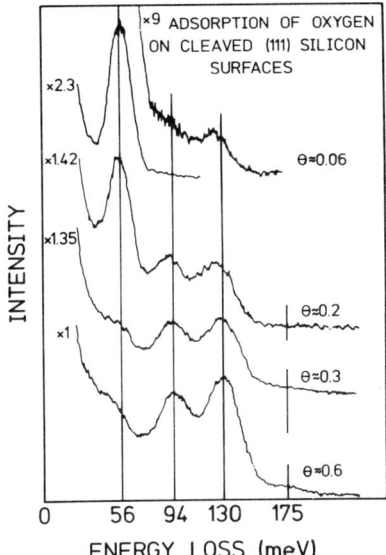

Fig. 6.17 Energy loss spectra of a Si(111) 2 × 1 surface with several coverages of oxygen

Ibach offered two possible configurations which are exhibited in Fig.6.18a and b. The two systems belong to the C_{2v} symmetry group having three A_1 modes [6.52] (only the A_2 modes have dipole moments parallel to the axis, i.e., perpendicular to the surface). In both models the highest frequency (175 meV) derives from the stretching vibration or the O-O quasimolecule. The small loss intensity may be explained in both cases by a strong coupling of the oxygen atoms. In the case of the bridge position the amount of the normal component of the dipole moment depends on the ratio of the binding energies of the oxygen atoms to each other and to the substrate as indicated in Fig.6.16. In the case of the stretched position (Fig.6.18b) a similar argument is valid if one considers that the free oxygen molecule is nonpolar. Thus a dipole moment may be induced by adsorption with a possibly related shift of charge distribution in the molecule which depends on the coupling to the substrate. A third possible model introduced by GODDARD et al. [6.53] is shown in Fig.6.18c. According to the ELS results one may conclude that oxygen adsorbs on Si(111) 2 × 1 surface in a molecular state with the three possible microstructures shown in Fig.6.18. Though a

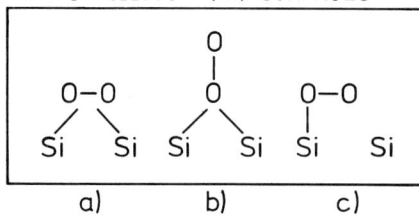

Fig. 6.18 Three possible models for adsorbed oxygen on silicon. The first and the second model were suggested by IBACH [6.47], the third has been presented by GODDARD et al. [6.53]

dissociative adsorption in an asymmetric position would have the same number of perpendicular components it seems quite unlikely with regard to the observed frequencies.

6.4.4 Adsorption of Hydrogen on Si(111) 2 × 1

The adsorption of hydrogen on Si(111) cleaved surfaces is an example of an atomic adsorption. This system was investigated by FROITZHEIM et al. [6.54]. Fig.6.19 shows the loss spectrum of a monolayer coverage. It exhibits only a single loss at 257 meV. Because of the high frequency and the large mass ratio of hydrogen and silicon it is possible to calculate the force constant of the Si-H bond (0.518 u.a.) which nearly coincides with the result of a first principle calculation of an on-top position carried out by APPELBAUM et al. [6.55].

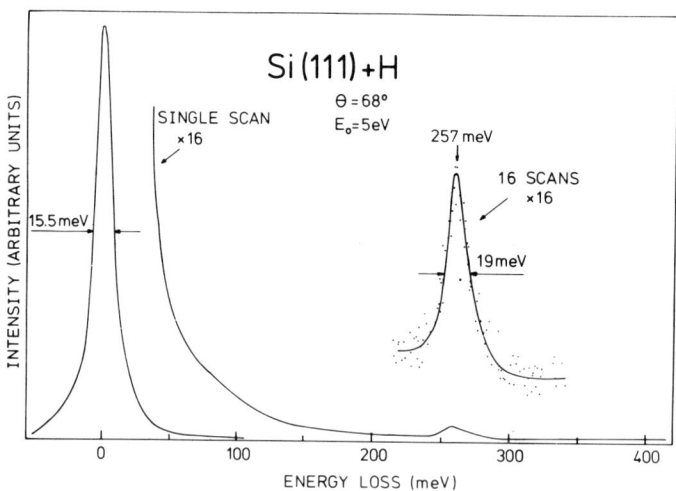

Fig. 6.19 Electron energy loss spectrum of a cleaved silicon (111) surface covered by a monolayer of atomic hydrogen. The loss at 257 meV corresponds to the Si-H stretching vibration

6.4.5 Hydrogen Adsorption on W(100)

ELS studies on tungsten (100) surfaces were performed for the first time by PROPST et al. [6.56]. They investigated vibrational states of adsorbed hydrogen, nitrogen, water and carbon monoxide. Unfortunately the resolution of their spectrometer was not quite sufficient and in addition there existed no detailed theoretical model for this type of spectroscopy at that time. A more recent study with resolution improved by at least a factor five was carried out by FROITZHEIM et al. [6.57].

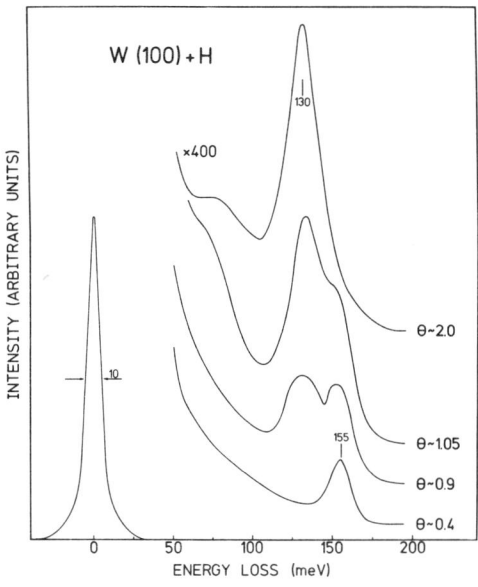

Fig. 6.20 Electron energy loss spectra of H on W (100). The two losses at 155 and 130 meV correspond to atomic hydrogen adsorbed in on-top and bridge sites, respectively. The peak around 70 meV at higher coverages is caused by a small CO contamination ($\theta \simeq 0.01$) during hydrogen exposure

Fig.6.20 shows the loss spectra for different exposures of hydrogen. Contrary to the study of Propst et al. for lower coverages at $\theta \simeq 0.4$ only a loss at 155 meV can be observed. At higher coverages a second loss at 130 meV appears which saturates at $\theta \simeq 2$ (two atoms per surface atom), while the 155 meV loss disappeared. (In the work of Propst et al. these losses were not resolved.) According to Fig.6.15 one has to interpret these losses as due to dissociative adsorption of hydrogen on symmetrical positions. Considering results established with other methods [6.58,59] it is assumed that the 155 meV loss corresponds to hydrogen absorbed in an on-top position and the 130 meV loss to a bridge position. The two stages are, however, not simply sequentially filled as previous models assumed but instead a reversible exchange and equilibrium between the relative occupation of the two sites exists as an analysis of the relative intensities of the two losses clearly demonstrates (Fig.6.21).

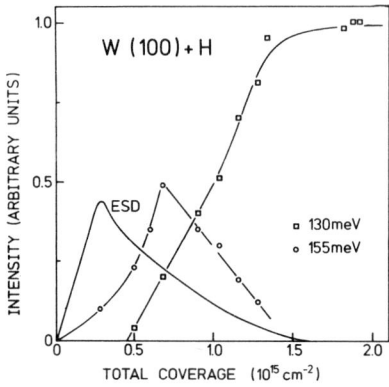

Fig. 6.21 Measured intensity of the losses at 155 and 130 meV, respectively, vs. coverage. The intensity of the electron stimulated desorption is shown for comparison

6.4.6 Adsorption of Oxygen on W(100)

Fig.6.22 represents a series of loss spectra of several coverages of oxygen on tungsten (100) [6.60] for room temperature adsorption. For small converages ($\theta \simeq 0.17$) only a single loss is observed at $\hbar\omega = 75$ meV. At $\theta \simeq 0.3$ a second loss of lower frequency appears at $\hbar\omega = 50$ meV, while the high-frequency loss shifts to 78 meV. For coverages around $\theta \simeq 0.5$ the loss spectrum consists of three losses. Four peaks can be observed at $\theta \simeq 0.85$ which coalesce to two broad losses at 90 and 65 meV when the coverage becomes about unity. At even higher coverages an additional loss appears at 125 meV. An interpretation of this rather complex situation was again

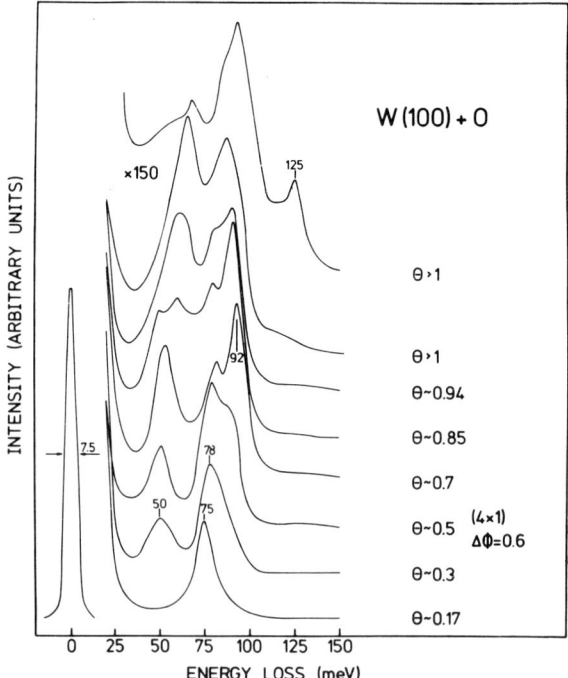

Fig. 6.22 Electron energy loss spectra of W(100) for several coverages of oxygen. The spectra are recorded at a primary energy of 5 eV with an angle of incidence $\theta_I = 70°$

possible only by correlating the above results with numerous data derived from other methods. For the interpretation of the vibration spectra the tungsten substrate was assumed to be a heavy rigid wall because of the large mass ratio of the tungsten and the oxygen atoms. In the following the main results of the interpretation shall be discussed.

a) The appearance of a single loss peak clearly indicates a dissociative absorption. Considering the open lattice structure of the tungsten (100) surface and the loss frequency, it is assumed that the oxygen prefers the fourfold coordination sites (Fig.6.23a)

b) At coverages around $\theta \simeq 0.25$ the two-domain $4 \times 1, 1 \times 4$ LEED pattern starts to develop which was explained by the formation of double rows previously [6.61] (Fig.6.23b). This interpretation is confirmed by the loss spectrum consisting of two loss peaks at this coverage. The formation of such rows is the only way whereby surface tungsten atoms can achieve a W^{4+} oxidation state and a nearest neighbor configuration similar to the rutile type lattice of WO_2. As estimations suggest the formation of such rows is energetically favorable since the repulsive energy between the double rows is smaller than the repulsive energy between isolated randomly or ordered adsorbed O-atoms.

c) At coverages higher than $\theta \simeq 0.5$ the LEED pattern shows a $c(2 \times 8)$ structure (exactly at $\theta = 5/8$ [6.62]). It was suggested and also confirmed by considerations concerning the binding energy and the frequencies observed, to interpret the loss spectra at this coverage range as resulting from the formation of triple and quadrupole rows of oxygen atoms on the surface (see Fig.6.23c and d).

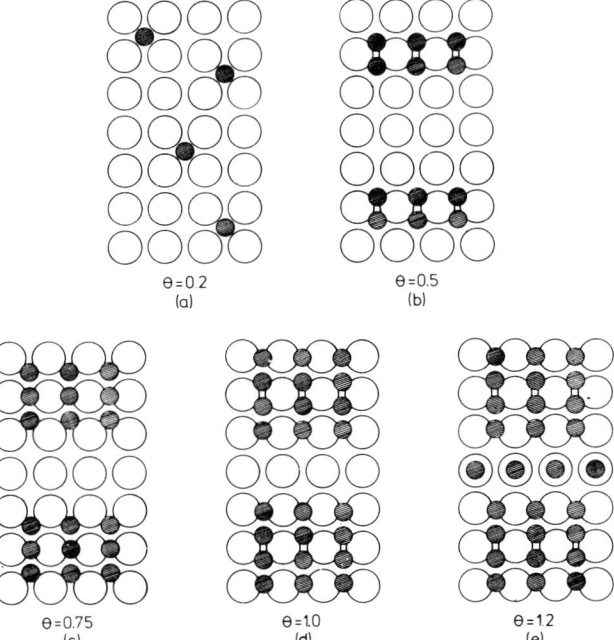

Fig. 6.23 a-e Models for the surface structure of oxygen on W(100) adsorbed at 300 K

d) The appearance of a 125 meV loss at coverages $\theta \simeq 1$ is explained as due to an oxygen atom in an on-top position. The frequency is in agreement with the vibrational frequency of WO of 130 meV. At about this coverage MADEY [6.63] found the state of high ESD yield which may also suggest an on-top position. It is therefore suggested that the 125 meV loss is identical with the β_1 state in ESD (Fig.6.23d).

6.4.7 The Adsorption of CO on Tungsten (100)

For small exposures to CO (< 1 L) two losses at $\simeq$ 70 meV and $\simeq$ 78 meV are observed in the loss spectra [6.64] (Fig.6.24). Both frequencies are identical to the vibrations of isolated carbon and oxygen atoms, respectively. The spectrum is therefore a clear evidence of dissociative adsorption of the β-CO state in agreement to previous XPS and isotope mixing studies [6.65,66]. With further exposures to CO two additional losses at 45 meV and 258 meV are observed (Fig.6.24). The intensity of these two losses depends on the CO partial pressure. The high frequency is close to the frequency of the free molecule vibration. It is therefore assumed that the two additional losses are related to the characteristic vibrations of the undissociated α-CO state corresponding to a molecule in upright position. Because of the small influence of the two adsorption states on each other it is assumed that at least at low pressures α-CO is adsorbed on top of the tungsten atoms. In agreement with other authors it was found that CO replaces adsorbed hydrogen on the surface. Coadsorption of CO and hydrogen is possible only if the β-CO state is not completely filled. In the presence of α-CO the adsorption of hydrogen is impossible.

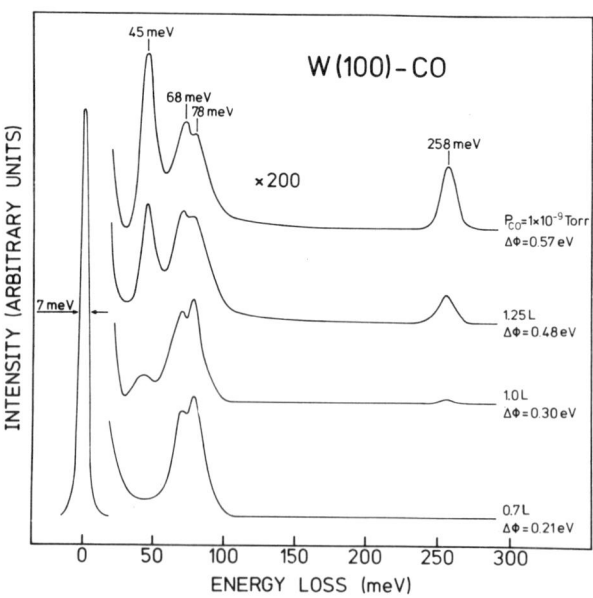

Fig. 6.24 Loss spectra of a W(100) surface for several exposures to CO. The values $\Delta\phi$ indicate the work function change due to the coverage. The spectrum indicated by $\Delta\phi = 0.57$ eV was measured at a partial pressure of CO of 10^{-9} Torr. The other spectra were measured at a total pressure below 10^{-10} Torr

6.5 Experimental Studies of Electronic Transitions

In this section some considerations of the problems which are related to the quantitative analysis of loss spectra shall be made. However, at first some remarks concerning the instrumental requirements for ELS of clean and gas-covered semiconductors and metals shall be given.

6.5.1 The Apparatus

The most important experimental condition for ELS studies of electronic transitions is again a high-resolution spectrometer where the energy resolution should be at least of the same order of magnitude as the energy losses to be expected, namely $\approx$ 100 meV. In addition the primary energy should be variable between 50 and 1000 eV. At present the most common method is to use commercially available Auger cylindrical mirror analyzers (see Chap.1) which will be driven in the first or second derivative mode in order to increase the sensitivity. This can be done by modulating the transmission energy and using a lock-in technique. However, it must be mentioned that differentiation annihilates valuable information on the unstructured loss intensity. It also overestimates sharp structures of low intensity because only the slope of the curves will be measured.

6.5.2 Relationship Between the Spectrometer and the Interpretation of the Loss Spectra

For a quantitative interpretation of the loss spectra taking into account also the loss intensity, a spectrometer is needed which has sufficient energy resolution without modulation techniques and which is capable of an angular resolution of at least 2°. These conditions must be fulfilled as the present available theory (see Sec.6.2) describes only small angle scattering. On the other hand, scattering into larger solid angles means larger wave vector transfer, and neglecting the wave vector dependence of the dielectric constant is no longer a useful approximation. Mathematically this means on one hand that the loss function becomes ϑ and ψ dependent and must be included in the integration over the solid angle which is in general impossible to carry out. In addition at large wave vector transfer the transverse dielectric constant may no longer be equal to the longitudinal dielectric constant which means that the measured spectrum is a mixture of two different loss functions. Even if this may be neglected, multiple scattering and small wave length effects (see Sec.6.6) may play an important role when the aperture angle of the spectrometer becomes too large. But also with an infinitely small aperture and $\theta_I = \theta_S$ there is a finite wave vector transfer due to the energy loss (6.10) with the characteristic measure $\vartheta_E = \hbar\omega/2E_0$. This measure is also important in another point of view. While in the case of surface scattering the angular half width of the intensity is about ϑ_E this is not the case in volume scattering [6.23]. In the latter the angular distribution of the scattered electrons peaks more weakly around the specular direction. Thus for larger values of ϑ_E the intensities of surface scattering (proportional to surface loss function) and volume scattering (proportional to bulk loss function) are of the same order of magnitude because the surface loss intensity is also proportional to ϑ_E^{-2} for $\vartheta \simeq 0$. Therefore at large values of ϑ_E the contribution of the volume scattering cannot be neglected. The ratio of the two contributions then becomes an unknown function of the angle of incidence and the primary energy which controls ϑ_E, the entrance depth of the electrons, the absorption coefficient and the extending length of the surface charge fluctuations. An estimation for the spectrometer presented in Fig.6.7 yields a ratio of volume to surface scattering of about 10^{-2} and 1 for phonons and plasmons, respectively. Another severe complication in the quantitative interpretation of loss spectra is the possibly strong energy dependence of the reflection coefficients R_I, R_S which means that the loss spectra of the diffraction-loss process (proportional to R_I) and the loss-diffraction process (proportional to R_S) become different, whereby this difference is energy dependent. It must therefore be determined whether the approximation $|R_I-R_S|/|R_I-R_S| \ll 1$ is sufficiently valid for every system under study. This can be done by checking for a possible energy dependence of the loss spectra. If such effects are observed the primary energy has to be increased (to decrease ϑ_E) but this may be limited by the fact that the cross section decreases proportional to $E_0^{-1/2}$.

6.5.3 Excitations of Electronic Transitions at Clean Silicon Surfaces

Spectra (not differentiated) of Si(111) cleaved and annealed surfaces were presented by FROITZHEIM et al. [6.40,41] for the first time. The measurements were taken with the spectrometer shown in Fig.6.7. The energy resolution was about 80 meV and the overall collecting aperture about 1.5° having nearly square shape. Fig.6.25 represents the typical spectra of a cleaved and an annealed surface, respectively.

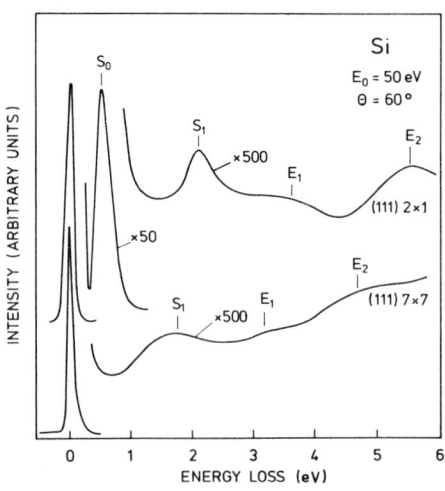

Fig. 6.25 Energy loss spectra of a Si(111) surface. The upper curve corresponds to a cleaved and the lower to an annealed surface

The cleaved surface exhibits a dominant maximum at 0.52 eV and three smaller ones at 2.25, 3.5 and 5.5 eV. Following previous papers [6.67] the loss peaks are indicated by S_0, S_1, E_1 and E_2 sequentially. In the spectra of the annealed surface S_0 has disappeared completely and S_1 is shifted to lower loss energy by about 0.5 eV. It can also be noticed that the whole structure is much weaker than is the spectra of the cleaved surface with a 2 × 1 reconstruction. It was the aim of the above cited work to derive the optical data of the surface, i.e., to calculate ε_1^S and ε_2^S (see Subsecs. 6.2.6 and 6.2.7). The validity of the approximation $|R_I - R_S|/|R_I + R_S| \ll 1$, was found to be correct for values of $\vartheta_E \leq 0.1$. In addition angle dependence measurements demonstrated that neglecting the volume scattering was a valid approximation (see Fig. 6.26).

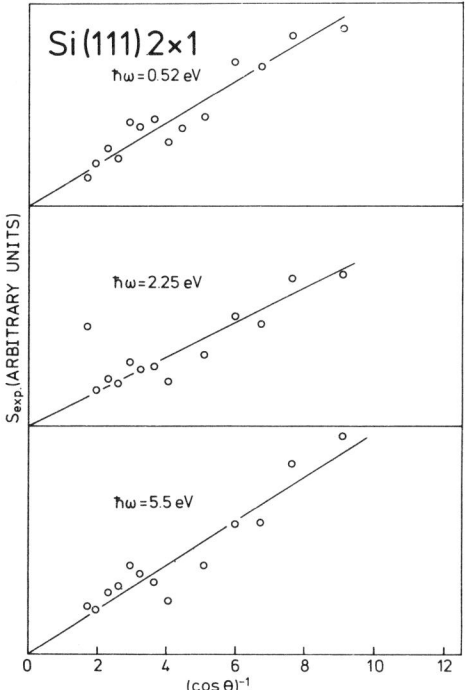

Fig. 6.26 Relative loss intensities of the losses indicated by S_0, S_1 and the E_2 in Fig. 6.25 vs. angle of incidence

With the help of the dielectric layer model (Subsec. 6.2.6) the authors tried to calculate the optical data of the cleaved surface. The results of this calculation are presented in Fig. 6.27. For a detailed discussion of these data and their comparison with those derived from optical methods the reader is referred to the original papers [6.40, 41], [6.68, 69].

In order to contrast high-resolution ELS with ELS using a cylindrical mirror analyzer it is worthwhile to discuss a paper by IBACH et al. [6.67] which deals also with the investigation of Si(111). The measurements were taken with normal

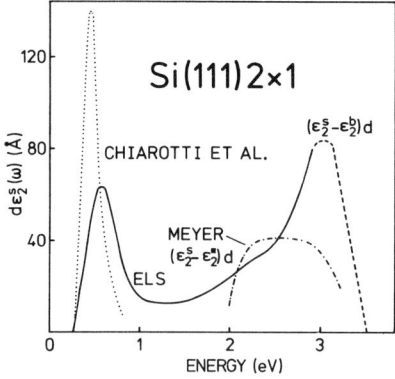

Fig. 6.27 Adsorption constant $d \cdot \varepsilon_2^s(\omega)$ of a surface active layer with thickness d for Si(111) 2 × 1. The dotted line and the dashed-dotted line correspond to the calculated values form optical data given in [6.68] and [6.69]

incidence because of the integrated coaxial gun. The spectra were measured in the second derivative mode yielding 0.7 eV FWHM. Fig.6.28 shows a series of loss spectra with different coverages of oxygen. The primary energy was 100 eV. As before the losses related to surface transitions are indicated by S while E indicates transitions between bulk states. The interesting information which has to be noticed in the present case is the vanishing of S_1 and S_3, the apparent increase of S_2 and the splitting of the surface plasmon which is proportional to the coverage. In order to understand this behavior, the same experiments were carried out for the (100) surface and the results are shown in Fig.6.29. The spectra exhibit here also a strong maximum at 7.2 eV while the two losses indicated by S_2 and S_3 appear only as weak

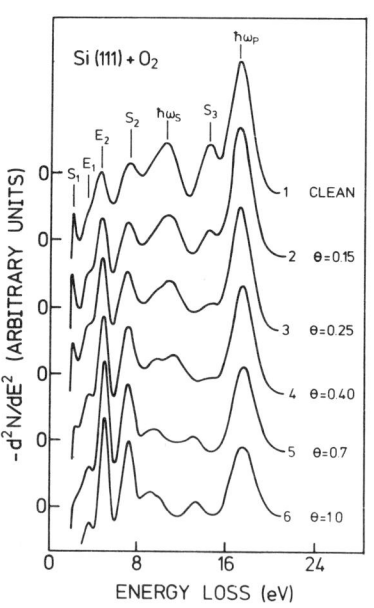

Fig. 6.28 Negative second derivatives of the loss spectra of silicon (111) 7 × 7 for several coverages of oxygen. Primary energy was 100 eV

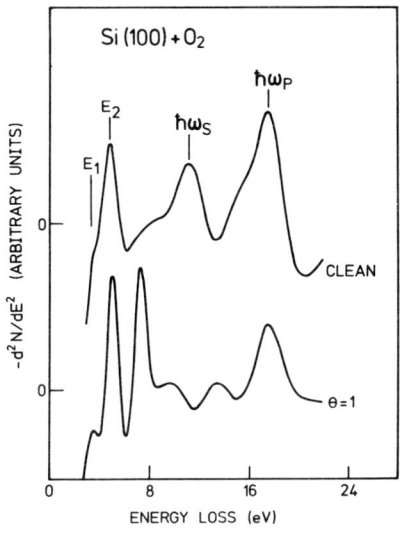

Fig. 6.29 Negative second derivatives of the loss spectra for the silicon (100) 2×1 surface. For monolayer coverage the spectra of the (111) and (100) surface agree within the limits of reproducibility

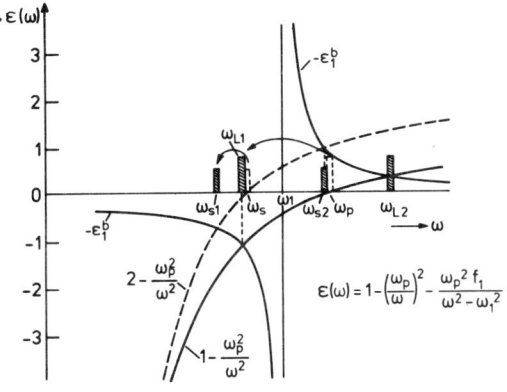

Fig. 6.30 Change of the possible plasmon modes ω_S and ω_P by a single adsorption at ω_1. The dielectric constant is composed of that of a free electron gas and a single oscillator with oscillator strength $\omega_p^2 f_1$. The new bulk modes are indicated by ω_{L1} and ω_{L2} (solutions of $-\varepsilon_1^b(\omega) = 1-(\frac{\omega_p}{\omega})^2$) and the new surface modes by ω_{S1} and ω_{S2} (solutions of $-\varepsilon_1^b(\omega) = 2-(\frac{\omega_p}{\omega})^2$)

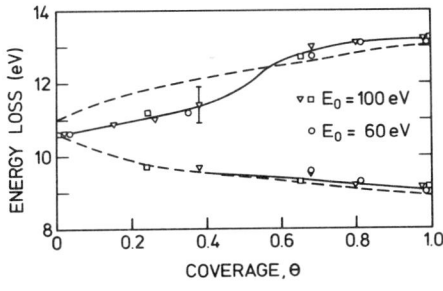

Fig. 6.31 Splitting of the surface-plasmon peak with coverage. The dotted line is the splitting calculated from a dielectric model (see Fig. 6.30)

shoulders in the spectrum of the clean surface. This suggests that the S_2 loss corresponds probably to transitions involving surface states as well as to oxygen related transitions. On the other hand, the interpretation of the S_1 and S_3 losses is straightforward as they disappeared after oxygen adsorption which indicates that the states involved in these transitions must be saturated. The splitting of the surface plasmon as well as the shift of the two resulting peaks could be explained nearly quantitatively by the model of a single oscillator discussed by WILSON [6.70] and is demonstrated in Fig. 6.30. A comparison of the calculated and measured shift versus coverage is plotted in Fig.6.31. It is noticed in the cited paper that this splitting and also the number and energetic positions of the loss peaks were reproducible and independent of the primary energy in the range of 40 to 100 eV [6.67,71]. It is emphasized in the paper that the spectra of the oxygen covered surface ($\theta = 1$) showed no energy dependence at all. This may be mainly due to the fact that the oxygen adsorption is disordered [6.72] and that therefore $q_{\shortparallel}$ is no larger a well-defined quantum number causing a broadening of the specular reflex and a smoothing of all angular and energy dependent effects.

6.5.4 Electronic Excitations at Ge(111) Surfaces

A similar situation as in the case of Si has been observed at Ge(111) cleaved surfaces by FROITZHEIM et al. [6.40-41] (Fig.6.32). Again the maxima S_0 and S_1 correspond to surface state transitions. The physical interpretation concerning the nature of the surface states is nearly the same as in the case of silicon. Despite this qualitative agreement there is a remarkable quantitative disagreement between Si and Ge especially on the annealed surfaces which is not fully understood yet.

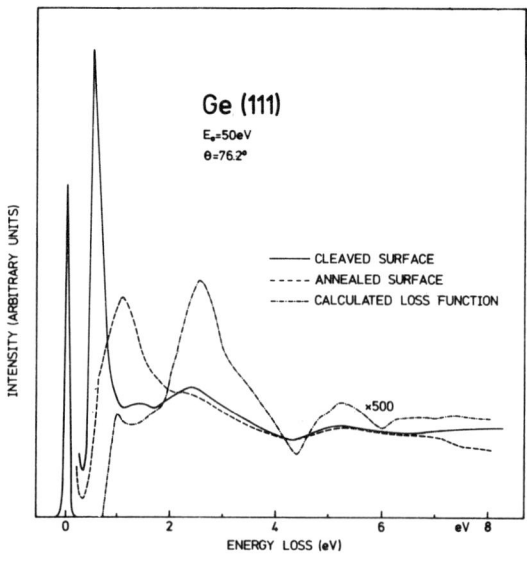

Fig. 6.32 Typical energy-loss spectrum for a cleaved (111) germanium surface. The solid line indicates the measured loss spectrum of a (111) 2 × 1 surface while the dashed line represents the loss spectrum of a (111) annealed surface. The dot-dashed line is the calculated spectrum from bulk optical data [6.79] after (6.21). The smoother shape of the experimental curve is due to the higher wave vector transfer in ELS discussed in Subsection 6.5.2

Another fruitful technique complementary to ELS is the spectroscopy of core-level excitations which was presented by LUDECKE et al. [6.73] and ROWE et al. [6.74] for the first time on Ge and is based on an earlier work done by GERLACH [6.75,76]. In the case of interband transitions only differences of the energetic levels of two states can be measured (whereby the absolute levels are generally unknown). However, by exciting transitions between a core-level and empty states above the Fermi level it is possible to determine the absolute energetic positions of these states if the energy of the filled core level is known.

Fig.6.33 shows the loss spectra of a Ge(111) face as presented in the paper of LUDEKE et al [6.73]. The upper curve represents a loss spectrum of an annealed surface and the lower curve that of an argon bombarded surface. The spectra were measured with a cylindrical mirror analyzer in the second derivative mode. The interest should here be concentrated on the energy range around 30 eV. The visible structure is related to transitions between the 3d-level and empty states above the Fermi level [6.73]. Because of the high surface sensitivity of the 29.3 eV peak to adsorption of carbon or oxygen it is assumed to correspond to a transition between a core level and an empty surface state, where the two other losses are transitions between the 3d core level and empty bulk states of the conduction band. A comparison with optical measurements [6.77] determined the energy of the empty surface state as 0.5 eV above the valence band edge.

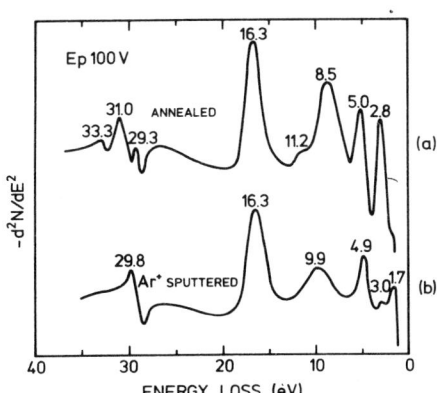

Fig. 6.33 a,b Second derivative energy-loss spectra of (111) Ge surfaces: a) Ar$^+$ sputtered and annealed; b) Ar$^+$ sputtered [6.73]

6.5.5 Gallium Arsenide

Fig.6.34a represents loss spectra of a clean cleaved (upper curve) and oxygen covered GaAs(110) surface (lower curves) measured with the spectrometer shown in Fig.6.7. Fig. 6.34b shows the comparison of a measured and a calculated loss spectrum [6.78]. As can be seen there is an excellent agreement between the measured and the calculated spectrum and there are no differences between the two curves which may hint at the existence of surface states. The calculation was carried out

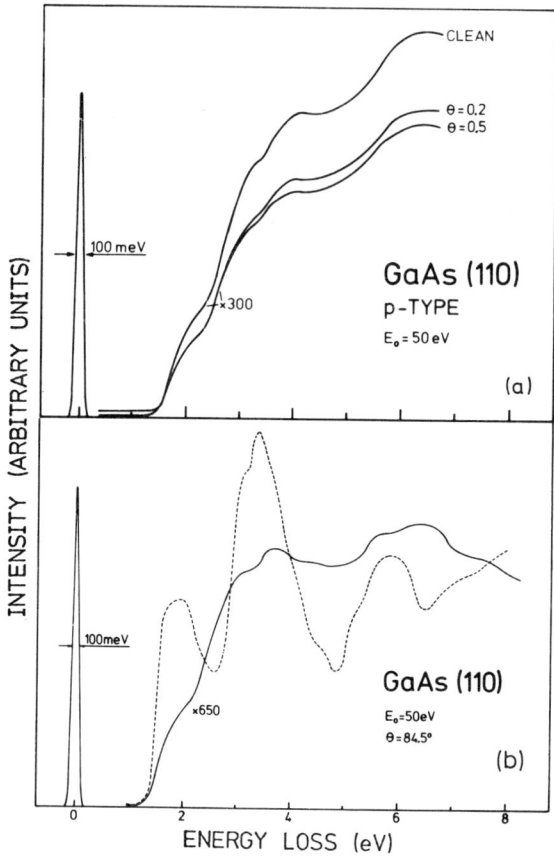

Fig. 6.34 a) Energy loss spectra of a clean and oxygen covered GaAs (110) cleaved surface; b) Comparison of a measured (solid line) and a calculated (dashed line) loss spectrum of a GaAs (110) cleaved surface

by means of optical data taken from EDEN [6.79]. While in this work with up to 8 eV loss energy no indication for the existence of surface states was found, LUDEKE et al. [6.73] observed transitions between 3d core levels and empty surface states at (100) and (111) faces (Fig.6.35) and later on at (110) faces [6.80]. This discrepancy existed already when DINAN et al. [6.81], who performed band bending measurements as a function of doping, could explain their results only by the existence of a surface state density of $2 \times 10^{13} cm^{-2} eV^{-1}$, 0.85 eV below the conduction band edge. This controversy seems to be solved now by work function measurements of HUIJSER et al. [6.82]. They demonstrated that the density of surface states may also be a function of the step density of the surface which may on its part also be a function of doping. This finding agrees with the results of LUDEKE et al. [6.73] because they performed their measurements with so-called "stabilized" surfaces. Since their crystal faces were prepared by molecular beam epitaxy, the resulting surfaces were either Ga-rich or As-rich as well as stoichiometric depending on beam composition. By this method the "intrinsic" properties of the surfaces of binary compounds may be manipulated without adsorbing foreign impurity atoms changing the extrinsic surface conditions in order to make surface sensitive effects visible.

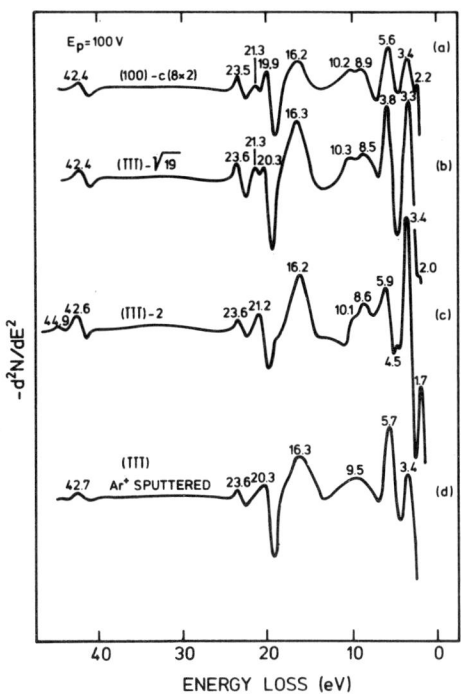

Fig. 6.35a-d Second-derivative energy-loss spectra of GaAs: a) (100) surface, Ga stabilized; b) (111) surface, Ga stabilized; c) ($\bar{1}\bar{1}\bar{1}$) surface, As stabilized; d) ($\bar{1}\bar{1}\bar{1}$) surface, sputtered [6.73]

The interesting loss range in the present discussion is the range around 20 and 2 eV [6.73]. The loss spectrum of the Ga stabilized (100) c(8×2) face (Fig. 6.35) exhibits two maxima at 2.2 and 19.9 eV which are surface sensitive. Therefore Ludeke interpreted these as transitions between the valence band and the Ga 3d core level and empty surface states near the bottom of the conduction band arising from Ga bonds at the surface. The spectra of the Ga stabilized (111) surface again exhibits structure at 20.3 eV in the region of the d-band transitions. This structure disappears as the surface becomes As stabilized, while a strong peak appears at 1.7 eV, which is attributed to transitions from filled surface states connected with As to empty conduction band states. From ESCA studies [6.77] the energy of the Ga 3d core levels with respect to the top of the valence band was quite accurately known. So it was possible to determine the relative positions of the filled As niveaus and the empty Ga niveaus within the bulk band gap (Fig.6.36).

The fact that FROITZHEIM et al. [6.78] did not observe any surface state transition may possibly be due to low step densities of the studied surfaces. Another explanation is that the overlap of the wave functions of the As surface states and Ga surface states may be too small for a measurable probability of electron excited transitions between these two states. Unfortunately in the paper of LUDEKE et al. [6.80] where core level excitations at GaAs (110) 1×1 surfaces are reported, no spectrum of the lower loss energy range is presented which could give additional information concerning this question.

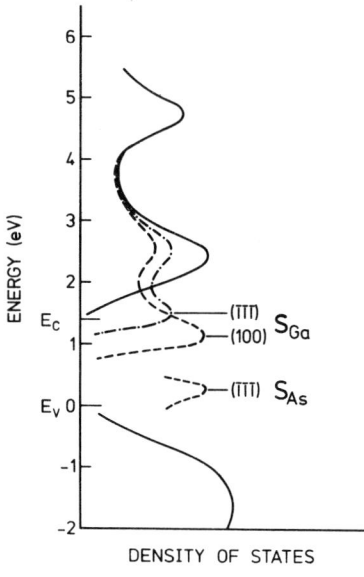

Fig. 6.36 Density of states of GaAs surfaces based on energy-loss spectra of Ga 3d-core level excitations: solid line, bulk like density of states for a ($\bar{1}\bar{1}\bar{1}$) As stabilized surface; dot-dashed line, for ($\bar{1}\bar{1}\bar{1}$) Ga stabilized surface; dashed line, for (100) Ga stabilized surface. E_v and E_c mark the valence and the conduction band edge, respectively. S_{Ga} and S_{As} indicate the estimated peak positions in the density of empty and filled surface states, respectively [6.73]

6.5.6 Selection Rule Effects Observed at Ge and GaAs

The already cited paper of LUDEKE et al. [6.80] is of great interest from the point of view of its interpretational spectrum. In this work a strong energy and surface orientation dependence of the loss spectra is reported. Fig.6.37 shows the loss spectra of GaAs and Ge for several surface orientations and primary energies, respectively. Only the loss region of interest is plotted. It is known that the d-core level is split into $d_{5/2}$ and $d_{3/2}$ levels with respect to the total angular momentum, with the $d_{5/2}$ about 0.6 eV above the $d_{3/2}$ for both materials. If, furthermore, the dangling bond states are sufficiently localized so that a description in terms of the total angular momentum is possible one would expect certain rules to be operative for the d-core state-to-surface-state transitions as long as the excitation is of dipole character or can be validly described by the first Born approximation (optical transitions). The fact that at primary energies above 100 eV only one of the two transitions (except at Ge(100) 2×2) can be observed, while at lower primary energies both transitions occur is interpreted as such a selection rule effect. The authors concluded that above 100 eV an optical description is adequate while below this value the appearance of the second loss results from a breakdown of the (dipole) selection rule $\Delta J = 0, \pm 1$ whereby quadrupole and spin-flip transitions must be considered then which allow the "forbidden" transition. If this interpretation should hold then ELS can be used to determine the symmetry of final states, information generally not available by other surface techniques.

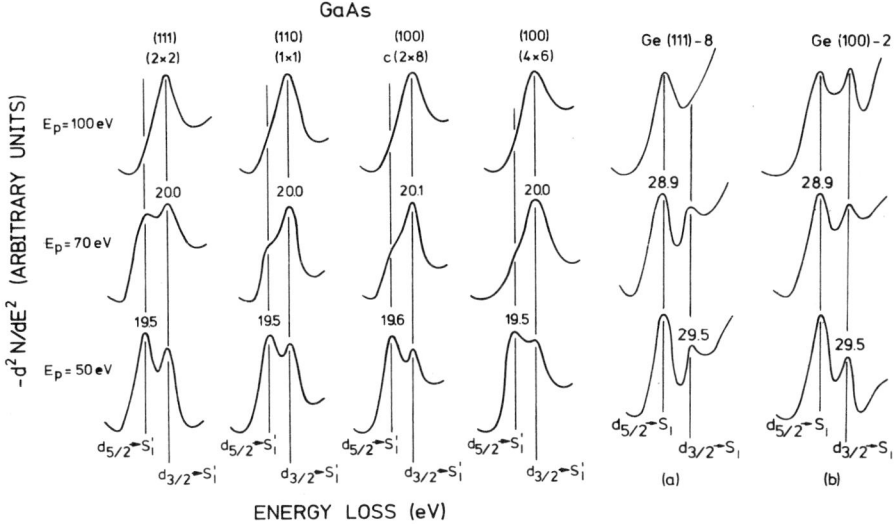

Fig. 6.37 Loss spectra due to transitions from d-core levels to empty surface states for different primary electron energies and various GaAs and Ge surfaces. For preparation of the surfaces see [6.80]

According to their results the authors summarized that the empty dangling bond states of GaAs and Ge(111) (8 × 8) exhibit dominant s- and p-like character, respectively, whereas that of the Ge(100) (2 × 2) surface consists of an admixture of these two.

6.5.7 Electronic Transitions at SiO and SiO$_2$

A very impressive example for the combination of ELS of interband transitions and core level excitations is the work done by KOMA et al. [6.83] on Si, SiO and SiO$_2$. These systems have already been dealt with in several papers by other authors [6.67,72]. Fig.6.38 shows the spectra of a Si(2p) excitation for several coverages of oxygen and the oxidized Si(100) 2 × 1 surface. The spectra of the oxygen covered and oxidized (100) and (111) surfaces are similar as is already reported in the paper of IBACH et al. [6.72]. The dominant maximum (Fig.6.38a) is very sensitive to surface contamination and it is interpreted therefore as being due to excitations into an empty surface state (dangling bond state). From knowledge of the energetic position (98.9 eV below the Fermi level) of the Si(2p) level, the position of the empty surface state coincides with the top of the valence band. The losses at 102.6 and 99.7 eV are attributed to transitions between a core level and the bulk conduction

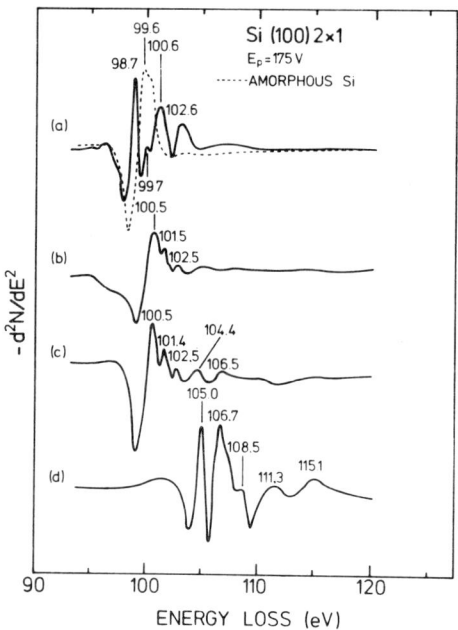

Fig. 6.38a-d Loss spectra due to excitations of Si(2p) core electrons for a Si surface: a) clean; b) with 0.6 monolayer oxygen coverage; c) monolayer coverage; d) covered with amorphous SiO_2 layer. The dashed curve in a) is the loss spectrum of the Ar^+-ion bombarded, amorphous surface

band. The 100.5 and the 101.5 eV losses of the oxygen-covered surface are believed to be oxygen related while those at 104.4 and 106.5 eV (Fig.6.38c) are believed to be due to the initial formation of SiO_2. In Fig.6.39a a comparison of the Si(2p) core level excitation spectrum and the loss spectrum due to interband transitions (dashed

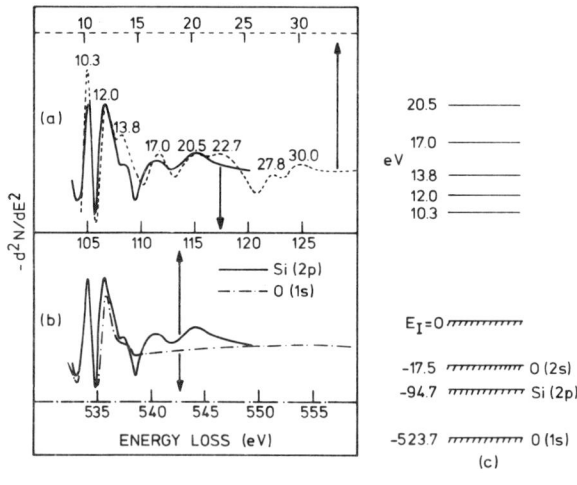

Fig. 6.39a-c Spectral data for amorphous SiO_2; a) comparison of the loss spectrum due to interband transitions (dashed curve) to that due to Si(2p) excitations into empty states above the Fermi level. b) Comparison of the spectra due to excitations of the Si(2p) excitations (solid line) and the O(1s) excitations (dashed-dotted line) into empty states above the Fermi level. The arrows indicate the corresponding abscissas. c) Energy level scheme deduced from the loss spectra

line) is represented. Up to $\hbar\omega$ = 20.5 eV there is excellent agreement in the structure of both spectra. With the assumption that the losses at 10.3 to 105 eV, respectively, have the same final state (this is suggested by results derived from soft X-ray emission measurements), the authors deduce the existence of final states above the Fermi level because of the good agreement of the two spectra (Fig.6.39c). In addition a fit of the Si(2p) and the O(1s) excitation spectra is shown in Fig. 6.39b. From the excellent match of the first three losses additional information is obtained concerning the localization of the final states. The missing of the two higher losses in the O(1s) excitation spectrum may be interpreted as a stronger localization of these states on the silicon atoms. In the interband loss spectrum the observed losses at 22.7, 27.8 and 30.8 eV are interpreted as plasmon excitations and transitions between the O(1s) core level and the two first final states (Fig.6.39c)

6.6 Conclusion

In the present chapter a concept has been presented of a method which is on one hand comparatively new but on the other hand well developed. The information and results obtainable by this method are closely related to those of optical methods but ELS is in many respects an easier experiment to carry out and gives more direct information. Another main advantage of this method arises from the fact that the "Dielectric Theory" can be easily applied as an underlying theory which is relatively clear.

This has been demonstrated especially by discussions and interpretations of vibrational spectra of adsorbed gases but it is also a supportive basis for the understanding of loss spectra arising from electronic transitions. We have tried to select the experimental examples presented in this chapter to explain and represent the possibilities of this method; this means it was not our aim to collect arbitrarily all available data. According to this we could show in Section 6.3 that the excitation of long wave length optical surface phonons is well understood in infrared as well as in noninfrared active material. It could also be demonstrated that ELS gives information on adsorption systems generally not available by other methods. This has been shown by means of the present available experimental data for gas-covered semiconductor and metal surfaces. Concerning ELS of electronic structure near the surface we presented only experiments dealing with semiconductor surfaces. The reason for this is that there is only little experimental material on metal surfaces. It should be mentioned, however, that there are three older papers [6.84,86] both dealing with clean and gas-covered Ni surfaces. Unfortunately these measurements were carried out with a LEED system with very small resolution. It is therefore difficult to interpret these measurements for reasons given in Subsection 6.5.2.

The most important experiments were made on Si, Ge and GaAs clean and gas-covered surfaces. These experiments are reported in Section 6.5. Especially the system Si plus oxygen has been discussed extensively in order to show that the combination of results arising from interband transitions and core level excitations may give information about filled and empty energy levels presently not available by any other method.

A progressive develeopment of ELS in the high resolution field can be seen in a further improvement of the spectrometers. Here it is desired to decrease the ratio of the background-to-signal intensity to values of the order of 10^{-4} or less and to increase the primary intensity in order to become able to investigate short wave length effects.

As present theoretical attempts [6.87] already show, experimental investigations of these effects especially studies on short wave length optical phonons would represent great progress in ELS. Up to now there exists no real experimental concept for such investigations. However, experimental results of these effects would be an important step forward in the understanding of the microstructure of solid surfaces because, contrary to long wave length effects, which are mainly controlled by bulk parameters, the short wave length effects strongly depend only on the characteristic microstructure of the surface region because of the small extending length of those waves into the solid. ROUNDY et al. [6.87] could show by model calculations, for instance, that the loss spectra of electrons, scattered from short wave length phonons, are strongly coupled to the lattice dynamical model of the surface described by the microstructure, force constants, etc.

Another aspect is the excitation of surface magnons, which could give an important contribution about the knowledge of magnetic properties of surfaces. This method could be further developed by the combination of ELS with spin-polarized electrons.

According to ELS of electronic structure, progress can be seen in the combination of ELS with electron microscopy [6.88] which enables one to investigate small regions of material with reference to their composition and their chemical state. This method may be especially interesting for studies more oriented to technological applications.

Acknowledgement

The author acknowledges gratefully helpful comments and critical reading the manuscript by Dr. H.P. Bonzel.

References

6.1   H. Raether: In *Springer Tracts in Modern Phyics* Vol. 38 (Springer, Berlin, Heidelberg, New York 1965) p. 85
6.2   H. Boersch, H. Miessner, W. Raith: Z. Phys. $\underline{168}$, 404 (1962)

6.3   J. Geiger: *Elektronen und Festkörper* Vol. 128 (Sammlung Vieweg, Braunschweig 1968)
6.4   D. Pines: *Elementary Excitations in Solids* (W.A. Benjamin, New York, Amsterdam 1963)
6.5   D. Bohm, D. Pines: Phys. Rev. $\underline{92}$, 609 (1953)
6.6   R.A. Ferrell: Phys. Rev. $\underline{101}$, 554 (1956); Phys. Rev. $\underline{107}$, 450 (1957)
6.7   H. Watanabe: J. Phys. Soc. Japan $\underline{16}$, 912 (1962)
6.8   R.H. Ritchie: Phys. Rev. $\underline{106}$, 874 (1957)
6.9   V.J. Davisson, B. Germer: Nature $\underline{119}$, 558 (1927)
6.10 C.B. Duke, G.E. Laramore: Phys. Rev. $\underline{B3}$, 3183 (1971); G.E. Laramore, G.B. Duke: Phys. Rev. $\underline{B3}$, 3198 (1971); C.B. Duke, U. Landmann: Phys. Rev. $\underline{B6}$, 2956 (1972)
6.11 W.H. Weber, M.B. Webb: Phys. Rev. $\underline{177}$, 1103 (1969)
6.12 E. Fermi: Phys. Rev. $\underline{57}$, 485 (1940)
6.13 J. Hubbard: Proc. Phys. Soc. (London) $\underline{A68}$, 976 (1955)
6.14 H. Fröhlich: *Max-Planck-Festschrift* (BEB Deutscher Verlage der Wissenschaften, Berlin 1958) p. 277
6.15 H. Fröhlich, H. Pelzer: Proc. Phys. Soc. (London) $\underline{A68}$, 525 (1955)
6.16 P. Schmüser: Z. Phys. $\underline{180}$, 105 (1964)
6.17 J. Daniels, C. v. Festenberg, H. Raether, K. Zeppenfeld: In *Springer Tracts in Modern Physics*. Vol. 54 (Springer, Berlin, Heidelberg, New York 1970) p. 76
6.18 A.A. Lucas, M. Sunjiec: *Progress in Surf. Sci.*, Vol. 2, ed. by S.G. Davison (Pergamon Press 1972) p. 2
6.19 H. Ibach: J. Vac. Sci. Technol. $\underline{9}$, 713 (1972)
6.20 E. Evans, D.L. Mills: Phys. Rev. $\underline{B7}$, 853 (1973)
6.21 E. Evans, D.L. Mills: Phys. Rev. $\underline{B5}$, 4126 (1972)
6.22 D.L. Mills: Surf. Sci. $\underline{48}$, 59 (1975)
6.23 D.L. Mills: "Interaction of low energy electrons with surface lattice vibrations", to be published in: *Surfaces and Interfaces*, Vol. I, ed. by L. Dobrzynski (Marcel Dekker, New York)
6.24 L. v. Hove: Phys. Rev. $\underline{95}$, 249, 1374 (1954)
6.25 R.F. Wallis: "Lattice Dynamics of Crystal Surfaces", *Progress in Surface Science*, Vol. 4, Part 3 (Pergamon Press); G. Leibfried: Gittertheory der mechanischen und thermischen Eigenschaften der Kirstalle. In *Handbuch der Physik* Vol. VII/1 (Springer, Berlin, Heidelberg, New York 1955) p. 104
      A.A. Maradudin: E.W. Montroll, G.H. Weiss: "Theory of lattice dynamics in the hormonic approximation", Solid State Phys. Supp. 3 (1963)
6.26 R. Fuchs, K.L. Kliewer: Phys. Rev. 140, 2076 (1965); Phys. Rev. $\underline{144}$, 495 (1966)

6.27 R. Engelmann, R. Ruppin: J. Phys. C: Proc. Phys. Soc. $\underline{1}$, 614 (1968);
R. Ruppin, R. Engelmann: J. Phys. C: Proc. Phys. Soc. $\underline{1}$, 630 (1968);
R. Engelmann, R. Ruppin: J. Phys. C: Proc. Phys. Soc. $\underline{1}$, 1515 (1968)

6.28 K. Huang: Proc. Roy. Soc. $\underline{A208}$, 352 (1951)

6.29 M. Born, K. Huang: Dynamical Theory of Crystal Lattices (Clarendon, Oxford 1968)

6.30 A.A. Lucas, E. Kartheuser: Phys. Rev. $\underline{B1}$, 3588 (1970)

6.31 T. Wolfram, R.E. DeWames, E.A. Krant: J. Vac. Sci. Technol. $\underline{9}$, 685 (1971)

6.32 W. Ludwig: *Springer Tracts in Modern Physics* Vol. 43 (Springer Berlin, Heidelberg, New York 1967) p. 206

6.33 R.E. Allen, G.P. Alldredge, F.W. de Wette: Phys. Rev. Lett. $\underline{23}$, 1285 (1969); $\underline{24}$, 301 (1970)

6.34 S.Y. Tsong, A.A. Maradudin: Phys. Rev. $\underline{187}$, 878 (1969)

6.35 R.E. Allen, G.P. Alldredge, W.F. de Wette: Phys. Rev. Abstr., 1st July 1971, p. 12

6.36 E. Fick: *Einf. in d. Grundl. d. Quantenmechanik* (Akadem. Verlagsgesellschaft Frankfurt/M. 1968) p. 457

6.37 D.L. Mills: private communication

6.38 J.D.E. McIntyre, D.E. Aspnes: Surf. Sci. $\underline{24}$, 417 (1971).

6.39 F. Meyer, E.E. de Kluizenaar, G.A. Bootsma: Surf. Sci. $\underline{27}$, 88 (1971)

6.40 H. Froitzheim, I. Ibach, D.L. Mills: Phys. Rev. $\underline{B11}$, 4980 (1975)

6.41 H. Froitzheim: Jül-Rep. 1179, März 1975

6.42 H. Froitzheim, I. Ibach: Z. Phys. $\underline{269}$, 17 (1974)

6.43 H. Ibach: Phys. Rev. Lett. $\underline{24}$, 1416 (1970)

6.44 H. Ibach: J. Vac. Sci. Technol. $\underline{9}$, 713 (1972)

6.45 H. Froitzheim: unpublished results

6.46 H. Ibach: Phys. Rev. Lett. $\underline{27}$, 253 (1971)

6.47 H. Ibach, K. Horn, R. Dorn, H. Lüth: Surf. Sci. $\underline{38}$, 433 (1973)

6.48 H. Froitzheim: unpublished results

6.49 S.E. Trullinger, S.L. Cunningham: Phys. Rev. Lett. $\underline{30}$, 913 (1973)

6.50 R.G. Greenler: J. Chem. Phys. $\underline{44}$, 310 (1966)

6.51 G. Herzberg: *Molecular Spectra and Molecular Structure* 2. Aufl. (van Nostrand, New York 1974)

6.52 K.F.W. Kohlrausch: Der Smekal Raman Effekt, Ergänzungsband (Springer 1938)

6.53 W.A. Goddard, A. Rendondo, T.C. McGill: to be published in Solid State Comm.

6.54 H. Froitzheim, H. Ibach, S. Lehwald: Phys. Lett. $\underline{55A}$, 247 (1975)

6.55 J.A. Appelbaum, D.R. Hamann: Phys. Rev. Lett. $\underline{34}$, 806 (1975)

6.56 F.M. Propst, T.C. Piper: J. Vac. Sci. Techn. $\underline{4}$, 53 (1967)

6.57 H. Froitzheim, H. Ibach, S. Lehwald: Rev. Sci. Instr. $\underline{46}$, 1325 (1975);
H. Froitzheim, H. Ibach, S. Lehwald: Phys. Rev. Lett. $\underline{36}$, 1549 (1976)

6.58 P.J. Estrup, J. Anderson: J. Chem. Phys. $\underline{45}$, 3354 (1966)

6.59   T.E. Madey: Surf. Sci. 29, 571 (1973)
6.60   H. Froitzheim, H. Ibach, S. Lehwald: Phys. Rev. B14, 1362 (1976)
6.61   C.A. Papageorgopoulos, J.M. Chen: Surf. Sci. 39, 313 (1973)
6.62   E. Bauer, H. Poppa, Y. Viswanath: Surf. Sci. 58, 517 (1976)
6.63   T.E. Madey: Surf. Sci. 33, 355 (1972)
6.64   H. Froitzheim, H. Ibach, S. Lehwald: Surf. Sci., to be published
6.65   J.T. Yates, T.E. Madey, N.E. Erickson: Surf. Sci. 43, 257 (1974)
6.66   L.W. Anders, R.S. Hansen: J. Chem. Phys. 62, 4652 (1975)
6.67   H. Ibach, J.E. Rowe: Phys. Rev. B9, 1951 (1974)
6.68   F. Meyer: Phys. Rev. B9, 3622 (1974)
6.69   G. Chiarotti, S. Nannarone, R. Pastore, P. Chiaradia: Phys. Rev. B4, 3398 (1971)
6.70   C.B. Wilson: Proc. Phys. Soc. 76, 481 (1960)
6.71   H. Ibach: private communication
6.72   H. Ibach, J.E. Rowe: Phys. Rev. B10, 710 (1974)
6.73   R. Ludeke, L. Esaki: Phys. Rev. Lett. 33, 653 (1974)
6.74   J.E. Rowe: Solid State Comm. 15, 1505 (1974)
6.75   R.L. Gerlach, J.E. Houston, R.L. Park: Appl. Phys. Lett. 16, 179 (1970)
6.76   R.L. Gerlach, A.R. Charme: Surf. Sci. 29, 317 (1972)
6.77   R.A. Pollak, R.F. McFeely, S.P. Kowalczyk, D.A. Shirley: Phys. Rev. B9, 600 (1974)
6.78   H. Froitzheim, H. Ibach: Surf. Sci. 47, 713 (1975)
6.79   R.C. Eden: Ph.D. Thesis, Stanford University, available as dissertation No. 67-17, 415, University Microfilms, Inc. Ann Arbor, Michigan 48106 (unpublished)
6.80   R. Ludeke, A. Koma: Phys. Rev. Lett. 34, 817 (1975)
6.81   J.H. Dinan, L.K. Galbraith, T.E. Fischer: Surf. Sci. 26, 587 (1971)
6.82   H. Huijser, J. van Laar: Surf. Sci. 52, 202 (1975)
6.83   A. Koma, R. Ludeke: Phys. Rev. Lett. 35, 107 (1975)
6.84   J. Küppers: Surf. Sci. 36, 53 (1973)
6.85   F. Steinrisser, E.N. Sickafus: Phys. Rev. Lett. 27, 992 (1971)
6.86   E. Sickafus, F. Steinrisser: Phys. Rev. B6, 3714 (1972)
6.87   V. Roundy, D.L. Mills: Phys. Rev. B5, 1347 (1972)
6.88   R.F. Egerton, J.G. Phillip, P.S. Turner, M.J. Whelan: J. Phys. E (Sci. Inst.) 8, 1033 (1975)
6.89   E.A. Stern, R.A. Ferrell: Phys. Rev. 120, 130 (1960)
6.90   J.E. Inglesfield, E. Wikborg: Solid State Comm. 14, 663 (1974)

Recent References Added in Proof

S. Anderson: Vibrational excitations and structure of CO adsorbed in Ni(100). Solid State Commun. 21, 75 (1977)

S. Anderson: Surface vibrational excitations of O in the p(2×2) O and c(2×2) O structures on Ni(100). Solid State Commun. 20, 229 (1976)

J.E. Rowe, G. Margaritondo, H. Ibach, H. Froitzheim: Oxidation of silicon: New electron spectroscopy results. Solid State Commun. 20, 277 (1976)

N.R. Avery: Chemical effects in the electron loss spectrum from a W(100) surface. Chem. Phys. Lett. 43, 250 (1976)

R. Ludeke, A. Koma: Low-energy-loss spectroscopy of Ge surfaces. Phys. Rev. B 13, 739 (1976)

Subject Index

Adiabatic approximation 65
Adsorbate, ELS 223
Alignment 30
Angle-resolved photoemission spectroscopy 155,183
Anisotropic effects, ELS 218
Anisotropy effects, AES 69
Angular acceptance 16
- distribution, inelastically scattered electrons 212
- emission pattern, photoemission 179
- half width, electron energy loss spectroscopy 213
Au, photoemission spectra 165
Auger electron appearance potential spectroscopy (AEAPS) 10,14,59,80
-- spectroscopy (AES) 10,14,59,92
- energies 67
- microanalysis 99
-/X-ray microanalysis 102

Ba, Auger appearance potential spectroscopy 32
Ba, soft X-ray appearance potential spectroscopy 85
Background, in electron diffraction 125
Backscattering factor 98
Be, differential ionization cross section 90
Begrenzungsterm 7

Binding energy of electrons 60
Bonding shifts in photoemission 173
Bremsstrahlung, resonances 86

C, differential ionization cross section 89
-, K-shell excitation 107
Characteristic energy losses 8
Chemical information, DAPS 79
- shifts, AES, 93
- shift, in photoemission spectroscopy 157
Cl, differential ionization cross sections 89
Cluster calculations 177
Coatings 30
Coefficient of quasielastic reflection 74
Coherence width in electron diffraction 121
Collective oscillations of surfaces 208
Comparison, AES and DAPS 109
-, AES and ILS 110
-, threshold spectroscopies versus ionization loss spectroscopies 107
-, threshold spectroscopies 104
Constant final state spectroscopy 154
- initial state spectroscopy 154,193
Continuum light sources 161
Core-level excitations 240
Corrugated electrodes 30,34

Coster-Kronig transitions 64
- yield 64
Cr, Auger electron-, soft X-ray appearance potential-, Auger appearance potential spectroscopy 104
-, soft X-ray appearance potential spectroscopy 84
Crossed-field deflection spectrometers 43
Cross section, in ELS 210
- transitions 93
Cu, differential ionization cross section 89
-, stepped surfaces 139
Cylindrical deflector analyzer 15,31
- mirror analyzer (CMA) 15
---, design 39
---, optimum transport current 23

Dangling bond states 244
Debye-Waller factor 127
Defect structure 117
Deflection energy analyzers 17
Delayed transitions 78
Derivative spectrum 28
Diagram lines 68
Dielectric constant 207
- layer model 236
- response function 4
- shift 238
- theory 207,210
Difference curves, photoemission spectroscopy 175
Diffraction pattern, of atoms with random displacements 128
--, simple defect structures 122
Diffraction, optical simulation 122
Diffusion, adsorbates 144
Dipole fields 208
- layer 214
Disappearance potential spectroscopy (DAPS) 10,69,74,79

Disorder, adsorbate lattices 143
Domains 141
Double Auger process 68
- filter analyzer 15,47
- pass CMA 41
Dynamical background subtraction 98

Effective charge 214,222
Electron diffraction 117
- energy loss spectroscopy (ELS) 10,205
- impact ionization 60
- microprobe analysis (EMA) 59
- optics 20
- sources 32
- spectrometers, comparison 43
--, design 13,19,23
- spectroscopies, requirements 13
- stimulated desorption 3
- trajectories 22
Electronic surface transitions 217
- transitions, in ELS 234
Empty surface states 240
Energy analysis, basis principles 16
- dispersion of analyzers 18
- resolution of analyzers 17
Escape depth in photoemission 153
Etendue of analyzers 13,21
EuO, spin polarized photoemission 195
Ewald construction 120,127
--, for irregular step arrays 135
--, for stepped surfaces 132
--, different disordered surfaces 147
Extra-atomic relaxation 69,158,172

Facets 141,145
-, small ones 146
Fast electrons 208
Field emission spectroscopy 8,14
- fluctuations 210
Figures of merit, analyzers 18

Filter lens analyzer 15
Final state wave functions 164
Fluorescence yield 63,64,65,104
Forbidden transitions 244
Fringing field 26
-- of CMA 42
Frozen orbitals 172
- orbital approximation 64

GaAs, constant initial state spectroscopy 193
-, electronic excitation of surfaces 240
-, source of spin polarized electrons 197
-, stepped surfaces 135
Ge, electronic excitations of surface states 239
-, partial yield spectroscopy 193
-, stepped surfaces 140
Ge+Sb, partial yield spectroscopy 193
Ghost peak, SXAP 83

Incidence energy modulation method 15
Ideal field boundary 26
Inelastic low-energy electron diffraction (ILEED) 9,14
- scattering, theory 207
Information depth, 4,103
--, DAPS and AEAPS 106
--, of ELS 10
Infrared spectroscopy 224
Initial state wave functions 164
Intra-atomic relaxation 67
Ion neutralization spectroscopy 8,14
Ionization cross section 62
- loss spectroscopy (ILS) 10,14,59,87
- loss spectrum 77

Kinematical approximation, electron diffraction 117

La, Bremsstrahlung 86

LEED optics analyzer 15
- pattern, calculation 119
LiF, photoemission 192
Light sources, photoemission 159
Line sources, photoemission 160
Loss function 208
Low-energy electron diffraction 14
Luminosity of analyzers 18

Magnetic deflection analyzer 15
- deflection spectrometers 43
- fields 29
Matrix elements, photoemission 163
Maxwell energy distribution 23
Mean free path 4,7,168
---, secondary electrons 192
Mg, differential ionization cross sections 89
Molecular orbital schemes 177
Möllenstedt lens analyzer 15
Mott scattering 194
Multichannel electron multipliers 28
Multiphonon excitation 214

N, differential ionization cross sections 89
Na, differential ionization cross sections 89
Ni + CO, density of states 177
Ni + hydrocarbons, photoemission 182
Ni, NiO, soft X-ray appearance potential spectroscopy 84
Nomenclature Auger transitions 66

O, differential ionization cross sections 89
Optical data of a surface layer 236
- surface phonons 213,214
Optimization principles of energy analyzers 22

Pair distribution function 136
Partial yield spectroscopy 192

Patterson function 121
Phonons 213
Photo-assisted field emission 156
Photoemission, adsorbates 174
- principles 152
- spectroscopy 151
--, basis processes of 156
--, parameters 154
--, theory 163
Plane mirror analyzer 15,38
Plasma waves 208
--, excitation 216
Point defects, electron diffraction 124
Poisson distribution 214
Pt, stepped surfaces 139
Pulse counting 28

Quantitative Auger analysis 95

Radiationless transitions 64
Radiative Auger effect 68
- transitions 64
Reciprocal space of surfaces with defects 125
Reconstruction of the surface 223
Relaxation 171
Resolving power 17
Resonance levels 153
Retarding potential analyzer 15,16
-- spectrometers 44

Scanning Auger electron spectroscopy 14
Secondary electrons 8
Segregation, of Si to the surface of Be 90
Selection rule, dipole scattering 215
-- effects, in electronic excitations by electrons 243
--, ELS 224

Self-energy of electrons 4
Sensitivity, of DAPS 78
- of analyzers 17
- of spectroscopies 10
Si, angle resolved photoemission 185
-, differential ionization cross section 89,90
-, electronic excitations of surface states 235
-, stepped surfaces 139
-, surface phonons 222
Si + H, photoemission 180
Si + H, surface vibrations 228
SiO, electronic transitions of surfaces 244
SiO_2, electronic transitions of surfaces 244
Si + O, surface vibrations 226
Signal-to-noise ratio, AEAPS 81
Soft X-ray appearance potential spectroscopy (SXAPS) 10,82,104
Space-charge limited electron beams 23
Spectra of clean vanadium in AEAPS and DAPS 105
Spectrometer, for AES and ILS 88
-, for ELS 219
Spherical deflector analyzer 15,34
---, optimum transport current 23
Spherical grid retarding potential analyzer 46
- mirror 41
Spin polarization in photoemission 155
Spin-polarized photoemission 194
Spline technique 99
Spot shapes, for irregular step arrays 137
- shape, in electron diffraction 123
- splitting, due to regular step arrays 131
Stainless steel, AES, DAPS 108
Step arrays, electron diffraction 130
- arrays, irregular 134

Step density 134
- edges, orientation 134
- height, electron diffraction 134
Stretching vibration 225
Surface dielectric constant 217
- effective charge 223
- magnons 247
- optical effect 153,170
- roughness effects, AES 95
- sensitivity, DAPS 79
- sensitivity, ILS 87
- state density 241
- states 240
Synchrotron radiation, photoemission spectroscopy 161
Sweeping modes for energy analysis 21

Tandem spherical deflector spectrometer 15
Terrace width, electron diffraction 134
Threshold spectroscopies 71,104
- spectroscopy, in photoemission 156
- yield measurements, photoemission 191
Ti_2O_3, disappearance potential spectroscopy 79
Time-of-flight analysis 15,48
Total yield spectrum, photoemission 191
Transmission of analyzers 18
Trajectories of electrons 22
Two step model 207

Ultraviolet photoemission spectroscopy, characteristics 159
--- (UPS) 10,14,156
UO_2, stepped surfaces 139

V, Auger electron spectroscopy 94
-, disappearance potential spectroscopy 76,105
Vectorial photoeffect 154
Vibrational modes 223

W, angle resolved photoemission 187
-, core excitations 110
W + C, ionization loss spectra 91
W + CO, surface vibrations 233
W + H, angle resolved photoemission 188
W + H, surface vibrations 229
W + O, surface vibrations 230
-, photoemission 182
-, stepped surfaces 148
Wien filter analyzer 15

X-ray adsorption spectrum 77
- appearance potential spectroscopy (SXAPS) 59
- excited electron appearance potential spectroscopy (XEAPS) 59
- photoemission spectroscopy (XPS) 10, 14,156
---, characteristics 158

Yield spectroscopies 189

ZnO, stepped surfaces 139
-, surface phonons 220
Zoom lens system 35

Interactions on Metal Surfaces

ed. by *R. Gomer*
Topics in Applied Physics, Vol.4
(1975) Pp.X+310
ISBN 3-540-07094-X

J.R. Smith (General Motors Technical Center)
Theory of Electronic Properties of Surfaces

S.K. Lyo, R. Gomer (University of Chicago)
Theory of Chemisorption

L.D. Schmidt (University of Minnesota)
Chemisorption: Aspects of the Experimental Situation

D. Menzel (Technische Universität München)
Desorption Phenomena

E.W. Plummer (University of Pennsylvania)
Photoemission and Field Emission Spectroscopy

E. Bauer (Technische Universität Clausthal)
Low Energy Electron Diffraction (LEED) and Auger Methods

M. Boudart (Stanford University)
Concepts in Heterogeneous Catalysis

Theory of Van der Waals Attraction

D. Langbein

Springer Tracts in Modern Physics, Vol.72
(1974) Pp.145
ISBN 3-540-06742-6

Surface Physics

Springer Tracts in Modern Physics, Vol.77
(1975) Pp.VI+125
ISBN 3-540-07501-1

P. Wissmann
The Electrical Resistivity of Pure and Gas Covered Metal Films

K. Müller
How Much Can Auger Electrons Tell Us about Solid Surfaces?

Light Scattering in Solids

ed. by *M. Cardona*
Topics in Applied Physics, Vol.8
(1975) Pp.XI+339
ISBN 3-540-07354-X

M. Cardona (MPI Festkörperforschung)
Introduction

A. Pinczuk (Argentinial Comisión Nacional de Energia Atómica); E. Burstein (University of Pennsylvania)
Fundamentals of Inelastic Light Scattering in Semiconductors and Insulators

R.M. Martin (Xerox Research Center); L.M. Falicov (University of California)
Resonant Raman Scattering

M.V. Klein (University of Illinois)
Electronic Raman Scattering

M.H. Brodsky (IBM Research Center)
Raman Scattering in Amorphous Semiconductors

A.S. Pine (MIT, Lincoln Laboratory)
Brillouin Scattering in Semiconductors

Y.-R. Shen (University of California)
Stimulated Raman Scattering

Springer-Verlag
Berlin Heidelberg New York

A monthly journal Applied Physics

Board of Editors	**S. Amelinckx,** Mol. · **V. P. Chebotayev,** Novosibirsk **R. Gomer,** Chicago, Ill. · **H. Ibach,** Jülich **V. S. Letokhov,** Moskau · **H. K. V. Lotsch,** Heidelberg **H. J. Queisser,** Stuttgart · **F. P. Schäfer,** Göttingen **A. Seeger,** Stuttgart · **K. Shimoda,** Tokyo **T. Tamir,** Brooklyn, N.Y. · **W. T. Welford,** London **H. P. J. Wijn,** Eindhoven
Coverage	application-oriented experimental and theoretical physics: *Solid-State Physics* *Quantum Electronics* *Surface Physics* *Laser Spectroscopy* *Chemisorption* *Photophysical Chemistry* *Microwave Acoustics* *Optical Physics* *Electrophysics* *Integrated Optics*
Special Features	**rapid** publication (3–4 months) **no** page charge for **concise** reports prepublication of titles and abstracts **microfiche** edition available as well
Languages	Mostly English
Articles	original reports, and short communications review and/or tutorial papers
Manuscripts	to Springer-Verlag (Attn. H. Lotsch), P.O. Box 105280 D-69 Heidelberg 1, F.R. Germany Place North-American orders with: Springer-Verlag New York Inc., 175 Fifth Avenue, New York. N.Y. 10010, USA

Springer-Verlag
Berlin Heidelberg New York

/539.72112E38>(1/